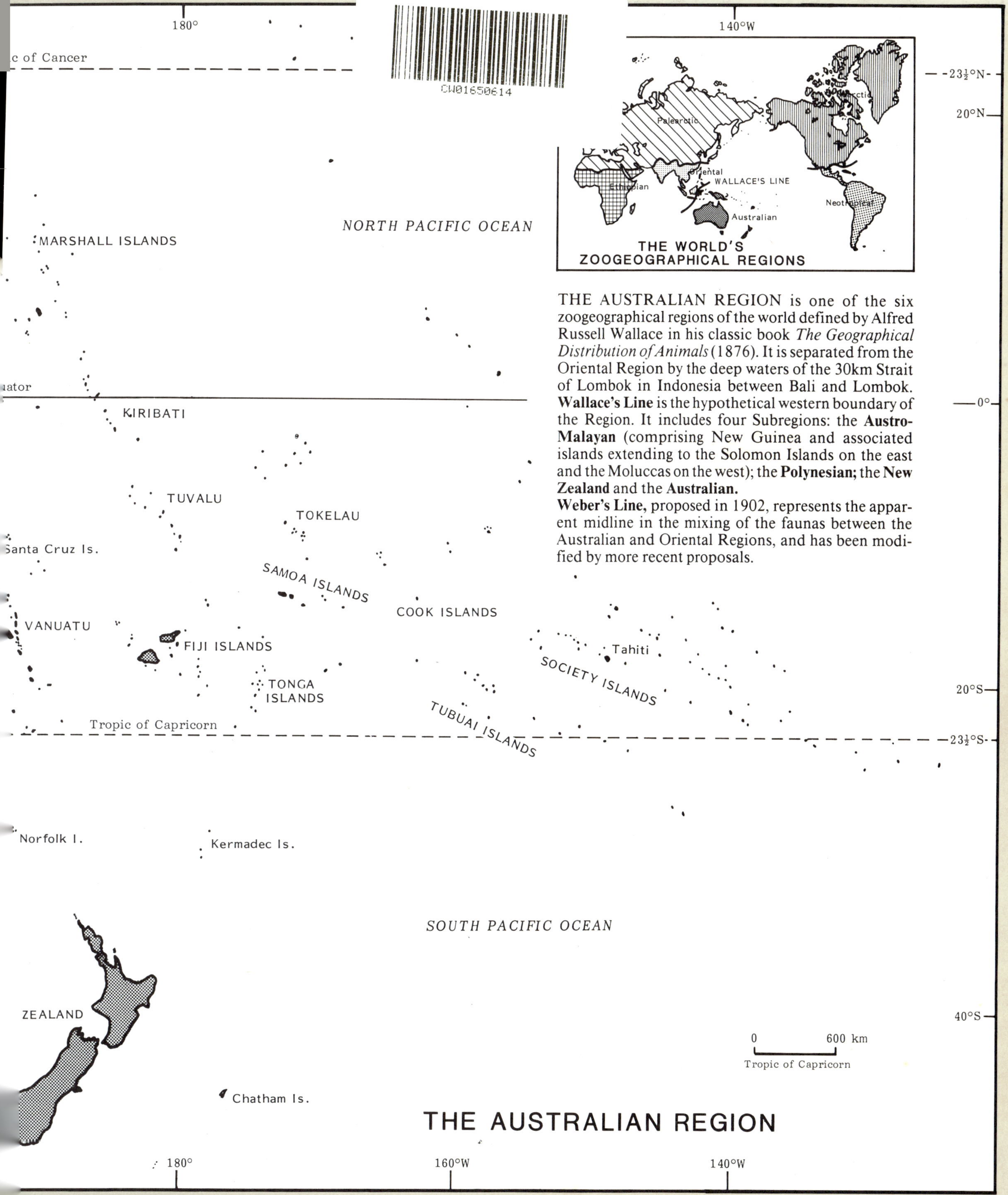

THE AUSTRALIAN REGION is one of the six zoogeographical regions of the world defined by Alfred Russell Wallace in his classic book *The Geographical Distribution of Animals* (1876). It is separated from the Oriental Region by the deep waters of the 30km Strait of Lombok in Indonesia between Bali and Lombok. **Wallace's Line** is the hypothetical western boundary of the Region. It includes four Subregions: the **Austro-Malayan** (comprising New Guinea and associated islands extending to the Solomon Islands on the east and the Moluccas on the west); the **Polynesian;** the **New Zealand** and the **Australian.**

Weber's Line, proposed in 1902, represents the apparent midline in the mixing of the faunas between the Australian and Oriental Regions, and has been modified by more recent proposals.

MELANESIA (from the Greek meaning black islands) is that part of Oceania lying in the west Pacific to the south-west of Polynesia and south of the Equator.

MOLUCCAS (or Spice Islands) are Indonesian islands of the Malay Archipelago bounded by Sulawesi (Celebes) on the west, New Guinea on the east, Timor and the Arafura Sea on the south, and the Philippines on the north.

AUSTRALASIA is that portion of Oceania lying between the Equator and 47°S and often used to bracket Australia and New Zealand.

AUSTRALIAN BIRDS

AUSTRALIAN BIRDS

DONALD & MOLLY TROUNSON

NATIONAL

Published by National Book Distributors
19A Roger Street, Brookvale, NSW, 2100
First published in 1987 as *Australia, Land of Birds*
by William Collins Pty Ltd, Sydney
Reprinted 1989, 1991, 1992 by National Book Distributors

Wholly devised and designed by Donald and Molly Trounson

Typeset by Asco Trade Typesetting Ltd. Hong Kong, and
Love Computer Typesetting Pty Ltd. Sydney
Printed in Singapore by Colourwork Press Pte Ltd

National Library of Australia Cataloguing-in-Publication data

Trounson, Donald, 1905–.
Australian Birds.

Includes index.
ISBN 1 875580 33 6

1. Birds—Australia—Identification.
I. Trounson, Molly, 1928– . II. Title.

598.2994

Photographs
Title-page spread: F. Spolc
Part-title spread Part I: D. & M. Trounson
Part-title spread Part II: P. Soleness

HOW TO FIND A BIRD

If you do not know the name of the bird

1. Consult the Contents List and narrow your search to the promising Sections.
2. Once you think you have found your bird, check the distribution map to confirm that it lives in the area, and consult the notes provided.

If you know the name of the bird

1. If you can remember the appropriate Section, the quickest way is to go straight to that Section. The chances are that you will at once recognise the bird from the photographs. Otherwise, consult the Section index.
2. If you can't remember the Section number, you can either find it in the Contents List, or look up the species directly in the general indexes at the end of the book.

If you are used to using the running order of the Systematic List

Consult the Key to the Distribution of Families Based on the Systematic List on page x.

CONTRIBUTORS

ILLUSTRATIONS

The Photographers
C. Avon
Barry M. Allwright, AFIAP, LAPS
Graham D. Anderson
Kathie Atkinson
Douglass Baglin
J. Bell
Hans & Judy Beste
Harry Butler, CBE
Chris Cameron
John Carnemolla
Norman Chaffer, OAM, FRZS
Graeme Chapman
M.A. Chappell
Brian Chudleigh
Brian Coates
Jack Cupper
John Dart
Ron Dencio
Douglas D. Dow
Foch Dowling
Alistair Drake
Derek England
Johnny Estbergs
Kenneth Fink
Graeme Folley
Alan Foster
Jim Frazier, FRPS (Mantis Wildlife Films)
Cliff Frith
Peter Fullagar, BSc, PhD
Tom & Pam Gardner
Ray Garstone
Colin Gill
Gloria Glass
Robert Graves
John Gray
Bob Green
Philip Green
Geoff Harrow
Andrew Henley, BSc, DPhil, ARPS
Tony Hertog
David Hollands
David Hosking, ARPS
Eric Hosking, OBE, Hon FRPS, FIIP
Keith Ireland, EFIAP, AAPS
C.H.H. Jerrard
Fred James
Gavin W. Johnstone
G.R. Jones
Pat Jordan
Ralph & Daphne Keller
Peter Klapste
C.R. Knights
Peter Knowles
Nick Kolichis
Bill Labbett
Wayne Lawler
Ronald Legge
Fred Lewitzka
T.R. Lindsey
Tom Lowe
Roy Mackay
Garth May, EFIAP, ARPS, APSA
Ian McCann
John McCann
Arnold R. McGill, OAM, MLO, FRZS, FRAOU
Ross McLennan
Malcolm McNaugton
David Millege
Michael & Irene Morcombe
Ian Morgan
Ian Morris
P.D. Munchenberg
Jim Napier
Penny & Jerry Olsen
Frank Park
Steve Parish
Bill Peckover
Alwyn Pepper, OAM, EFIAP
Trevor Pescott
Harold J. Pollock
Murray Price
Jack Purnell
Keith & Beryl Richards
Paul Roach
Graham Robertson
Len Robinson
Alan E.F. Rogers
G. Rogerson
Lyndon F. Schick
Vincent Serventy, AM
Michael Seyfort
Ederic Slater, BEM
E.C.J. Smith, LAPS
Peter Soleness
Michael F. Soper
Chris Spiker
Frank Spolc
Gary Steer
Ron & Valerie Taylor
Athal Tolhurst
K. Vagg
Richard Vaughan
Trevor Waite
John Warham, Phd, FRPS
Brendan Watkins
John Waterhouse
Gary Weber, FRPS, EFIAP
Cyril Webster, EFIAP
Wardene Weisser
Bert Wells, FRPS, EFIAP, AAPS and
Babs Wells
Edgar Whitbourn
Michael Wright
Eric Zillmann

Artist
T.R. Lindsey

TEXT

The Editorial Team
Rae Henson, BA, Dip.Ed
Writer and researcher

Dariel Larkins
Ornithologist

N.W. Longmore
Department of Ornithology
Queensland Museum

Arnold McGill, OAM, MLO, FRZS, FRAOU Ornithologist

Special Contributors
Richard Donaghey, PhD
Ornithologist
(Section 15 Bowerbirds)

Penny Olsen, BSc (Hons) and
Jerry Olsen, BSc, M.Ed
Raptor biologists
(Section 2 Diurnal Birds of Prey)

Referees
Walter Boles, BSE (Hons)
Ornithologist
The Australian Museum, Sydney

H. John de S. Disney, MA
Former Curator of Birds
The Australian Museum, Sydney

Graeme Chapman
Ornithologist and
Wildlife photographer
CSIRO, Canberra

N.W. Longmore

T.R. Lindsey
Ornithologist and Book Editor
National Photographic Index of Australian Wildlife

Harold J. Pollock
Ornithologist

Proof Reading
N.W. Longmore
Zoë and Gerald Middleton

BIRD CALLS

Ernest S. Hoskin
Life Member, NSW Field Ornithologists Club

CARTOGRAPHY

Arnold McGill, OAM, MLO, FRZS, FRAOU Ornithologist
(Distribution maps)

John Roberts
Cartographer
Sydney University
(End paper maps)

Irene and Bruce Young
Hornsby Commercial Art Services
(Art work on the distribution maps)

PREFACE

The enjoyment of birds . . . depends neither upon intensive study nor academic qualifications

Roger Tory Peterson (1954)

The idea for this book was first floated in May 1966. It was to have been called the *National Photographic Index of the Birds of Australia.* The concept was simple: there were to be twelve photographs on the right-hand page and a set of matching captions on the left-hand page. A prospectus was produced, illustrated with a selection of our first bird photographs. The proposal, however, proved to be ahead of its time, for we quickly found that some 200 of the species had never been photographed.

The seed, however, was sown for another idea—to lay the foundations for a permanent national collection of bird photographs in which all the birds of Australia would in time be represented. We believed that participation in its creation would appeal to all bird photographers and that it would become a valuable asset for the community. A wide measure of support from many quarters followed, and the project became operational as the National Photographic Index of Australian Birds (later Wildlife) on 3 June 1969 under the wing of the Australian Museum, Sydney, in association with the National Library of Australia, Canberra. By the early 1980s, many thousands of photographs of Australia's birds had been obtained, and today only three species remain to be photographed: the elusive Night Parrot, the secretive Buff-Breasted Button-quail, and the visiting Black-winged Monarch from New Guinea.

With such a wealth of photographs to choose from, our 1966 concept of a book became a viable proposition and in 1982, after retiring from the Index, we decided to embark on its production. Thanks to the cooperation of the Index staff and the contributions of 118 photographers, the 864 photographs reproduced in the book represent the most extensive photographic coverage of the birds of this country ever presented in a single volume. Particularly notable is the generous allocation given to photographs exemplifying the great diversity of plumage variations related to sex, race, growth and season.

In planning the organisation of the book, account was taken of the needs of the many people who are interested in birds but do not have the expert knowledge to identify them. In 1931, Neville W. Cayley, in his *What Bird Is That?*, responded to this widespread problem by departing from the technical classifications of the Systematic List and grouping the birds in a series of chapters according to habitat—to achieve, in his own words, 'the simplest means of identification'. Our own response was to divide the book into a number of sections in which the birds are grouped according to common attributes that are at once recognisable in everyday terms.

A further aid to identification lies in the assembly of photographs in the form of composite plates so that the birds can be directly compared. This proved to be no mean task as it involved drawing every bird by hand to a realistic scale in the exact position required to provide a matrix for the guidance of the plate-maker. An exceptional degree of colour control was also exercised during the proofing stages to ensure that the printed image reproduced the plumage of the bird as faithfully as possible.

For us, the book is the culmination of an association of over twenty years with bird photography. Its making has taken five years, and has involved the cooperation of many people. Our debt to them is profound.

Canberra
June 1989

INTRODUCTION

The Birds

The Australian Region Australia forms part of the Australian Region — together with New Guinea, the Moluccus, New Zealand and Polynesia — and was so named by Alfred Russell Wallace in 1876 in his classic book *The Geographical Distribution of Animals* as one of the six zoogeographical regions of the world (defined inside the front cover). In this great remote region of the earth, the flora and fauna has evolved in isolation, and the wildlife is so distinctive that it has the character of a separate animal kingdom. Numerous groups of its birds are found nowhere else, and many are of outstanding beauty and interest. They include, for instance, the Emu, the cassowaries, the mound-builders, the Plains-wanderer, the cockatoos and lorikeets, the owlet-nightjars, the lyrebirds, the scrub-birds, the fairy-wrens, scrubwrens and thornbills, the Australian chats, the sittellas and tree-creepers, the butcherbirds and allies, the birds-of-Paradise and the bowerbirds. But no group characterises the region more than the brush-tongued honeyeaters, whose abundance and variety is linked with the profusion of blossom-bearing trees that yield the nectar on which many of these birds feed. More numerous than in other regions are its pigeons and doves, frogmouths, kingfishers, cuckoo-shrikes, trillers, Australian robins, whistlers, flycatchers and woodswallows.

Australia Nineteen of the world's twenty-five orders, comprising eighty-one families, are represented in Australia. Estimates of the number of species varies but, on the basis of H. T. Condon's *Checklist of the Birds of Australia, Part 1 Non-Passerines* (1975) and R. Schodde's *Interim List of Australian Songbirds, Passerines* (1975), there are 726 species, including twenty-six introduced species. Of these, 393 are non-passerines and 323 passerines. Adding thirty-two species listed as uncommon and vagrant species in *The Atlas of Australian Birds* published by the Royal Australasian Ornithologists Union (1984) and not included in the Condon and Schodde lists, the number of species recorded in Australia and surrounding seas totals 758 — of a world total estimated at 8650 species. The 600 resident species have adapted to a great variety of climatic conditions in this vast continent, from the luxuriant rainforests in the wet monsoon belt in the tropical north to the harsh desert areas inland where rain may not fall for long periods and where temperatures may soar to 45°C or more. Nearly 60 per cent of them are endemic.

The Book

The species Represented in this book are all the birds that breed in Australia (600) and the most regularly recorded visitors (77) — total 677 species. Also included are over 150 examples of plumage diversity relating to race, sex, growth and season.

The photographs There are photographs for all the species except for the Night Parrot, the Black-winged Monarch, the Buff-breasted Button-Quail and the Lesser Sooty Owl. Paintings are used for the first two; the quail and the owl are paired with their close relatives, the Chestnut-backed Button-Quail and the Sooty Owl respectively. In all, 864 photographs are employed in the make-up of this book and they are the pick of many thousands viewed. They were selected primarily for their value in showing the diagnostic features of each bird, but always with an eye to their aesthetic appeal. Each Section is identified by one or more full-page photographs typifying the birds in the Section.

Layout of the illustrated plates Photographs of look-alikes are placed together for comparison whenever possible and, to preserve a sense of orientation, western and eastern birds are placed left and right respectively and the northern and southern birds at the top and bottom. The photographs are scaled, within the limits available, to a size relationship between the birds on a given page.

The text The Species Notes are paired with the photographs on the opposite page and are written to a pattern throughout the book so that data complementary to the illustration is always in the same place. Thus, size will be found at the top right, with information on sex, the immature bird and voice in a coda at the bottom in italics. The text is in the form of a brief narrative about the bird and its life style, together with any points of interest, including historical references. When known, the sex or maturity of the bird in the photograph is given in bold type in the bottom left-hand corner. The primary purpose of the Introduction to each Section is to provide a descriptive framework to the families represented in the Section.

Distribution maps The distribution maps were originally produced by Arnold McGill before publication of *The Atlas of Australian Birds* (RAOU) in 1984, but were brought into harmony with the *Atlas* before going to press. Stippling, notably in Section 2, represents areas of occasional sightings.

Names The English names given follow the precept quoted and adopted in *A Field Guide to the Birds of Britain and Europe* (1954) by Roger Tory Peterson 'that the names in popular usage must always be the ultimate criterion'. In cases where two names vie for popularity, both are given, the alternative in brackets — with precedence given to names that are descriptive, short and in English (the name Jabiru excepted as it has been in use since 1790). When, on occasion, this results in a name recommended by the Royal Australasian Ornithologists Union being treated as an alternative, it is distinguished in square brackets. The scientific names, for the most part, are those currently used in the Union's literature.

Voice Descriptions of voice in the Species Notes perforce are limited by space to brief clues. These have been based partly on personal experience, Ernest Hoskin's contributions for the birds of New South Wales, and certain abbreviations used in books such as *A Field Guide to the Birds of Australia* by Graham Pizzey and the Readers Digest's *Complete Book of Australian Birds* that appealed, for which acknowledgement is readily given.

The voices of Australian birds are full of contrasts, varying from loud calls of the cockatoos, kookaburras and the Channel-billed Cuckoo, to the elfin trills of the fairy-wrens. The song of some is as lovely as those heard anywhere — especially notable being that of the Grey Shrike-thrush, the Pied Butcherbird, the Clamorous Reed Warbler and the Brown Honeyeater. Over fifty Australian birds have been recorded as mimics, notably the two lyrebirds whose gift of mimicry has no equal. Some too are clever ventriloquists, such as the Crested Bellbird and the Golden-headed Fantail Warbler.

The enjoyment of birds is immeasurably enhanced by familiarity with their calls, and once a good ear for them has been cultivated, voice alone can become a decisive aid to identification. There is no short cut, but using a tape recorder can help. Prerecorded tapes and discs are available from a number of sources some of which are listed on page 225.

Organisation The book is organised in two main divisions — Part I covering the land birds and Part II the aquatic birds — and is subdivided into 21 Sections. These are autonomous — each with its own introduction, index and credits to the photographers. Their make-up is easily identifiable from the self-explanatory titles. Families are mostly intact within the Sections, and the conventional running order of the Systematic List is observed wherever possible in the first part of the book.

Indexes The species indexes of English names in the Sections are complemented by two general indexes at the end of the book, one giving English names, and the other scientific names. The Contents list defines the groups of birds encompassed by each Section. A key is provided to the distribution of family groups among the Sections on page ix.

KEY TO THE DISTRIBUTION OF FAMILIES BASED ON THE SYSTEMATIC LIST

Passerines and Non-passerines. The order Passeriformes represents about three-fifths of all living birds. Its members are known as passerines or perching birds. They occupy the top of the evolutionary tree, and are characterised by a number of distinctive anatomical features, especially by their four-toed feet in which the hind toe is the longest and strongest, and which curls forward to meet the front toes to secure a firm grip when perching. (Non-passerines have a variety of very different feet.) Passerines also have unique muscular arrangements of the voice box (syrinx) which have endowed many of them with the gift of song. All the world's songbirds are passerines, but some are songless and others are possessed of anything but a musical voice.

CONTENTS

Part I
LAND BIRDS
Birds of the bush, forest and desert

Part II
AQUATIC BIRDS
Birds of the inland waters, shore and ocean

Overleaf: Little Corellas

PART ONE

LAND BIRDS

BIRDS OF THE BUSH, FOREST AND DESERT

1 Flightless Birds

Of the four families of flightless birds comprising the order Struthioniformes, two – the Emu and the Cassowary – are represented naturally in Australia, the only country where there are two families of still-living giant flightless birds. The Ostrich was introduced into South Australia from Africa in the 1880s for plume farming and small numbers of feral birds are still seen as a reminder of this unsuccessful venture. The Rhea of South America is unrepresented. These birds, together with New Zealand's dwarf Kiwi and its probable twenty-five species of extinct Moas, are known as ratites – so named because of their flat breastbones which have no keels. The feathers of the Emu and the Cassowary are double, with a secondary feather (or aftershaft) nearly as long as the main feather sprouting from the base of the quill. Their structure is also loose, since they lack the interlocking hooks (barbicels) which stiffen the feathers of most birds.

EMU Family Dromaiidae

The Emu is the world's tallest living bird after the Ostrich. It graces Australia's coat-of-arms in company with the Kangaroo. The first specimen was shot in February 1788 near where Sydney's Central Railway Station now stands, and was exhibited as a stuffed specimen in John Hunter's Museum in Leicester Square, London. The Emu owes its name to Oliver Goldsmith's *History of the Earth and Animated Nature* (1774) in which a similar bird, the Rhea of South America, is called 'Emu' (derived from the Portuguese 'Ema' then applied to large birds of ostrich-like appearance). This book was apparently carried with the First Fleet, for Captain Watkin Tench quotes 'Goldsmith's account' when describing the new bird in his *Narrative* of 1788. He refers to it as 'a species of ostrich, approaching nearer to the emu of South America than any other we know of', and says it was 'shot at a considerable distance, with a single ball, by a convict employed for that purpose by the Governor'.

Populations of the Emu in Tasmania and the Dwarf Emu of King Island and Kangaroo Island became early casualties of European settlement. They were recorded in vast flocks on Kangaroo Island in 1803 on Baudin's expedition by the zoologist Francois Péron. Fossil remains have also been found of three other Emu species.

Emus can do great damage by trampling down crops, and prodigious numbers were destroyed and eggs taken early this century to keep the numbers in check. In 1932, the Army, equipped with Lewis machine guns and 10 000 rounds of ammunition, was sent into the wheatbelt of Western Australia to deal with the situation. The birds scattered at the first burst of fire and the so-called 'Emu War' is remembered as an hilarious fiasco. However, to their credit, they eat great numbers of harmful insects; some 3 000 large caterpillars have been found in the stomach of a single bird. Today, the Emu is protected in most areas. 1 and 3

CASSOWARY Family Casuariidae

The stronghold of the heavy-bodied Cassowary lies in New Guinea and its adjacent islands where there are three species and many different forms. The range of one of these species (the Two-wattled Cassowary) extends southwards into north-eastern Queensland. 2 and 4

PHOTOGRAPHERS:
C.B. Frith 2;
P. Knowles (PLA) 3;
H.J. Pollock 1, 4

1, 3 EMU

Dromaius novaehollandiae 2 m

The Emu is widespread wherever there is woodland and scrub near water but it avoids settled areas and rainforest. It is highly nomadic, and flocks of up to 500 may migrate over great distances in times of drought. An Emu's strong 3-toed legs can carry it in bursts approaching 50 km/h. It feeds largely on plant food, native fruits, and insects when available especially during grasshopper plagues. It is insatiably curious and, though shy, can be lured by displaying unfamiliar objects. Breeds usually in winter. Nests on ground. Clutches average 5–11 eggs, sometimes as many as 20, but possibly as the result of two females laying in the same nest. Newly-laid eggs have a blue patina but this rubs off during incubation to expose a deep green shell. The chicks can be heard 'cheeping' in their shell within 1 or 2 days of hatching. The male incubates the eggs for about 8 weeks and may lose 4–8 kg in weight. It also rears the chicks and may stay with the brood for up to 18 months. Meanwhile, the female may take another mate. **Sexes**: *similar but female often larger*. **Immatures**: *greyer and paler*. **Voice**: *male: guttural grunts; female: resonant drummings*.

Below and top right

2, 4 [Southern] CASSOWARY

Casuarius casuarius 1.5 m

In Australia, the Cassowary is confined to two isolated populations in the rainforests at all altitudes of north-east Queensland – with occasional visits to canefields and adjoining parks and gardens. It tends to be solitary and no more than six birds have been seen together. It feeds mainly on fallen native fruits and berries. It raids orchards and will take fruit from people. It is wary and can retreat at prodigious speeds even in thick bush – through which it charges probably using its bony helmet or casque as a battering ram. It sometimes swims. It fights by leaping feet first at an enemy slashing it with the vicious sharp spikes on the inner claw of its 3-toed feet. It has been known to kill people when provoked. The male incubates the eggs and rears the chicks while the female roams in search of another mate, laying a clutch of 3–5 eggs two or three times in a season. Breeds June–October. Nests are lined with a few leaves in a ground scrape and are hard to find. **Female**: *larger*. **Immatures**: *brown, head and neck duller*. **Voice**: *low rumbles, booms, hisses – and loud roars when fighting*.

Below and bottom right

1

2

3

4

2

Diurnal Birds of Prey

Hawks, eagles and their kin comprise the order Falconiformes and are found throughout the world, except Antarctica. All have sharply hooked down-curved bills for tearing flesh and powerful taloned feet for grasping prey. Collectively they are known as raptors. They are also referred to as 'diurnal' (active in daylight) to distinguish them from the owls, although the Letter-winged Kite is a truly nocturnal hunter and many other birds in this group, like the Little Falcon, are also crepuscular. The female of most raptors is larger than the male, some considerably so like the Collared Sparrowhawk where the female is almost double the weight of the male. Some females also have proportionately larger beaks and feet. Of a world total of nearly 300 species, twenty-four occur in Australia. Three of the five families are represented; not present are the Secretary Birds and the Vultures.

EAGLES, OSPREY, KITES, GOSHAWKS AND HARRIERS Family Accipitridae

This diverse family of eighteen Australian members, from a world total of over 200, divides conveniently into seven groups.

EAGLES

This diverse and heterogeneous group of ninety species divided among twenty-nine genera is represented in Australia by only three species.These are the Wedge-tailed Eagle and the Little Eagle — which have fully feathered, or booted, legs and characterise the 'true eagles'— and the White-bellied Sea-Eagle which has bare legs. The Wedge-tailed Eagle is Australia's only member of a group of nine large powerful eagles belonging to the genus *Aquila*, the name given by the Romans to the famous Golden Eagle adopted bythem as a symbol of their power. It was popularly believed that Wedge-tailed Eagles were responsible for serious lamb losses and State Governments once offered bounties for their destruction.Today it is known that they rarely kill healthy lambs, and that losses from this cause are more than offset by the benefits derived from the eagles' large scale predation of rabbits. Wedge-tailed Eagles are now protected in all States.The Little Eagle is the Australian representative of the widespread *Hieraaetus* genus.It captures more birds than its larger relative the Wedge-tailed Eagle. The powerful White-bellied Sea-eagle (genus *Haliaeetus* is a great pirate, stealing from other raptors, but is also a capable and opportunistic hunter.
1–4, 6–7

OSPREY

The world's most widely distributed raptor after the Peregrine Falcon. It has anatomical features which set it apart from other raptors (see Species Note) and is placed in a family of its own. 5

SOARING KITES

The kites in this group are medium to large in size and soar effortlessly on rising air currents. There are five Australian species. 8–9, 12–17

HOVERING KITES

Elanid kites are small grey and white raptors that gracefully hover before parachuting down to catch the rodents on which they specialise. Of a world total of four species, Australia has two, the Black-shouldered Kite and the Letter-winged Kite. 10–11

GOSHAWKS AND SPARROWHAWKS

The genus *Accipiter*, comprising about fifty species worldwide, is the most widely distributed of all raptorine genera. Three are represented in Australia – the Brown Goshawk, the Grey Goshawk and the Sparrowhawk. Typically, they have rounded short wings, long tail and legs and often hunt from concealment, dashing out at their unsuspecting victims. The rare Red Goshawk (genus *Erythrotriorchis*) is typified by long wings and small erectile crest; it is also sufficiently different in other respects to be in a genus of its own. 18–22

HARRIERS

These long-legged, long-winged raptors typically fly low, quartering the ground over swamps and grasslands. They have a characteristic owl-like facial ruff which collects sounds from the ground and helps them locate prey, unlike other raptors which depend more on vision. Of seventeen species worldwide, Australia has two – the Swamp Harrier and the Spotted Harrier. 23–24

BAZAS

The local example of this genus is the Crested Hawk. It is also known as the Pacific Baza and the Cuckoo Falcon. Baza is a latinized Hindu word meaning goshawk. There are several mainly tropical Bazas. 27

FALCONS Family Falconidae

Of a world total of sixty-one species of this widely distributed family, six are represented in Australia. These range from the swift-flying Peregrine Falcon – probably the world's most famous bird of prey and the most widely distributed of all birds – to the small, hovering Kestrel. Typically, falcons are powerful hunters with long pointed wings – they outfly and stoop (dive) at their prey, often reaching very high speeds. A hobby, one of the smaller falcons, has been seen to overtake and seize a swift at probably close to 160 km/h. In England in the fifteenth century when falconry was the sport of the nobility, the penalty for taking a falcon was two years imprisonment or a year and a day for collecting an egg. 25–32

Opposite: Peregrine Falcon

PHOTOGRAPHERS: D. Baglin 4; H. & J. Beste 8, 12, 19; W.H. Butler 9; C.A.C. Cameron 16, 27; G. Chapman 17, 22; J. Cupper (BOC) 18, 21, 31; T. & P. Gardner 5, 13, 23, 24; A.L. Hertog 15; D.G.W. Hollands 28; K. Ireland 30, 32; G. May 20; M. & I. Morcombe 3; P. & J. Olsen 29; H.J. Pollock 2; J. Purnell 26; G. Steer (NPI) 1; M. Wright 11; Authors p. 7; 6, 7, 10, 14, 25

1, 2 WEDGE-TAILED EAGLE

Aquila audax ♂ 90cm ♀ 100cm

The noble Wedge-tailed Eagle is Australia's largest bird of prey, with a wing-span of up to 2.5 m. When soaring on upswept 'fingered' wings, it is easily recognised by its distinctive wedge-shaped tail, as illustrated. It is often seen at great heights of 2 000 m or more riding thermal currents. Singly, in pairs or loose groups, it frequents a variety of habitats from wooded mountain slopes to treeless plains, and hunts mostly in open areas. Its preferred diet is rabbits, but it also takes young kangaroos, wallabies, reptiles, birds and carrion, including dead lambs. Sedentary; immatures may be nomadic. Also occurs in southern New Guinea. Breeds May–November. Nests on sometimes huge platform of sticks, usually in a tree, occasionally on a cliff or on the ground. **Sexes**: *similar*. **Immatures**: *usually paler*. **Voice**: *thin, high whistling notes and yelps.*

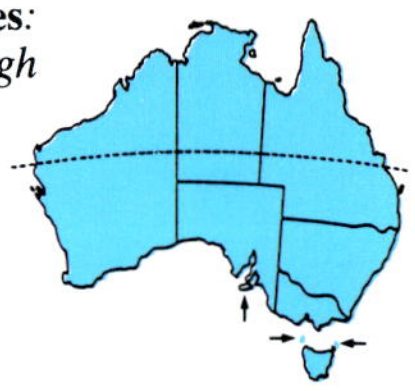

1

2

3

3, 4 WHITE-BELLIED SEA-EAGLE

Haliaeetus leucogaster ♂ 75cm ♀ 85cm

This magnificent fish-eating eagle is seen singly, in pairs or family parties along inland rivers, lagoons, lakes and reservoirs, in coastal areas and on nearby islands. It soars high, its broad rounded wings and short rounded tail giving it a distinctive silhouette. Besides fish, which it snatches from near the surface of water, it will take large birds, turtles, flying foxes and rabbits and eats carrion such as dead lambs. It likes perching on an exposed tree or headland. Mostly sedentary; immatures nomadic. Also occurs from New Guinea to India. Breeds June–December. Nests in huge stick nest in a tree, on rocks, or on the ground. **Sexes**: *similar*. **Immatures**: *mottled brown with whitish bulls-eye in wing*. **Voice**: *goose-like honks*.

4

Adult above, immature right

2

5 OSPREY

Pandion haliaetus ♂ 55 cm ♀ 65 cm

Named after Pandion the mythical king of Athens who was turned into an Osprey. The most specialised for fishing of all raptors, the Osprey has a number of anatomical peculiarities which assist it in grasping its slippery prey. These include a reversible outer toe as in the owls, long needle-sharp talons, and spines on the soles of its feet. A spectacular hunter, it patrols a stretch of water, hovers briefly, then plunges into the water feet first, sometimes disappearing beneath the surface to burst upwards again with a fish in its talons. So vice-like is the bird's grip on a fish that it sometimes has difficulty in releasing it, and there have been a number of instances of an Osprey being dragged under the water and drowned by a fish too heavy to be carried off. In Australia, it frequents the shores, estuaries and islands particularly of the north, singly, in pairs or small family parties, and is occasionally seen inland along large rivers. It has characteristically angled, narrow wings and a conspicuous dark line from its beak through its eyes to its nape. Often perches prominently. Adults sedentary. Breeds April-September. Nests coastally in a sometimes massive stick nest on cliff face, rocky shore, dead tree, or man-made structure. The eggs — cream, boldly blotched and spotted with rich brown and chocolate — are among the most beautiful of all eggs. They were greatly prized by egg collectors in former days before robbing birds' eggs was outlawed. **Sexes**: *similar.* **Immatures**: *more streaking; heavier chest band.* **Voice**: *plaintive whistles.*

Female left, male right

6,7 LITTLE EAGLE

Hieraaetus morphnoides ♂ 48 cm ♀ 55 cm

The Little Eagle frequents open forest and woodland, mostly singly or in pairs. It may be distinguished from other similar birds of prey by its flat, broad wings, short square tail, and feathered legs. There are two forms: a light and a dark morph, the former the more common. Its head feathers are erectile (as illustrated in the right photograph). It will, on occasion, eat carrion, but prefers live rabbits and other small animals, birds, reptiles and insects. Sedentary or partly nomadic. Breeds August-December. Nests 5-45 m high. Builds its own nest or adds to an old nest of another species. Also in New Guinea. **Sexes**: *similar.* **Immatures**: *mostly chestnut.* **Voice**: *double or treble high-pitched whistle.*

Adult left, juvenile right

8, 9 BRAHMINY KITE (Red-backed Sea-Eagle)

Haliastur indus ♂ 48 cm ♀ 51 cm

In Australia, this striking, rich chestnut and white plumaged kite frequents northern coastal areas, particularly estuaries, harbours and offshore islands. Often perches conspicuously in a tree. Tolerant of approach. Sometimes flocks with other birds of prey, especially the Whistling Kite. It is a scavenger, congregating to feed on scraps, rubbish and dead fish. It also sweeps low in slow wheeling flight, snatching fish, small animals and crabs; or insects from the air, foliage or the ground. Adults usually sedentary. Breeds April—October. Bulky stick nest in woodland or mangrove tree. Range extends through New Guinea and South East Asia to India. **Sexes:** *similar* **Immatures:** *mottled brown* **Voice:** *repeated 'pee-ah-h-h'.*

Adult left, juvenile right

5

6

7

8

9

2

10 BLACK-SHOULDERED KITE

Elanus axillaris ♂ 36cm ♀ 38cm

Frequents open woodland, grassland or cultivated areas, perching conspicuously, singly, in pairs or loose groups. It soars and glides with raised wings but is most often seen hovering waiting to drop swiftly on a mouse, insect or lizard. Distinguished from the Letter-winged Kite by yellow legs and less owl-like facial features. Mostly nomadic, some seasonal movement, very dependent on mouse numbers. Sometimes breeds twice a year March-May and September–November. Nests often high in thick foliage. Endemic. **Sexes**: *alike.* **Immatures**: *golden tan streaking on breast, head and back.* **Voice**: *harsh 'scree' or 'chip'.*

11 LETTER-WINGED KITE

Elanus scriptus ♂ 36cm ♀ 38cm

The black M or W marking on the underwing gives this kite its name. Distinguished from the Black-shouldered Kite by horn coloured legs and sometimes grey crown. Frequents grasslands and wooded creeks inland and feeds largely on rats and mice. Large flocks are drawn to exploit Long-haired Rat plagues. Nocturnal in habits, its large eyes adapt it for hunting through the night. Often roosts in flocks in daytime and is the only Australian raptor to breed in close colonies. Endemic. **Sexes**: *alike.* **Immatures**: *heavy golden tan streakings on chest, head and back.* **Voice**: *similar to Black-shouldered Kite.*

12 WHISTLING KITE (Eagle)

Haliastur sphenurus ♂ 53cm ♀ 58cm

A scavenger of carrion, sometimes in large numbers around refuse dumps. May rob feeding waterbirds, and hunts low after grass fires, often with other birds. Recognised by somewhat dropped wings and long, round-tipped tail, slow wheeling flight and effortless soaring. Perches alone or in family parties. Quiet and unaggressive. Nomadic during drought. Usually breeds June–November. Bulky stick nest in tall tree. Range extends from New Guinea to New Caledonia. **Sexes**: *similar.* **Immatures**: *more streaked and spotted.* **Voice**: *drawnout descending whistle followed by ascending burst of notes.*

13, 14 BLACK (Fork-tailed) KITE

Milvus migrans ♂ 52cm ♀ 55cm

This gregarious kite inhabits open areas and, as its scientific name suggests, it is a wanderer. Is often seen scavenging in large flocks, sometimes numbering hundreds, around slaughterhouses, refuse tips, stockyards and airfields. It also takes mice, rats and insects (appearing in force when plagues of these pests occur), and nestling birds. It will spiral above bush fires looking for victims. It characteristically twists its forked tail in flight and soars effortlessly. Nomadic. Usually breeds July–November. Nests are flattish and untidy, built in trees lining watercourses. A wide-ranging species in the Old World. **Sexes**: *similar.* **Immatures**: *paler.* **Voice**: *plaintive mewing or whining.*

15 BLACK-BREASTED BUZZARD

Hamirostra melanosternon ♂ 52cm ♀ 60cm

This robust bird is identifiable by its large white wing spots and very short tail. Singly, in pairs sometimes in family groups, it frequents open woodland in the tropical north and timbered watercourses in arid areas. It soars high and glides effortlessly over the tree-tops. Feeds mainly on live animals, such as small mammals and birds. Said to drop heavy objects into the nests of emus and bustards to break open the eggs. Breeds mostly June–December. Nests often on a dead branch near water. Endemic. **Sexes**: *similar.* **Immatures**: *more chestnut.* **Voice**: *hoarse, repeated yelp, harsh whining.*

16, 17 SQUARE-TAILED KITE

Lophoictinia isura ♂ 53cm ♀ 56cm

Unlike the Black Kite and Whistling Kite, this species is seldom seen in flocks, and is altogether a rarer and less well-known bird. It inhabits riverine forests and well-wooded areas near open country and feeds on a wide variety of insects and small animals, especially bird nestlings. It characteristically soars and glides just above the tree tops with its wings held in a V. Like the Black-breasted Buzzard, white spots are exhibited under its wings in flight. Seasonal movements and possibly nomadic. Breeds August–November. Bulky loose nests in high tree. Endemic. **Sexes**: *alike.* **Immatures**: *more rufous head, less streaked.* **Voice**: *hoarse quavering yelp, twitter, soft whistle.*

18 RED GOSHAWK

Erythrotriorchis radiatus ♂ 51cm ♀ 61cm

This rare and powerful Goshawk was sketched near Sydney by the convict artist Thomas Watling, but is now confined to northern Australia. Singly or in pairs, it frequents coastal rainforest and woodland. It soars high with 'fingered' wings spread wide. Long conspicuously barred underwing and tail are characteristic features. It preys on birds as large as cockatoos, kookaburras and ducks. Also takes reptiles and small mammals. Breeds April–November. Nests high in a tree. Endemic. **Sexes**: *similar.* **Immatures**: *unknown, perhaps more rufous.* **Voice**: *high-pitched chatter.*

19 BROWN GOSHAWK

Accipiter fasciatus ♂ 45cm ♀ 55cm

This widespread predator, seen singly or in pairs, frequents all types of woodland, forest or watercourse vegetation. On rounded wings, it spirals over tree tops hunting, and takes birds, rabbits, snakes and large insects. It also dives on victims from tree cover. Northern birds are paler and smaller; south-west birds more rufous. Breeds June–January. Nests often high in trees. Also occurs in New Guinea, Christmas Island and much of Indonesia. **Sexes**: *similar.* **Immatures**: *upperparts dark brown; underparts off white with bold streaking on breast.* **Voice**: *high-pitched chatter, more relaxed 'seeep, seeep, seeep'.*

20, 21 GREY (White) GOSHAWK

Accipiter novaehollandiae ♂ 42cm ♀ 54cm

This species can be either grey or white; the latter sometimes with a light grey wash. The two forms interbreed. Northern birds are smaller; only the white form occurs in Tasmania and predominates in the Kimberley region of WA. The White Goshawk, especially in flight, bears some resemblance to the Sulphur-crested Cockatoo, and could be confused with it. The species lives in woodland and heavily forested areas, particularly along watercourse vegetation. It feeds on small to medium-sized birds, rabbits and a variety of insects. It perches in ambush poised to dash suddenly after prey. When soaring, the rounded wings and tail are obvious, the latter somewhat shorter than that of the Brown Goshawk. Adults sedentary; immatures nomadic. Breeds June–December. Nests high in trees. Also in New Guinea, Melanesia and Indonesia. **Sexes**: *similar.* **Immatures**: *white form-white; grey form-wider barring, slight brown wash. They have yellow eyes which deepen to red in adults.* **Voice**: *loud, piercing 'queet', shrill chatter.*

Grey form left, white form right

10

11

12

13

14

15

16

17

18

19

20

21

2

22 COLLARED SPARROWHAWK

Accipiter cirrhocephalus ♂ 32cm ♀ 38cm

This agile hunter attacks at lightning speed, sometimes taking prey larger than itself, but predominantly birds such as finches and sparrows. Singly or in pairs, it frequents forest, woodland, mallee scrub, watercourse margins and even suburban parkland. Distinguishable from the Brown Goshawk by its slighter build and square tail. Sedentary; immatures nomadic. Breeds August–January. Nests usually high in tree. Also in New Guinea and adjacent islands. **Sexes**: *similar.* **Immatures**: *upperparts grey-brown, underparts off white, broad vertical streaks on breast.* **Voice**: *shrill chatter.*

23 SWAMP (Marsh) HARRIER

Circus approximans ♂ 57cm ♀ 61cm

This long-winged species frequents the wetter coastal areas amid swampy grass, reeds and croplands. Wary at its nest. It hunts low often with legs dangling, quartering the ground in search of small animals, eggs and carrion. Usually solitary or in pairs. Roosts with other harriers. Migratory in Tasmania; partially elsewhere. Also in New Zealand and from the New Guinea region to Fiji. Breeds September–January. Nest mainly on ground. **Recognition**: *Adults have a distinctive white rump.* **Female**: *slightly browner.* **Immatures**: *darker.* **Voice**: *high-pitched 'see-ah' and descending 'kee-aa'.*

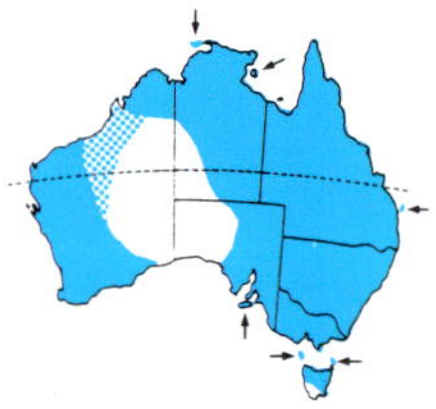

Immature Male

24 SPOTTED HARRIER

Circus assimilis ♂ 55cm ♀ 60cm

A bird of the inland plains, usually seen singly or in pairs. It is found mainly in the more arid areas in grassland, lightly timbered country, and cropland; only occasionally in wetter areas. It weaves lazily low over the ground, often with legs trailing, hunting a wide variety of small mammals, reptiles and ground birds, rarely eating carrion. Nomadic and partially migratory. Breeds June–November. Nests in low-medium trees (the only harrier to nest in trees). Also nests occasionally on telephone poles. Range extends to Indonesia. **Sexes**: *similar.* **Immatures**: *rufous brown.* **Voice**: *'kikiki' in nest defence.*

25 BROWN FALCON

Falco berigora ♂ 45cm ♀ 50cm

Sometimes known as the 'Cackling Hawk' because of its call and noisy habits. One of the few Australian birds where an Aboriginal name, *berigora*, forms part of the scientific name. Perhaps the most common and widespread of the falcons in Australia. Its plumage is variable in colour from light to dark and sometimes red, even in the same area. It frequents mostly lightly timbered country but is found almost anywhere except in closed forest. On upswept wings, it soars high, occasionally hovering before dropping onto prey. It is a generalist, eating reptiles, mice, young rabbits, small birds, insects and carrion. Adults sedentary, immatures nomadic, sometimes in loose groups when prey is abundant. Breeds June–November. Mainly uses stick nests of other species, especially corvids, very occasionally a broken off tree stump. Also in New Guinea. **Sexes**: *similar.* **Immatures**: *buff collar and edges to feathers, legs blue-grey.* **Voice**: *raucous cackles and squawks.*

Light form left, dark form right

26, 29 PEREGRINE FALCON

Falco peregrinus ♂ 42cm ♀ 50cm

The Peregrine Falcon occurs almost throughout the world. Its remarkable speed is legendary and it is said to exceed 250 km/h when stooping at prey. Victims are usually birds taken in mid-air. Its very presence causes universal panic among other birds. Favoured habitat is rocky hills, mountains and cliffs, but it may appear almost anywhere, even in urban areas. Sedentary; immatures nomadic. Breeds July–December. Nests on ledges, in tree hollows or stick nests of other birds. In Australia, the Peregrine Falcon has not suffered as grievously from the effects of pesticides as it has elsewhere in the world. It is especially vulnerable to DDT which causes thin eggshells and a serious decline in breeding success. **Female**:*heavier beak and thicker legs and toes.* **Immatures**: *browner than adults with dense black vertical streaking on underparts and sandy brown edges to feathers.* **Voice**: *loud screaming chatters 'kek-kek-kek' and whines.*

Male above, female below.

27 CRESTED HAWK [Pacific Baza]

Aviceda subcristata ♂ 40cm ♀ 45cm

The range of this, the only prominently crested hawk in Australia, extends to New Guinea and adjacent islands and the Solomon Islands. Seen alone, in pairs, family parties or flocks, it lives quietly and unobtrusively in well-wooded areas. It soars just above the tree tops and occasionally indulges in impressive tumbling display flights. Feeds on frogs, insects and small reptiles often gleaned or flushed from foliage. Sedentary and nomadic. Breeds September–December. Nests usually in a high tree. **Sexes**: *alike.* **Immatures**: *browner.* **Voice**: *mellow 'wee-choo', 'toc-toc-toc', other whistles and trills.*

28 BLACK FALCON

Falco subniger ♂ 50cm ♀ 55cm

Comparatively little known, it can be mistaken for the dark form of the Brown Falcon but, unlike the latter, which often carries its more rounded wings at a slight upward angle in flight, it glides with more pointed wings held horizontally. It prefers dry open country and only breeds there. A feared predator, it feeds on rabbits and ground birds, and will strike prey like galahs in the air. Quail are also part of its diet and it follows their movement. Breeds in spring. Usually re-uses nests of crows or other hawks. Endemic. **Sexes**: *similar.* **Immatures**: *legs, cere and eye ring blue-grey.* **Voice**: *'gak-gak'.*

30 LITTLE FALCON [Australian Hobby]

Falco longipennis ♂ 32cm ♀ 35cm

Hobbies, widespread in Eurasia, Africa and the Australian Region, look like miniature Peregrines and are noted for their extraordinary aerobatics and remarkable speed in level flight. Small birds, bats and insects are taken on the wing. They live in open wooded country, parks and gardens. Scarce in Tasmania. Populations in the inland arid areas are paler. Breeds September–January. Nests in high tree or pylon. **Adults**: *blue-grey* **Voice**: *high-pitched chatter 'kekekek' and twitters.*

Immature

31 GREY FALCON

Falco hypoleucos ♂ 36cm ♀ 43cm

This rare and elegantly-plumaged falcon is a bird of the vast plains of the arid interior, where it frequents open scrub and timber, especially tree-lined watercourses. It occasionally appears in the more humid areas. It soars high to drop on prey, or makes low-level dashes at its mainly bird prey. Sedentary; seasonal movement or nomadic. Breeds June–November. Nests in structures of other hawks and corvids. Endemic. **Sexes**: *similar.* **Immatures**: *upperparts darker, underparts more streaked.* **Voice**: *similar to Peregrine Falcon.*

32 NANKEEN [Aust] KESTREL

Falco cenchroides ♂ 33cm ♀ 35cm

Traditionally called 'Nankeen' from the colour of a buff-brown cloth once imported from Nanking. This small falcon mainly hunts by patiently hovering in mid-air ready to drop on an unsuspecting mouse, small bird or insect. Singly, in pairs or family parties, it frequents most habitats, particularly open woodland and cultivated areas. Ranges to New Guinea and Indonesia. Scarce in Tasmania. Breeds August–November. Nests variously on cliffs, buildings and hollow trees. **Female**: *tail and head chestnut.* **Immatures**: *like adult female.* **Voice**: *chatters, whines.*

Male

22

23

24

25

26

27

28

29

30

31

32

3

Nocturnal Birds of Prey

OWLS

The owl has fascinated man from time immemorial and it has both been regarded with superstitious dread and venerated for its supposed wisdom. The hypnotic stare of an owl and its habit of blinking only with its upper lids gives it an air of superiority and we still subscribe to the notion of its wisdom in the popular rhyme:

A wise old owl sat on an oak
The more he saw the less he spoke:
The less he spoke the more he heard;
Why aren't we like that wise old bird?

Edward Hersey Richards

Their superb gifts as night hunters set owls apart from all other birds and are due to their extraordinary vision and hearing, and to their utterly silent flight, achieved by a velvety pile covering their feathers. Their eyes are forward-facing, like ours, and give them stereoscopic vision which enables them to pin-point their prey with deadly accuracy. Their bills even point downwards so as not to obstruct their line of sight. The eyes themselves are marvellously constructed to collect the maximum amount of light – estimated at more than thirty-five times better than ours. They have super-sensitive retinas and the huge eyes are so specialised that they cannot swivel. Owls, therefore, have to turn their entire heads to shift their gaze and, being endowed with 'rubber' necks, they can wind their heads backwards in either direction through an arc of up to 270°. An old saying has it that if you walk around an owl on a post it will twist its head off! An owl is able to scan for sounds by moving its large ear flaps and the edges of its facial discs. The exceptionally large ear openings are unequal in size and one is a little higher than the other. These adaptions give owls such acute hearing that their ability to locate prey in the dark is astonishing – the squeak of a mouse or rustle of a leaf can bring a Masked Owl down on its victim in absolute darkness, a time when owls cannot see, despite popular belief.

HAWK OWLS Family Strigidae

Of a world total of 134 species, only four occur in Australia. All belong to the genus *Ninox*, a group of sixteen species, collectively known as Hawk Owls, which extends to India, Japan and New Zealand. Vision is most acutely developed in the hawk owls and their hunting technique is to quarter the ground systematically like the hawks searching for prey. For this reason, they are usually seen abroad earlier than the Masked Owls. Feet are feathered. Females are larger than males. 1–7

MASKED OWLS Family Tytonidae

Of a world total of twelve species, five occur in Australia. They are characterised by their large heart-shaped facial masks and dark eyes, and are sometimes known as 'monkey-faced owls'. They have short tails, long feathered legs and a serrated grip under the middle claw. Their eyes tend to be smaller and perhaps catch less light than those of the Hawk Owls, but their hearing is more acute and is highly directional. Masked Owls have voracious appetites and expel large coated pellets. It has been estimated that a Barn Owl can eat over 1 000 mice a year and is worth twelve cats in a farmyard. 8–12

NIGHTJARS AND FROGMOUTHS

All this group of night birds (order Caprimulgiformes) have huge mouths, small legs and weak feet. Three of its five families are represented in Australia.

NIGHTJARS Family Caprimulgidae

Nightjars were known in America in the 18th century as 'Goatsuckers'. This stems from an old belief that they milked goats at night and made them blind, and the legend persists in the scientific name which is based on the Latin word *caprimulgus*, milker of goats. Nightjars are noted for their loud persistent, monotonous calls which acutally do jar. They fly open-mouthed sieving the air for prey – aided by their facial cat-like whiskers. By day they squat quite openly among ground litter relying on their superb camouflage and will only flush when almost trodden on. The suddenness of their take-off can be quite startling as they burst upwards ghost-like in swift zigzag noiseless flight. When breeding they avoid detection by laying eggs that blend with the ground and, as there is no nest, these are sometimes moved during the incubation period. The remarkable reflection from a nightjar's eyes has been described as 'like twin coals of fire'. There are seventy-six species worldwide; three occur in Australia. 13–15

OWLET NIGHTJARS Family Aegothelidae

With huge forward-facing dark eyes, Owlet-nightjars are superficially like long-tailed owls. They also have characteristics both of the Frogmouth and the Nightjar, and are sometimes called 'Moth Owls'. They are found both in New Guinea, where there are six species, and in Australia where there is one species. Here they are the smallest of the night birds. 18

FROGMOUTHS Family Podargidae

These grotesque birds with their large frog-like mouths are masters of camouflage enabling them to spend their whole lives fully exposed to view and roost where the tree branches match their cryptic plumage. They feed mostly on the ground. They are poor fliers and have weak feet – hence their scientific name *Podargus* which is a Latinized word from the French for gouty foot. The bright yellow linings of their mouths are displayed menacingly when under threat. Of twelve species, which range through New Guinea to Malaysia, three occur in Australia, including the two largest. 16, 17, 19–21

Opposite: Barking Owl

PHOTOGRAPHERS: H. & J. Beste 2, 10, 11, 14; J. Frazier 3, 12, 19; T. & P. Gardner 13, 18; L. Robinson 1, 16; A.H. Tolhurst 5; B.A. & A.G. Wells 4; E.E. Zillmann p. 17; 6, 7, 15; Authors 8, 9, 17, 20, 21.

3

1, 2, 4, 5 BOOBOOK OWL [Southern Boobook]

Ninox novaeseelandiae ♂ 23–36cm ♀ 27–36 cm

The Boobook, smallest and most familiar of the owls in Australia, also occurs extensively in other territories of the Australian Region. The Australian populations vary greatly in plumage and size both geographically and ecologically. As many as 10 races were recognised in earlier days, but Schodde and Mason, in their *Nocturnal Birds of Australia* (1980), reduced them to four, comprising two main populations extending over the continent divided on a line following the western footslopes of the Great Dividing Range with the nominate race *boobook* (A) to the east and race *ocellata* (B) to the west. They note that the western birds are of a more sandy red hue than the eastern birds living in similar country; also that the western birds are smaller and paler in the north (in accordance with Bergmann's Rule), while the trend in the east is reversed. However, an isolated population is identified in the dark and humid rainforest of the high country of north-eastern Queensland, race *lurida* (C), which is smaller. It is also darker (in accordance with Glober's Rule). The Tasmanian birds, race *leucopsis* (D), are even smaller and conspicuously spotted. They have golden-yellow eyes and pale yellow toes like their New Zealand cousins, and some migrate to the mainland as far as Sydney to winter. The Boobook lives in a wide variety of habitat but is particularly partial to eucalypt forest and woodlands. It hunts mostly at nightfall and before dawn, and preys on mice, sparrows and other small birds. It can sometimes be seen diving after moths and beetles under street lights in suburban areas. It draws in its feathers under threat and turns sideways to present a thin profile. Small birds will mob a Boobook unmercifully if it becomes conspicuous during its daytime rest. It may attack passers-by when caring for young, but can also be playful. Breeds August/September—December/January. Nests in the hollow of a dead tree 3—25 m above ground. **Sexes:** *females slightly larger and more richly coloured.* **Immatures:** *paler than adults.* **Voice:** *its leisurely two-syllable call which, though soft in tone, can carry for several kilometres, has been variously described as 'mo-poke', 'boo-book', 'buk-buk', 'wook-wook' and 'moop-oop'.*

Illustrations: Race (A) 2 (taken in Victoria) and 5 (taken in NSW); Race (B) 1 (taken in the Kimberley) and 4 (taken further south). [Races (C) and (D) not illustrated]

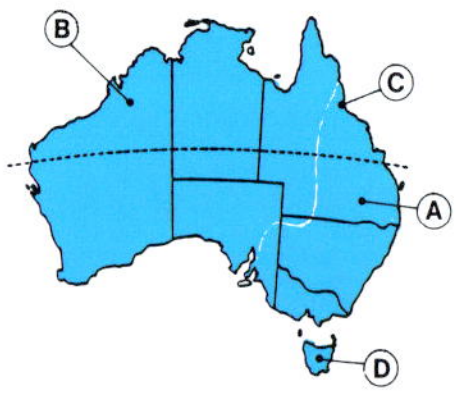

3 RUFOUS OWL

Ninox rufa 50 cm

The Rufous Owl, widespread in New Guinea, is uncommon in Australia and restricted to the tropical north. Size and plumage tonings vary slightly, the northern birds being the palest. All live in rainforest and its fringes. Sometimes seen in the quieter garden suburbs of Darwin. It is a wary unobtrusive bird, roosting by day singly or in pairs in its 'home' tree, and feeding on tree-dwelling mammals such as flying foxes, sugar gliders, small birds and insects: Sedentary. Breeds winter and spring. Nests in tree hollows 8–40 m high. **Female**: *slightly smaller and darker.* **Immatures**: *like adults.* **Voice**: *soft low 'woo-hoo'.*

6 BARKING OWL

Ninox connivens 43 cm

Also called 'Winking Owl' (but all owls wink). (A) race *connivens* is the darker; (B) race *peninsularis* is more variable. The most hawklike of Australian owls, relying almost wholly on sight for hunting. Frequents woodland and open forest. Prey includes mammals, quite large birds, and insects taken on the wing. Pairs and family parties use several established roosting trees. Sedentary. Breeds July–October. Nests in tree hollow to 30 m high. Also from New Guinea to the Moluccas. **Sexes**: *similar.* **Immatures**: *like adults.* **Voice**: *loud barking 'wook-wook'; and blood-curdling 'screaming woman' calls.*

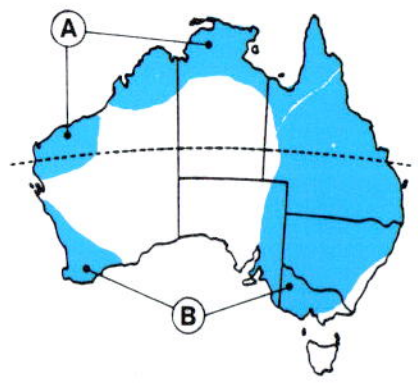

7 POWERFUL OWL

Ninox strenua ♂ 65 cm ♀ 54 cm

Australia's largest owl, with a wingspan of over 1 m. Its favoured habitat is the dense eucalypt forests of the high country. It uses a series of roosting trees within a selected area in which it hunts. It preys on possums, gliders, quite large birds such as kookaburras and currawongs, and a large variety of other mainly arboreal creatures. Most hunting is by evening and dawn light. It lives in the same territory year after year. Breeds in winter from May. Nests in a tree hole above 15 m. Endemic. **Sexes**: *similar.* **Immatures**: *back paler, smaller chevrons on breast.* **Voice**: *drawn-out double-note 'whooooo-hoooo'.*

8, 9 MASKED OWL

Tyto novaehollandiae

Mainland races: ♂ 33–41 cm ♀ 38–46 cm
Tasmanian race: ♂ 35–42 cm ♀ 43–57 cm

(A) race *novaehollandiae*; (B) race *kimberli*; (C) race *castanops*; (D) race *melvillensis*. The Masked Owl is the largest and most powerful representative of the genus *Tyto* in Australia and the female of the Tasmanian race is the largest in the world. The dark Tasmanian birds are still reasonably common, but their mainland relatives have declined alarmingly. In general, the northern race is the palest but there are also plumage variations from dark, through intermediate to pale within each race. The Masked Owl feeds mainly on terrestrial mammals up to the size of rabbits. The giant Tasmanian bird has a voracious appetite, and throws out huge pellets. When on the defensive, it spreads its wings downwards, rocks from side to side, snaps its beak, hisses, fixes the intruder with a hostile glare – and makes a truly fearsome spectacle. The Melville Island birds are very small. Also in southern New Guinea. The hybrid of Lord Howe Island is from introductions of the Tasmanian and mainland races. Breeds variably, mostly March–May and October–November. Nests in tree hollow or cave. **Sexes**: *similar apart from size.* **Immatures**: *like adults.* **Voice**: *drawn-out, rasping or hissing screech.*

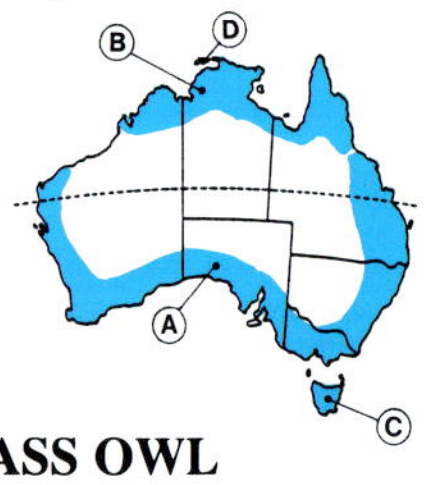

Tasmanian race left, mainland race right

10 (A) SOOTY OWL

Tyto tenebricosa ♂ 37—43 cm ♀ 44—51 cm

(B) LESSER SOOTY OWL

Tyto multipunctata ♂ 31—35 cm ♀ 35-38 cm

These elusive thick-set and robust owls have the largest eyes of any masked owl. They live in dense gloomy habitats, and are on the wing at dusk hunting for terrestrial and arboreal prey, including small birds. They breed opportunistically and nest very high in tree hollows. **Sexes:** *feet and bill of (A)'s female larger, plumages similar.* **Voices:** *weird swooping whistles — northern birds higher pitch.*

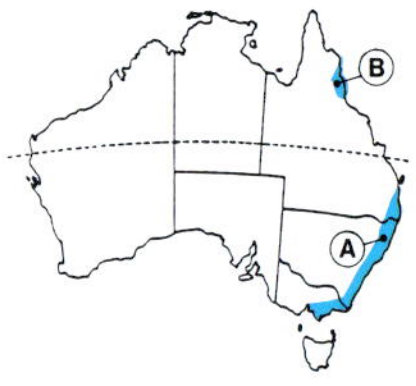

11 BARN OWL

Tyto alba 30–39 cm

This pale, nearly worldwide, owl is closely associated with man through its traditional use in the Old World of barn lofts and church steeples as nesting sites. It hunts harrier-like for ground-dwelling prey during the hours of darkness, using its remarkable gifts of hearing rather than sight. It is highly nomadic and populations fluctuate dramatically during periods of mice plagues. It can eat over 1 000 mice a year. By day it roosts unobtrusively in secluded places. Breeds variably. Nests in tree hollow, cave or old building. **Sexes**: *similar.* **Immatures**: *spotted below.* **Voice**: *eerie drawn-out screeching.*

12 EASTERN GRASS OWL

Tyto longimembris ♂ 32–36 cm ♀ 35–38 cm

This rarely-seen long-legged owl lives out its whole life on the ground and frequents grassland, swampy heath, canegrass or reeds. It feeds mainly on terrestrial rodents. Coastal birds are sedentary, but their distribution inland varies greatly with food supply – especially when there are plagues of native rats. Normally abroad at dusk, but will hunt by day when food is scarce. Breeds mostly March–June. Nests under dense deep tussocks. Ranges to Africa, southern Asia and western Pacific. **Sexes**: *similar apart from size.* **Voice**: *like Barn Owl but softer, deeper and less vocal.*

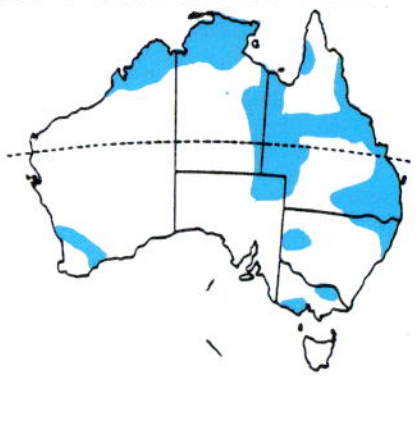

3

1, 2, 4, 5 BOOBOOK OWL [Southern Boobook]

Ninox novaeseelandiae ♂ 23—36cm ♀ 27—36 cm

The Boobook, smallest and most familiar of the owls in Australia, also occurs extensively in other territories of the Australian Region. The Australian populations vary greatly in plumage and size both geographically and ecologically. As many as 10 races were recognised in earlier days, but Schodde and Mason, in their *Nocturnal Birds of Australia* (1980), reduced them to four, comprising two main populations extending over the continent divided on a line following the western footslopes of the Great Dividing Range with the nominate race *boobook* (A) to the east and race *ocellata* (B) to the west. They note that the western birds are of a more sandy red hue than the eastern birds living in similar country; also that the western birds are smaller and paler in the north (in accordance with Bergmann's Rule), while the trend in the east is reversed. However, an isolated population is identified in the dark and humid rainforest of the high country of north-eastern Queensland, race *lurida* (C), which is smaller. It is also darker (in accordance with Glober's Rule). The Tasmanian birds, race *leucopsis* (D), are even smaller and conspicuously spotted. They have golden-yellow eyes and pale yellow toes like their New Zealand cousins, and some migrate to the mainland as far as Sydney to winter. The Boobook lives in a wide variety of habitat but is particularly partial to eucalypt forest and woodlands. It hunts mostly at nightfall and before dawn, and preys on mice, sparrows and other small birds. It can sometimes be seen diving after moths and beetles under street lights in suburban areas. It draws in its feathers under threat and turns sideways to present a thin profile. Small birds will mob a Boobook unmercifully if it becomes conspicuous during its daytime rest. It may attack passers-by when caring for young, but can also be playful. Breeds August/September—December/January. Nests in the hollow of a dead tree 3—25 m above ground. **Sexes:** *females slightly larger and more richly coloured.* **Immatures:** *paler than adults.* **Voice:** *its leisurely two-syllable call which, though soft in tone, can carry for several kilometres, has been variously described as 'mo-poke', 'boo-book', 'buk-buk', 'wook-wook' and 'moop-oop'.*

Illustrations: Race (A) 2 (taken in Victoria) and 5 (taken in NSW); Race (B) 1 (taken in the Kimberley) and 4 (taken further south). [Races (C) and (D) not illustrated]

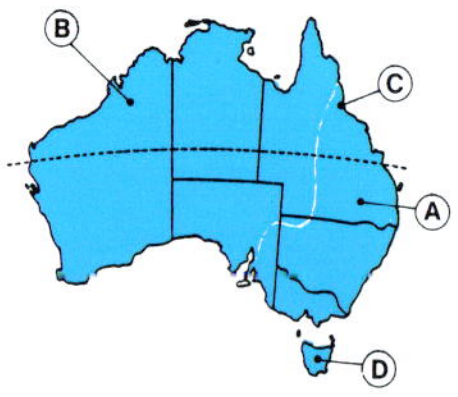

3 RUFOUS OWL

Ninox rufa 50 cm

The Rufous Owl, widespread in New Guinea, is uncommon in Australia and restricted to the tropical north. Size and plumage tonings vary slightly, the northern birds being the palest. All live in rainforest and its fringes. Sometimes seen in the quieter garden suburbs of Darwin. It is a wary unobtrusive bird, roosting by day singly or in pairs in its 'home' tree, and feeding on tree-dwelling mammals such as flying foxes, sugar gliders, small birds and insects: Sedentary. Breeds winter and spring. Nests in tree hollows 8–40 m high. **Female**: *slightly smaller and darker.* **Immatures**: *like adults.* **Voice**: *soft low 'woo-hoo'.*

6 BARKING OWL

Ninox connivens 43 cm

Also called 'Winking Owl' (but all owls wink). (A) race *connivens* is the darker; (B) race *peninsularis* is more variable. The most hawklike of Australian owls, relying almost wholly on sight for hunting. Frequents woodland and open forest. Prey includes mammals, quite large birds, and insects taken on the wing. Pairs and family parties use several established roosting trees. Sedentary. Breeds July–October. Nests in tree hollow to 30 m high. Also from New Guinea to the Moluccas. **Sexes**: *similar.* **Immatures**: *like adults.* **Voice**: *loud barking 'wook-wook'; and blood-curdling 'screaming woman' calls.*

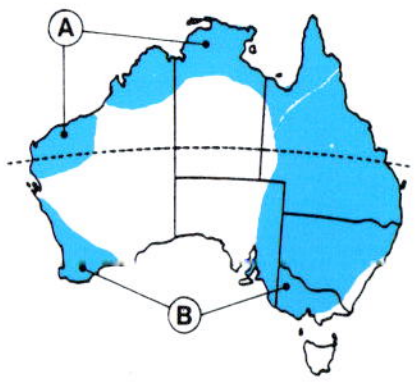

7 POWERFUL OWL

Ninox strenua ♂ 65 cm ♀ 54 cm

Australia's largest owl, with a wingspan of over 1 m. Its favoured habitat is the dense eucalypt forests of the high country. It uses a series of roosting trees within a selected area in which it hunts. It preys on possums, gliders, quite large birds such as kookaburras and currawongs, and a large variety of other mainly arboreal creatures. Most hunting is by evening and dawn light. It lives in the same territory year after year. Breeds in winter from May. Nests in a tree hole above 15 m. Endemic. **Sexes**: *similar.* **Immatures**: *back paler, smaller chevrons on breast.* **Voice**: *drawn-out double-note 'whooooo-hoooo'.*

8, 9 MASKED OWL

Tyto novaehollandiae

Mainland races: ♂ 33–41 cm ♀ 38–46 cm
Tasmanian race: ♂ 35–42 cm ♀ 43–57 cm

(A) race *novaehollandiae*; (B) race *kimberli*; (C) race *castanops*; (D) race *melvillensis*. The Masked Owl is the largest and most powerful representative of the genus *Tyto* in Australia and the female of the Tasmanian race is the largest in the world. The dark Tasmanian birds are still reasonably common, but their mainland relatives have declined alarmingly. In general, the northern race is the palest but there are also plumage variations from dark, through intermediate to pale within each race. The Masked Owl feeds mainly on terrestrial mammals up to the size of rabbits. The giant Tasmanian bird has a voracious appetite, and throws out huge pellets. When on the defensive, it spreads its wings downwards, rocks from side to side, snaps its beak, hisses, fixes the intruder with a hostile glare – and makes a truly fearsome spectacle. The Melville Island birds are very small. Also in southern New Guinea. The hybrid of Lord Howe Island is from introductions of the Tasmanian and mainland races. Breeds variably, mostly March–May and October–November. Nests in tree hollow or cave. **Sexes**: *similar apart from size.* **Immatures**: *like adults.* **Voice**: *drawn-out, rasping or hissing screech.*

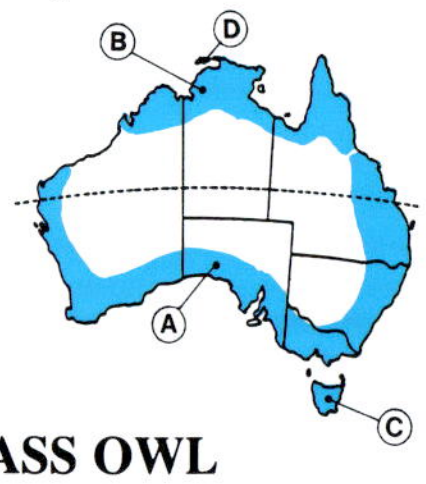

Tasmanian race left, mainland race right

10 (A) SOOTY OWL

Tyto tenebricosa ♂ 37–43 cm ♀ 44–51 cm

(B) LESSER SOOTY OWL

Tyto multipunctata ♂ 31–35 cm ♀ 35-38 cm

These elusive thick-set and robust owls have the largest eyes of any masked owl. They live in dense gloomy habitats, and are on the wing at dusk hunting for terrestrial and arboreal prey, including small birds. They breed opportunistically and nest very high in tree hollows. **Sexes:** *feet and bill of (A)'s female larger, plumages similar.* **Voices:** *weird swooping whistles — northern birds higher pitch.*

11 BARN OWL

Tyto alba 30–39 cm

This pale, nearly worldwide, owl is closely associated with man through its traditional use in the Old World of barn lofts and church steeples as nesting sites. It hunts harrier-like for ground-dwelling prey during the hours of darkness, using its remarkable gifts of hearing rather than sight. It is highly nomadic and populations fluctuate dramatically during periods of mice plagues. It can eat over 1000 mice a year. By day it roosts unobtrusively in secluded places. Breeds variably. Nests in tree hollow, cave or old building. **Sexes**: *similar.* **Immatures**: *spotted below.* **Voice**: *eerie drawn-out screeching.*

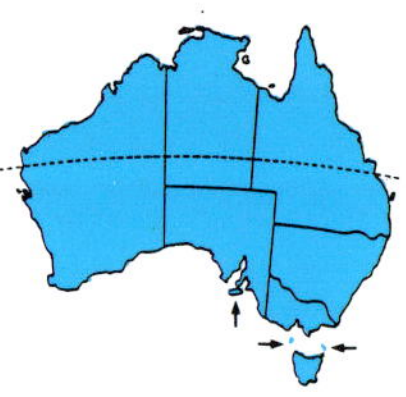

12 EASTERN GRASS OWL

Tyto longimembris ♂ 32–36 cm ♀ 35–38 cm

This rarely-seen long-legged owl lives out its whole life on the ground and frequents grassland, swampy heath, canegrass or reeds. It feeds mainly on terrestrial rodents. Coastal birds are sedentary, but their distribution inland varies greatly with food supply – especially when there are plagues of native rats. Normally abroad at dusk, but will hunt by day when food is scarce. Breeds mostly March–June. Nests under dense deep tussocks. Ranges to Africa, southern Asia and western Pacific. **Sexes**: *similar apart from size.* **Voice**: *like Barn Owl but softer, deeper and less vocal.*

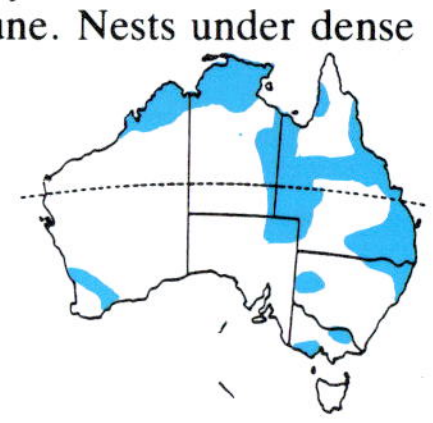

1

2

3

4

5

6

7

8

9

10

11

12

3

13 SPOTTED NIGHTJAR

Eurostopodus argus 32 cm

Where the range of the White-throated Nightjar ends west of the Great Dividing Range, this widespread and paler nightjar takes over. It frequents a variety of habitat including the sparse vegetation of the arid inland. It roosts by day on ground which blends with its plumage and squats in shade motionless except for turning its back constantly towards the sun. It hawks for flying insects after dark in bat-like flight. Possibly migratory. Breeds from September to January (south) and to December (north). Nests on the ground. **Sexes**: *alike.* **Immatures**: *redder.* **Voice**: *musical 'caw-caw-caw-gobble-gobble-gobble'.*

14 LARGE-TAILED NIGHTJAR

Caprimulgus macrurus 28 cm

Utters bouts of axe-like 'chops' between catching insects, varying from a few to maybe 100. Stories of wagers being placed on the next number of 'chops' by bushmen are legendary – hence the popular name 'Betting Bird'. It lives along edges of humid rainforest and dense monsoon scrub. It hunts exclusively on the wing, early in the evening and before dawn, scooping flying insects into its huge mouth while weaving amongst the trees like a large moth. Breeds August January. Nests amongst ground litter. Ranges to India and S. E. Asia. **Recognition**: *white in tail.* **Sexes**: *alike.* **Immatures**: *slightly duller.* **Voice**: *a succession of 'chops'.*

15 WHITE-THROATED NIGHTJAR

Eurostopodus mystacalis 35 cm

One of the world's largest nightjars and the biggest in Australia. It frequents woodlands and stony ridges in the south-east and the fringes of rainforest in the north wherever there is ample dry leaf litter in which to roost by day and nest. It feeds exclusively on nocturnal flying insects, mostly moths. Seasonally migratory from south to north. Breeds September – November in the north and later in the south. Nests on ground. Also in New Guinea and Solomon Islands. **Sexes**: *alike.* **Immatures**: *duller and redder.* **Voice**: *weird ascending musical whistles and laughter.*

16, 19 TAWNY FROGMOUTH

Podargus strigoides

(A) ♂ 34–40 cm ♀ 35–40 cm
(B) ♂ 45–53 cm ♀ 45–49 cm

(A) race *strigoides*; (B) race *phalaenoides*. The Tawny Frogmouth is sometimes erroneously called a 'Mopoke', which is an alternative name for the Boobook Owl. When sleeping by day, it makes no attempt to hide like the owls. It mostly chooses a bare branch to harmonise with its plumage along which it sits motionless throughout the day with feathers tightly compressed pretending to be part of the tree. It also sleeps on the ground in shade. Plumage and size vary to a bewildering degree but, in general, the northern birds (illustrated above) are paler and can be half the size of the southern; the birds east of the Great Dividing Range are larger and more boldly marked than the greyer birds of the west (the latter illustrated below); and those of the inland are small and rufous. There are also sexual and phase differences to add to the confusion. The species frequents a widely varied habitat, including forests, woodlands and inland mulga. It becomes active soon after sundown and scouts for prey from a low vantage point – ready to glide silently to ground to snatch terrestrial prey, including beetles, grasshoppers, scorpions, snails and sometimes small mammals and birds. Frogmouths often get killed on the roads as a result of their ground feeding habits. Is often seen in family groups. Sedentary. Breeds August–December. Nests 2–10 m high on bare horizontal branch. Endemic. **Male**: *heavier bill and darker.* **Immatures**: *like adults.* **Voice**: *bursts of low ventriloquial 'ooms' and 'oods' and a variety of other sounds including growling like a dog when threatened.*

17, 20 MARBLED FROGMOUTH

Podargus ocellatus (A) 37–41 cm
(B) 44–48 cm

(A) race *marmoratus*; (B) race *plumiferus*. In Australia, race *marmoratus* lives in the rainforests of Cape York. The once widespread southerly race *plumiferus*, known as the Plumed Frogmouth, has suffered grievously from over exploitation of its habitat, and is now restricted to small widely separated patches of coastal forest in northern NSW and southern Queensland. Marbled Frogmouths are active from dusk into mid-evening and towards dawn and prey on ground-dwelling insects from low perches. When threatened, they elongate and stiffen their bodies drawing their feathers closely in to present a thin profile in a cryptic pose. (The photographs opposite are of the same bird and are eloquent of the change in attitude that takes place). Breeds August–October. Nests on horizontal branch 3–15 m high. Ranges beyond Australia from New Guinea to the Solomon Islands. **Female**: *plainer and more chestnut.* **Immatures**: *like adults.* **Voice**: *'koo-loo' repeated in short bursts.*

18 AUSTRALIAN OWLET-NIGHTJAR

Aegotheles cristatus 23 cm

(A) race *cristatus*; (B) race *tasmanicus*. This diminutive and agile 'monkey-faced' bird lives in all types of timbered country. It hunts after dark on the wing for flying insects, or pounces from a perch for ground insects when aerial prey is scarce. By day it hides in bolt holes in trees or cliffs. Plumage becomes paler inland and northwards. The Tasmanian race is darker and smaller. Breeds August–December. Nests at bottom of tree hollow or cliff hole. Also in New Guinea. **Sexes**: *alike.* **Immatures**: *like adults.* **Voice**: *high-pitched rattling 'churrs'.*

21 PAPUAN FROGMOUTH

Podargus papuensis 50–60 cm

(A) race *rogersi*; (B) race *baileyi*. The Papuan Frogmouth, whose stronghold is New Guinea, has an enormous bill, and is the largest frogmouth in Australia. Race *rogersi* is the larger and darker of the two Australian populations. The species frequents the edges of tropical rainforest, dense woodland and mangrove. It is most active during early evening and the last hours before dawn, and takes larger prey than the other frogmouths. This includes mainly large beetles, grasshoppers and other terrestrial and arboreal insects which it hunts from low vantage points – gliding down on silent wings. During the day it retires within the forest to roost, often in small family groups, usually on high bare branches overhung with dense foliage. It 'freezes' in typical frogmouth fashion when approached. Sedentary. Breeds August–January. Nests in tree 6–12 m high. **Sexes**: *male heavier and darker.* **Immatures**: *paler, eye yellow.* **Voice**: *mainly drumming low booming 'woos' and 'uums'; sometimes ghostly laughing sounds.*

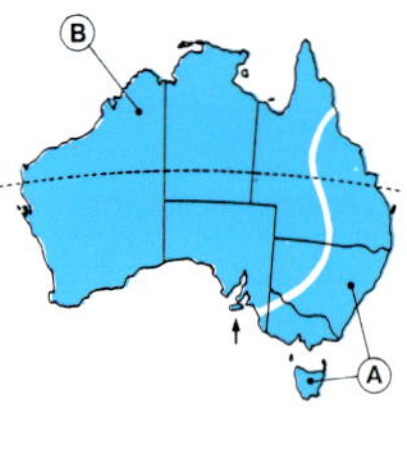

(B) race *phalaenoides*: above northern form below southern form

(A) race *mamoratus*

(A) race *rogersi*

13

14

15

16

17

18

19

20

21

4

Ground-Frequenting Birds

The birds in this Section feed entirely on the ground or are largely terrestrial in behaviour and belong mostly to small families unrepresented in the other Sections. They are grouped according to common attributes under four self-explanatory headings. Other ground-frequenting birds, such as the scrubwrens, grass warblers, certain parrots and pigeons have been left in their respective Sections.

BIRDS OF THE OPEN COUNTRY

CHATS Family Ephthianuridae

Australia's five species of chats make up an entire endemic family. Europe's Winchats and Stonechats belong to the thrush family and are unrelated. Chats have brush tongues like those of honeyeaters and their eggs are also similar in colour and markings. 1–6

PIPITS AND WAGTAILS Family Motacillidae

Wagtails There are eleven species of wagtails which occur widely throughout the Old World. Of these, the Yellow Wagtail (illustrated), occurs as a regular visitor, the Grey, White and Yellow-headed (Citrine) species only rarely. (The unrelated Willie Wagtail is a flycatcher – see Section 10.)

Pipits Of over forty species of pipits worldwide, only one occurs in Australia. Scientifically named from a specimen collected in New Zealand, it is there appropriately called the New Zealand Pipit, but known elsewhere as Richard's Pipit. Pipits are so-called from their twittering voices. 7, 10

LARK-LIKE BIRDS

Family Alaudidae This widespread family of about seventy-five species is represented in Australia only by the Singing Bushlark and the introduced Skylark. Most larks live in open country and are usually so well camouflaged that their presence is often unsuspected until flushed from cover when almost upon them. They are accomplished songsters and the spectacular soaring song flight of the Skylark has inspired poets and composers throughout the ages.

Family Sylviidae (Part) Australia's two endemic songlarks are both vociferous songsters when breeding. 8, 9, 11, 12

BIRDS OF THE LOW SCRUB AND FOREST FLOOR

FAMILY ORTHONYCHIDAE

Quail-thrushes There are five species of quail-thrushes, one endemic to New Guinea and four endemic to Australia. They are so named because of their quail-like habits and thrush-like appearance.

Whipbirds The endemic whipbirds live a furtive existence in thick undergrowth and maintain contact with each other by their loud distinctive calls. The name 'whipbird' derives from the remarkable swelling whip-crack calls of the eastern species.

Wedgebills The two superficially-alike wedgebills of the same genus as the whipbirds are distinguished mainly on voice.

Logrunners These denizens of the forest floor are sometimes called spinetails because of the stiff spiny-tipped shafts of their tail feathers, but it is a misconception that they mainly run about on logs. Of the two species, the northern one is endemic, while there are similar races in New Guinea to the southern one. 13–21, 26, 27, 29–32

APOSTLEBIRD Family Corcoracidae

This endemic mud-nester, though related to the White-winged Chough (Section 16), is placed here next to the babblers because of their similar behaviour and the frequent popular use of the same names for these birds such as 'Twelve Apostles' and 'Happy Family'. 22

BABBLERS Family Timaliidae

Four species of babbler occur in Australia and are endemic, except for a small population of the Grey-crowned Babbler which extends to southern New Guinea. They belong to a highly diverse Old World family of about 280 species which extends to Asia, Africa and parts of Europe. Babblers are highly sociable and do everything together in small tightly-knit groups to the accompaniment of incessant chattering. When disturbed, they fly off noisily in follow-my-leader order. 23–25, 28

FAMILY ACANTHIZIDAE (Part)

Pilotbird This endemic bird of the forest floor gets its name because of the company it sometimes keeps with the Lyrebird when feeding. It is a near relative of the Rock Warbler.

Bristlebirds Australia's endemic, rare bristlebirds, of which there are three species, are so named because of their unique pairs of strong decurved bristles which protrude from either side of their gapes. They are found in three, widely separated pockets in coastal heathland, where they are at great risk from bushfires. 33–36

SCRUB-BIRDS Family Atrichornithidae

The two species of this very rare Australian family possess extremely loud far-carrying voices. They are characterised anatomically by sturdy legs and feet, the absence of a wishbone, and similarities of their voice boxes, or syrinxes. They also both live furtively in dense undergrowth. Although isolated by the width of the continent, their common characteristics indicate that they have survived as remnants of a formerly-linked common ancestral population. (An article on the Noisy Scrub-bird is at page 36). 37, 38, 60

PITTAS Family Pittidae

Of the twenty-four species of this family of brightly coloured birds, which are spread throughout the tropical forests of Africa, south-east Asia, and the territories southwards to northern Australia, three breed here. The Blue-winged Pitta of Malaysia is a rare vagrant (not illustrated). 39–41

Top left: Orange Chat
Top right: Eastern Whipbird
Bottom left: Spotted Quail-thrush
Bottom right: Black breasted Button-Quail

4

FOWL-LIKE BIRDS

The two families in this group represent the order Galliformes in Australia, a worldwide order consisting of six well-marked families characterised by three toes in front and a shorter toe behind.

MOUND-BUILDERS Family Megapodiidae

The thirteen species of this unique family of megapodes, meaning great-footed, range from Australia to the Nicobar Islands, the Philippines and Samoa. Three occur in Australia, two of which, the Mallee fowl and the Brush-turkey, are endemic. Megapodes are the only group of vertebrate animals other than amphibians and reptiles that do not use their body heat to incubate their eggs or young; instead they replace it by burying their eggs in mounds and using the heat of rotting vegetation or sometimes the sun. The incubation can be forty or more days, during which the parents remain in daily attendance checking the internal temperature of the mound and using their tongues or heads as a thermometer and adjusting the top cover to ensure a constant level of 33°C. On hatching, the chicks dig themselves out unaided and are independent of their parents from the day of their birth. 42–44

PHEASANTS, FOWLS AND TRUE QUAILS Family Phasianidae

Taking its name from the pheasant, this large and diverse family of 189 species which is distributed in the temperate and tropical areas of the world, includes pheasants, quails, partridges and the domestic fowl. The family is poorly represented in Australia.

Pheasant and allies The Common Pheasant and the Peafowl are represented in Australia from periodic introductions. Few have prospered in the wild. The domestic fowl, which is a pheasant, is the most widespread of all birds. Its ancestors were probably one of the species of wild jungle fowl inhabiting India and South-East Asia which is thought to have been first domesticated by Bronze Age man as far back as 4000 BC.

True (or Typical) Quails In Australia, the true quails are represented by the introduced California Quail and three species of the genus *Coturnix*. The *Coturnix* quails are sexually dimorphic – the females of the Stubble Quail and Brown Quail being the larger and the more boldly marked, while the reverse applies in the King Quail where the male is much the brighter bird. They are strong fliers and migratory in habit. The presence of the characteristic small hind toe of the order to which they belong distinguishes them from the button-quails. The endemic Stubble Quail most closely represents the widespread migratory Common Quail *Coturnix coturnix* of Europe, Africa and Asia which, from time immemorial, migrated in great numbers annually from their breeding grounds in Europe to overwinter in North Africa. Until well into this century, they continued to be netted by the million along the shores of the Mediterranean for fattening for the table. The prodigious scale of these migrations in bygone days is attested in the Old Testament account of the exodus of the Children of Israel from Egypt in which we are told (Numbers 11:31) that, during their sojourn in the Sinai Desert, Moses' flock of 600 000 gathered quail for two days and a night. In his book *Birds of the World* (1961), O.L. Austin cites a researcher who estimated that the Children of Israel collected nine million quail on that famous occasion. In Palestine during March they still make their appearance in huge numbers in the course of a single night. 45–50

BUTTON-QUAILS, PLAINS-WANDERER AND BUSTARD

Members of the three families in this group, together with two in Section 19 – Rails (family Rallidae) and Cranes (family Gruidae) – represent the order Gruiformes in Australia. Worldwide, the order consists of twelve mostly small families.

BUTTON-QUAILS (OR HEMIPODES) Family Turnicidae

The fifteen species of this family (also known as Bustard-quails) are distributed across the southern parts of the Eastern Hemisphere. Of these, seven members of the genus *Turnix* occur in Australia, all being endemic except for the Red-backed Button-quail whose range extends to New Guinea and the island chain to the Solomons. Superficially, button-quails resemble the true quails and behave like them in some ways, but there are fundamental differences: like the bustards, they have no hind toe, their bill is more slender and weaker, their underparts are mainly plain and pale and the normal duties of the sexes are also reversed, with the male being left to rear the young; the females are also polyandrous and may take several mates in a season. They are weak fliers and do not flock. 51–56

PLAINS-WANDERER Family Pedionomidae

This unique quail-like bird has characteristics that are so unusual that it is classified in a family of its own. Its numbers have declined alarmingly and it is now rare and endangered. The World Wildlife Fund Australia has sponsored a four-year programme to formulate a land management plan to help it survive. 57, 58

BUSTARDS Family Otididae

Of about twenty-four species widespread in the Old World, particularly in Africa where sixteen species occur, the Australian Bustard (or Plains Turkey) which also lives in New Guinea, is the sole representative here. Despite human depredation, the Australian Bustard still roams across most of the continent, though depleted in numbers in most areas. 59

Opposite: Malleefowl

PHOTOGRAPHERS: C. Avon (Ardea) 45; H. & J. Beste 42; C.A.C. Cameron 15, 18; N. Chaffer 33; G. Chapman p. 23 (top left); 2, 3, 12, 13, 16, 17, 29, 34, 60; D.D. Dow 24; J.A. Estbergs, 39, 53; G. Folley p. 37 (plaque); A.T. Foster p. 23 (bottom right); 54; J. Frazier 4, 5; T. & P. Gardner 1, 31, 41, 44; R. Garstone 26; P.N. Green p. 23 (bottom left); 20, 21; E. & D. Hosking 7, 9, 46; W.J. Labbett 30; F.E. Lewitzka 55; I.R. McCann (ANT) 51; R. Miller 6; M. & I. Morcombe 38; W.S. Peckover 40; A.Y. Pepper 50; H.J. Pollock p. 25; J. Purnell 19, 35, 36, 56; L. Robinson 14, 32, 52; E.C. Slater 57, 58; G. Weber 22; H. Webster (NPI) 37; B.A. & A.G. Wells 8, 23, 28; Authors: p. 23 (top right); 10, 11, 25, 27, 43, 47, 48, 49, 59.

4

1 WHITE-FRONTED CHAT

Ephthianura albifrons 12 cm

The most commonly seen chat in southern Australia and Tasmania's only chat. In pairs and small parties, sometimes in large flocks during the non-breeding season, it frequents low dense cover, heath and moist vegetation near water. It is partially nomadic according to season. Like most chats, it perches on the tops of low bushes and tussocks. Its diet consists of insects, possibly also nectar. Breeds July–January. Nests near ground in low scrub such as samphire. Endemic. **Female**: *duller, more brown.* **Immatures**: *like females.* **Voice**: *metallic sounds like 'tang'.*

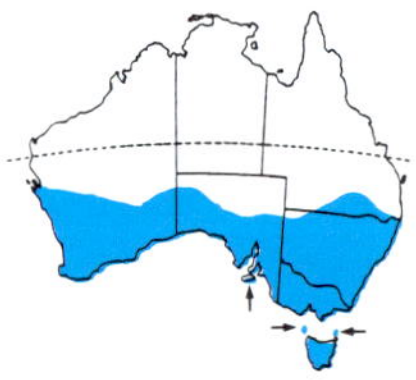

Male

2 ORANGE CHAT

Ephthianura aurifrons 12 cm

The range of the Orange Chat roughly corresponds with that of the Crimson Chat, but it is more thinly spread except for the arid inland where it is very much at home. In general, the habits of the two species are similar, but the Orange Chat is less nomadic and feeds only on insects. Breeds August–September. Nests in low cover in colonies, often in areas shared with the Crimson Chat. Endemic. **Female**: *lacks black on face (can be confused with Gibberbird but yellow rump of Orange Chat is distinctive).* **Immatures**: *like females.* **Voice**: *repeated musical notes in flight.*

Male

3 YELLOW CHAT

Ephthianura crocea 11 cm

This little known chat is found only in isolated colonies of up to 50 birds. It mostly inhabits swamps, often related to bore drains, and frequents dense vegetation such as lignum and sedge. Less terrestrial than the Crimson and Orange Chats, feeding also in bushes and trees. Diet is mainly insects, often flies. Breeds November–January. Nests near ground in thick cover. Endemic. **Female**: *paler and lacks black breastband.* **Immatures**: *grey-brown above.* **Voice**: *pretty three-note call.*

Male

4, 5 CRIMSON CHAT

Ephthianura tricolor 12 cm

Widespread west of the Great Dividing Range. Has adapted well to desert conditions through its highly nomadic movements and its versatile diet, which includes nectar, such as from Sturt's Desert Pea, as well as its main diet of insects. In pairs and loose flocks, sometimes with other chats, it frequents a wide variety of habitat including grasslands, the low vegetation of open plains, scrub and tree savannah. Breeds mainly August–October; variably in north. Nests close to ground. Males may lapse into eclipse plumage after breeding. Endemic. **Immatures**: *like females.* **Voice**: *repeated high-pitched tinkling notes.*

Male left, female right

6 GIBBERBIRD

Ashbyia lovensis 13 cm

The Gibberbird lives only among the sparse vegetation of the arid stony gibber plains. It is seen singly and in pairs, sometimes in small flocks. It has an erect stance and long legs and perches on stones rather than bushes. Possibly nomadic. During the breeding season it performs a song flight unique among chats. Rising almost vertically to over 30 m in stages it utters a piercing *weet weet weet* at each step, then plummets almost to its point of departure. Breeds generally May–October but may vary with conditions. Nests on ground in a depression. Endemic. **Sexes**: *similar.* **Immatures**: *browner.* **Voice**: *musical chatter.*

7 YELLOW WAGTAIL

Motacilla flava 19 cm

Unrelated to Australia's Willie Wagtail, which is a flycatcher, this species, a regular visitor from Siberia and Alaska, arrives in northern Australia from November to February. Occasionally strays southwards and has been recorded as far south as the Hawkesbury River area in NSW. Terrestrial. Singly or in small parties, it walks with long steps, head rocking back and forth, and pumping its tail up and down. Feeds on insects and leaps into air for prey. **Sexes**: *alike.* **Immatures**: *browner.* **Voice**: *shrill 'tsweep' or brief trill.*

8 SINGING BUSHLARK

Mirafra javanica 13 cm

In Australia, the Singing Bushlark, found also in India and Africa, lives in open country including pastures and is seen singly and in small groups, occasionally in large flocks. Feeds mostly on seed, also insects. Its plumage varies with the colour of the soil where it lives. Like its introduced relative the Skylark, it sings as it soars aloft, not so loudly but with a more varied song. Also sings from ground on perch by day and night. Breeds September–January in south; with onset of the wet in north. Nests among tussocks. **Sexes**: *alike.* **Immatures**: *paler.* **Voice**: *sweet melodious notes reminiscent of skylark; accomplished mimic.*

9 SKYLARK

Alauda arvensis 19 cm

Native to Eurasia and northern Africa, the Skylark was introduced to Australia progressively from the 1850s. Well established in the southeast and is spreading slowly. Famous for its remarkable song flight, it rises vertically from ground cover pouring out its song while soaring and hovering, then plummets silently to ground on closed wings. Singly or in parties, it lives in heath and other open country. Feeds on seeds, shoots and insects. Breeds September–January. Nests under tussocks. **Recognition**: *bigger than Singing Bushlark, small crest and does not bob tail.* **Sexes**: *alike.* **Immatures**: *paler.* **Voice**: *continuous joyous song.*

10 RICHARD'S PIPIT

Anthus novaeseelandiae 17 cm

There are many plumage variations in Australia of this common species which ranges widely from New Zealand to Africa, Europe and Asia – even to Siberia. It runs rapidly in short spurts and characteristically bobs its tail. Singly, in pairs or loose flocks, it mostly frequents open country and cleared land. Usually seen on ground, but also uses low perches such as stumps, rocks or fences. Eats seeds and insects. Breeds August–January. Nests in depression under tussocks. **Sexes**: *alike.* **Immatures**: *like adults.* **Voice**: *rasping 'shreep'; male in display flight repetitious trills.*

11 RUFOUS SONGLARK

Cincloramphus mathewsi 18 cm

The southern population of this widespread endemic species is migratory, arriving in the south to breed in the spring and departing to overwinter in the north in autumn. Sings continuously during the breeding season from conspicuous high perch and in display flight and is one of Australia's most vocal birds. Singly and small parties, it frequents grassy woodlands and open scrub. Feeds on ground on insects and seeds. Breeds September–February. Nests on ground in well hidden hollows. **Sexes**: *alike, but female smaller.* **Immatures**: *paler.* **Voice**: *loud 'witchy-weedle' monotonously repeated.*

12 BROWN SONGLARK

Cincloramphus cruralis ♂ 24 cm ♀ 18 cm

This endemic species is more often seen on treeless plains than the Rufous Songlark. Carries its long tail cocked and sits upright when perching. The male takes off singing on its fluttering nuptual display flight with legs dangling and drops back to the same spot. Southern birds migrate northwards or inland after breeding. Singly and in loose groups, it is a ground feeder taking insects and seeds. Breeds mainly September–February. Nests on ground under tussocks. **Female**: *paler and smaller.* **Immatures**: *like females.* **Voice**: *distinctive metallic 'skit-scot-a-wheeler' often repeated.*

Male

1

2

3

4

5

6

7

8

9

10

11

12

4

13 CHESTNUT-BREASTED QUAIL-THRUSH

Cinclosoma castaneothorax 24 cm
Race *marginatum* (western)

The western race lives among mulga scrub in stony hilly country. It is more boldly marked than the eastern race, but is similar in habits. Males will sometimes call soon after sunrise from a dead branch or other vantage point with two or three other males often replying. Breeds usually July–September but any time after heavy rain. Nests in a scrape under bush. Endemic. **Female**: *plainer: no black on underparts; no black markings on head or breast.* **Immatures**: *like females.* **Voice**: *repeated very high-pitched whistles.*

Male

16, 19 CHESTNUT QUAIL-THRUSH

Cinclosoma castanotum 23–26 cm

The Chestnut Quail-thrush is the most widely spread of the quail-thrushes and is found in a variety of habitat ranging from the timbered areas of semi-arid southern Australia to the spinifex and other low scrub of the sandy inland desert. Its range extensively overlaps that of the western population of the Chestnut-breasted Quail-thrush, but they tend to separate into their preferred habitats. There is some evidence of interbreeding between them. The chestnut on the back is more extensive and brighter in the desert birds. It rambles in pairs and small families, searching for insects and seeds and is probably nomadic in the desert. It is wary and takes off with whirring wings to a safe distance when alarmed. Breeds mainly July–September, but August–December in the desert areas or following rain. Nests in a scape under cover. Endemic. **Immatures**: *like females.* **Voice**: *high-pitched sibilant whistles, especially at dawn.*

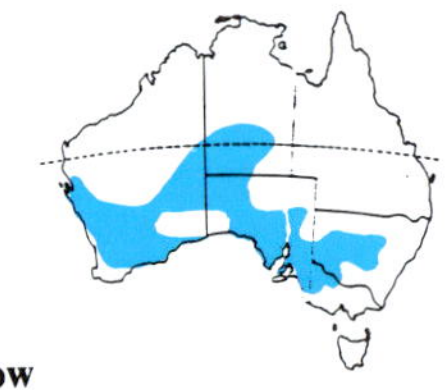

Male above, female below

14, 17 CINNAMON QUAIL-THRUSH

Cinclosoma cinnamomeum 19–22 cm
(A) Race *cinnamomeum*
(B) Race *alisteri* 18–21 cm
(Nullarbor Quail-thrush)

These small, short-tailed quail-thrushes, live under extremely harsh conditions on the sparsely vegetated stony plains of the central deserts. Race *cinnamomeum* is fairly numerous in the Lake Eyre basin. Race *alisteri*, isolated on the Nullarbor Plain, is, however, rare and its survival is threatened by land clearance; it is the only bird species found on the Nullabor Plain and nowhere else. The Cinnamon Quail-thrush interbreeds with the Chestnut-breasted Quail-thrush in some areas where their ranges meet. The species is wary and skulks in cover if approached. When alarmed, it runs away swiftly or bursts from cover to fly off in a quail-like flurry of wings. It flicks its tail when walking, and wanders singly, in pairs, and sometimes in small loose parties, searching for seeds and insects, often turning over small stones with its bill. Breeding: (A) any time after rain, but mainly July–September; (B) winter and spring, possibly autumn after rain. Nests consist of a ground scrape under a bush. Endemic. **Female**: *plainer; no black on underparts.* **Immatures**: *like females.* **Voice**: *repeated very high-pitched thin whistles.*

(A) male above, (B) immature male below

15, 18 CHESTNUT-BREASTED QUAIL-THRUSH

Cinclosoma castaneothorax 22–24 cm
Race *castaneothorax* (eastern)

The eastern race of the Chestnut-breasted Quail-thrush lives among dense mulga in low hilly country restricted to central eastern Australia. While its territory meets that of the Cinnamon Quail-thrush, each maintains a distinct identity with few reports of hybridisation. Though fairly common, it is rather shy and difficult to see. If approached, it will disappear under a bush or run away to cover – flicking its tail. If flushed, it will break cover in a quail-like burst and fly off in a flurry of wings. Singly, in pairs or small parties, it wanders on the ground foraging for seeds, insects and spiders – sometimes turning over stones. Chris Cameron, whose photographs are opposite, has occasionally noted a male perched on the top of a dead tree or exposed twig singing a prolonged and beautiful song. Breeds July–September and almost any time after good rain. Nests on ground well hidden under cover. Endemic. **Immatures**: *like females but duller with dusky flecks.* **Voice**: *thin, high-pitched whistle.*

Male above, female below

20, 21 SPOTTED QUAIL-THRUSH

Cinclosoma punctatum 26–28 cm

This, the largest of the quail-thrushes, lives in the high country of south-eastern Australia and is Tasmania's only quail-thrush. No other quail-thrush is found east of the Great Dividing Range. It favours stony hills and ridges afforested with dry eucalypt with a ground cover of litter, or an understorey of shrubs and other vegetation at lower altitudes. It is very wary and elusive, and takes off in explosive quail-like flight when alarmed. Singly, in pairs and family parties, it feeds on seeds and a variety of insects, caterpillars and sometimes small lizards. The male defends a breeding territory by singing from a prominent perch. Its double whistles distinguish it from other quail-thrushes. Breeds July–February. Nests in depression under cover. Endemic. **Immatures**: *like females.* **Voice**: *sibilant high-pitched notes and series of musical double whistles.*

Male left, female right

22 APOSTLEBIRD

Struthidea cinerea 33 cm

A highly sociable species which lives in close-knit groups of from 6—20 birds — often in twelves as suggested by the name. In the non-breeding season, flocks of hundreds may be seen. Never far from water in open woodland and scrub; also near country homesteads where it is relatively tame. Jumps around in restless groups making a great commotion, often with tails cocked. Feeds on the ground on insects and seeds. Will catch and kill house mice during plagues. Breeds August—February. Nest is a large bowl made of mud and grass. Endemic. **Sexes**: *alike.* **Immatures**: *smaller.* **Voice**: *continuous harsh chattering.*

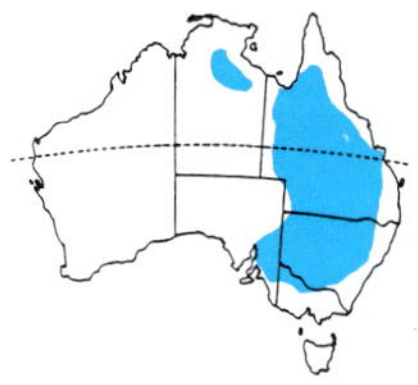

23 WHITE-BROWED BABBLER

Pomatostomus superciliosus 20 cm

The White-browed Babbler is common in the dry woodlands and semi-desert country of southern Australia, and keeps together in restless groups of from 3–10 birds. Intrusion sets parties off in an indignant chorus of scolding notes. Feeds mostly on the ground foraging for seeds and insects. Breeds July–December. Nests communally in clusters of domed-shaped nests with spout-like entrances. These structures also serve as dormitories for roosting. Endemic. **Sexes**: *alike.* **Immatures**: *almost similar.* **Voice**: *range of loud chattering notes and whistles.*

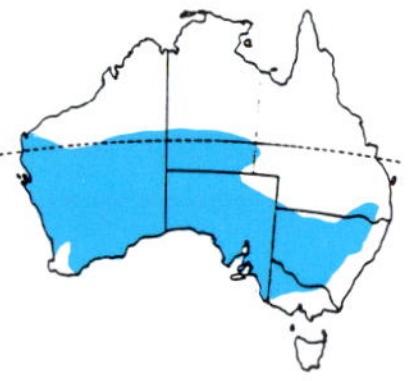

24 HALL'S BABBLER

Pomatostomus halli 23 cm

Until 1964, when a small babbler collected on the Harold Hall Expedition the previous year was found to differ from the widespread White-browed Babbler, this bird had not been recognised as a separate species – although the late Monty Schrader knew that babblers like this, long familiar to him at Eulo Qld, were different. It lives in lightly timbered mulga through which it moves constantly in noisy chattering groups. A ground feeder. Breeds communally September–June. Nests up to 10 m high. Endemic. **Recognition**: *small white bib; dark breast and belly.* **Sexes**: *similar.* **Voice**: *chuckles growls and liquid notes.*

13

14

15

16

17

18

19

20

21

22

23

24

4

25 CHESTNUT-CROWNED BABBLER

Pomatostomus ruficeps 22 cm

In its relatively restricted inland range, this babbler is fairly common. It frequents open scrub and woodlands and keeps together in groups of from 12–15 birds. Specific sites are selected for socialising and for use as playgrounds. Here they will chase each other around a bush so often in an apparent game of follow-my-leader that they make well defined circular tracks. Shy and elusive. Retreats noisily if disturbed. Feeds on insects and seeds. Breeds July–December. Nests 5–10 m high in a tree. Endemic. **Sexes**: *alike.* **Immatures**: *duller.* **Voice**: *penetrating whistles and chattering.*

26 WESTERN WHIPBIRD

Psophodes nigrogularis 25 cm

This rare bird is now restricted to five small and isolated opulations in coastal sandhills, heath and mallee scrub. Being strongly territorial it is very vulnerable to loss of habitat, and its former range has shrunk accordingly. The eastern birds are paler. Excessively shy, it lives in dense cover where it forages for insects and small invertebrates. Breeds July–November. Nests in dense scrub within 1 m of ground. Endemic. **Sexes**: *similar.* **Immatures**: *white and black throat markings absent.* **Voice**: *male four syllable 'lets-scratch-teacher' answered by female 'pick-it-up'.*

27 EASTERN WHIPBIRD

Psophodes olivaceus 28 cm

A bird of the gullies and wetter wooded areas east of the Great Dividing Range. An isolated northerly population has brown not white undertail tips. Solitary and in small parties, it hops chuckling through thick undergrowth grubbing for insects and small invertebrates under litter. Also probes fallen logs for prey. Shy but inquisitive. Sedentary. Breeds July–January. Nests near ground up to 3 m in thick undergrowth. Endemic. **Sexes**: *female smaller.* **Immatures**: *white on throat and neck absent.* **Voice**: *male's swelling note ending in explosive whip-crack often answered immediately by 'choo-choo' from female.*

28 GREY-CROWNED BABBLER

Pomatostomus temporalis 26 cm

The largest and most widespread of the babblers. It frequents a variety of woodland in groups of from 5–10 birds. Feeds mostly on the ground for insects, spiders and even lizards. Plumage and size varies geographically e.g. some western birds have a pale rufous breast; southern birds are larger and darker. Breeds July–February. Nests 3–6 m high in shrub or sapling; nest is dome shaped. Also in New Guinea. **Recognition**: *chestnut wing patch in flight.* **Sexes**: *similar.* **Immatures**: *slightly paler.* **Voice**: *clear 'yahoo'.*

29 CHIMING WEDGEBILL

Psophodes occidentalis 22 cm

The Chiming Wedgebill is almost identical in appearance to the Chirruping Wedgebill of the east, but differs markedly in voice and some habits. In pairs and parties of 10–20 birds, it is common to the west of its range and lives in thickets of low scrub and arid tree savannah. Feeds on ground on insects and seeds. Wary and runs quickly to cover when approached. Endemic. Breeds variably depending on rains. Nests low in thick bushes. **Sexes**: *alike.* **Immatures**: *like adults but paler bill and legs.* **Voice**: *descending chiming sweet tones 'did-you-get-drunk' repeated monotonously, sometimes by moonlight.*

30 CHIRRUPING WEDGEBILL

Psophodes cristatus 20 cm

The ranges of the two wedgebills meet in the Simpson Desert and both species have been heard near Oodnadatta. The Chirruping Wedgebill mainly lives in lignum swamps and spinifex of semidesert country and feeds on insects and seeds. Gregarious, sometimes in quite large parties. Less shy than its relative and more often seen. Breeds March–November, except midwinter. Nests below 3 m in low shrub. Endemic. **Recognition**: *faintly streaked on breast (the Chiming Wedgebill's is plain).* **Sexes**: *alike.* **Immatures**: *crest undeveloped.* **Voice**: *creaky song; female answers male's 'chip-chip-cher-oo' with 'chin-up'.*

31 (Southern) LOGRUNNER

Orthonyx temminckii 20 cm

A bird of the dank deep shades of semi-tropical and temperate forest. In pairs and small parties, it runs and hops over fallen debris, seldom flying. Rakes litter and leaf mould with its strong legs, using its stiff spine-shafted tail as a prop. Circular scrapings of about 15 cm indicate where it has been feeding. Breeds April–August. Domed nest on ground against rock or tree. Several races in New Guinea. **Male**: *white throat.* **Immatures**: *like females but plainer.* **Voice**: *excited bursts of 'kweek-kweek-kweek'.*

Female

32 CHOWCHILLA (Northern Logrunner)

Orthonyx spaldingii 31 cm

Common in the Atherton Tablelands. Its remarkable voice, reflected in its name, rings through the tropical rainforest at dawn and dominates the early morning chorus. Stopping suddenly, it remains relatively silent during the day. In pairs and small parties, it runs and hops over fallen debris and rarely takes wing. Rakes floor litter for insects and grubs with its strong legs. Breeds May–November. Domed nest up to 4 m from ground. Endemic. **Male**: *throat and foreneck white.* **Immatures**: *cinnamon-brown.* **Voice**: *resonant repeated 'chow-chowchilla'.*

Female

33 PILOTBIRD

Pycnoptilus floccosus 17 cm

This unique small bird lives in moist eucalypt forest and woodland. It sometimes accompanies the Superb Lyrebird as it fossicks in ground litter taking insects, earthworms and other small invertebrates disturbed by its large companion – hence its name. Solitary and fearless. A weak flier, it hops briskly and runs rapidly when alarmed; tail often carried high and flicked. Breeds August–January. Domed nest on or near ground. Endemic. **Sexes** and **immatures**: *similar.* **Voice**: *strong, sweet and varied; includes call of 'guinea-a-week'.*

34 WESTERN BRISTLEBIRD

Dasyornis longirostris 19 cm

This small bristlebird has become one of Australia's rarest birds. Once found along the coastal areas from Perth to Albany, the only population now known – estimated at about 100 pairs – is confined to the Two Peoples Bay area east to Albany where the Noisy Scrubbird was rediscovered in 1961. It lives in deep dense heath, is remarkably shy and rarely shows itself, even singing hidden from view. Runs swiftly; feeble flight. Breeds September –January. Domed nest near ground. Endemic. **Sexes**: *alike.* **Immatures**: *undescribed.* **Voice**: *variable, including male's 'chip-pee' answered by female's 'quick-more-beer'.*

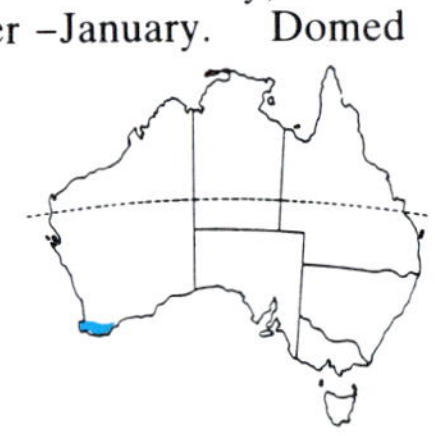

35 RUFOUS BRISTLEBIRD

Dasyornis broadbenti 27 cm

This, the largest of the bristlebirds, lives in heavy scrub and coastal thickets. Usually in pairs, it is a ground feeder on grubs, insects, seeds and berries. With raised and spread longish tail, it runs swiftly through undergrowth and seldom flies. Shy and elusive. Coorong, SA birds are pale grey on chin and centre of belly. A race formerly of the south-west corner of WA was last reliably reported in 1906 and is believed extinct. Breeds August–December. Domed nest up to 2 m from ground. Endemic. **Sexes**: *similar.* **Immatures**: *undescribed.* **Voice**: *penetrating and squeaky calls with responses from female.*

36 EASTERN BRISTLEBIRD

Dasyornis brachypterus 22 cm

Restricted to a series of isolated pockets in coastal and mountain areas from southeast Queensland to Victoria, the Eastern Bristlebird has become rare following the impact of settlement. It skulks in rank, dense undergrowth where it feeds on insects and seeds. It is reluctant to fly and runs with tail partly cocked and sometimes fanned. Solitary and shy. Sings from bushes and is easy to locate from its loud distinctive calls. Breeds August–December. Domed nest on or near ground. Endemic. **Sexes**: *alike.* **Immatures**: *undescribed.* **Voice**: *melodious, penetrating and variable, including 'it-wood-weet-sip'.*

25

26

27

28

29

30

31

32

33

34

35

36

4

37 NOISY SCRUB-BIRD

Atrichornis clamosus 24 cm

The small surviving colony of Noisy Scrub-birds discovered in the Two Peoples Bay/Mt Gardner area in 1961 and those released on Mt Many Peaks in 1983 (see article at page 36) live mainly in deep and densely vegetated mountain gullies. Shy and very difficult to see. Feeds among ground litter on insects and small invertebrates. Strong legs carry it swiftly through undergrowth. Winter breeder May–September. Nest spherical with side entrance usually low near running water. Endemic. **Male**: *black chest markings.* **Immatures**: *like females.* **Voice**: *very loud piercing staccato notes.*

Female

38 RUFOUS SCRUB-BIRD

Atrichornis rufescens 16 cm

The species lives in isolated pockets on the highlands of the Great Dividing Range among deep tangled understorey in moist forest. Rarely flying, it runs fast with tail cocked, and wriggles through litter more like a mouse than a bird. The male has one of the loudest voices in the bush; it is also ventriloquial and a gifted mimic. The female is mostly silent, also the male when not breeding. Breeds September–January. Domed ground nest lined with 'card-board' of chewed wood. Endemic. **Sexes**: *similar, but male has dark breast mark.* **Immatures**: *underparts chestnut.* **Voice**: *repeated loud 'chipping' phrases.*

39 RAINBOW PITTA

Pitta iris 18 cm

This strikingly marked pitta of the Kimberley and the Northern Territory lives in a varied habitat of monsoon rainforest, bamboo thickets and mangrove, and hops among moist ground litter like the Noisy Pitta. It eats insects, grubs and small snails, breaking their shells on any hard object. Solitary, shy and retiring. Breeds December–March. Nests vary from partly open in the Kimberley to the characteristic domed construction in the monsoon forests of the far north. Endemic. **Sexes**: *alike.* **Immatures**: *unrecorded.* **Voice**: *loud and clear 'want-a-whip'.*

40 BLUE-BREASTED [Red-bellied] PITTA

Pitta erythrogaster 19 cm

This breeding migrant from New Guinea arrives in the rainforests of Cape York Peninsula from October and returns March–April. A member of the most varied species of the pitta family with 26 recognised races ranging widely through New Guinea, Indonesia and the Philippines. Like the Noisy Pitta, it smashes snails on an 'anvil'. Hops about when feeding and flies erratically from one foraging spot to another. Often calls from trees up to 25 m from ground. **Sexes**: *similar, but chin, upper half of throat and cheeks of female sooty brown.* **Immatures**: *duller.* **Voice**: *mournful rasping notes.*

41 NOISY PITTA

Pitta versicolor 21 cm

A stocky brightly-coloured dweller of tropical and semi-tropical rainforest. Fossicks for grubs and insects; also eats berries and large forest snails whose shells are smashed on an 'anvil' of stone or wood. These are sometimes worn smooth with constant use and heaps of broken shells betray their location. Hops with upright stance, tail flicking, and flies swiftly. Shy. Breeds mainly October–January. Bulky domed nest on or near ground with 'doormat' of mammal dung. **Sexes**: *alike.* **Immatures**: *duller.* **Voice**: *resonant notes including 'or-rip-tor-ear'; also at night.*

42 [Orange-footed] SCRUBFOWL

Megapodius reinwardt 40 cm

This Scrubfowl, which also ranges through New Guinea to India, is found in Australia mostly in the dense wet coastal rainforests of the tropics. Solitary and in scattered colonies, fossicks for seeds, roots and grubs on the forest floor. It flies heavily to trees when disturbed. Sedentary. Its incubating mounds of soil and rotting litter are usually 3 m wide and 1.5 m high, but they can become enormous, growing even to 15 m wide and 4 m high. Up to 15 eggs are laid in holes to 2 m deep, then sealed. Breeds August–March. **Sexes**: *alike.* **Chicks**: *downy.* **Voice**: *clucking and squawking like Domestic Fowl.*

43 [Australian] BRUSH-TURKEY

Alectura lathami 70 cm

Endemic to Australia, this mound-builder lives in coastal rainforest and the drier high country scrub. Mainly a ground feeder on native fruits, seeds and insects. Sedentary. The egg-incubating mound, usually about 4 m wide and 1.5 m high, is built by the male, who also controls the mound's temperature at 33°C during the 50-day incubation period. The female lays her 18–24 eggs in holes at intervals according to conditions. Cape York birds have bluish-white wattles. Breeds August–December. **Sexes**: *alike except for male's breeding wattles.* **Voice**: *loud clucking and grunting.*

44 MALLEEFOWL

Leipoa ocellata 60 cm

Malleefowls usually excavate their mounds in poor sandy soil in mallee scrub and open woodland. Mounds can grow to 4 m across and 1.5 m high and their construction and maintenance of the egg chamber at 33°C involves shifting tonnes of soil and fermenting vegetable matter in a season. From 5–33 eggs are laid one at a time over 2–17 days with the male preparing the cavity to the female's satisfaction on each occasion. The species is wary. Food includes buds, flowers, fruits and spilled grain. Breeds September–April. Endemic. **Sexes**: *alike.* **Chicks**: *downy with wing feathers.* **Voice**: *loud booming and grunts.*

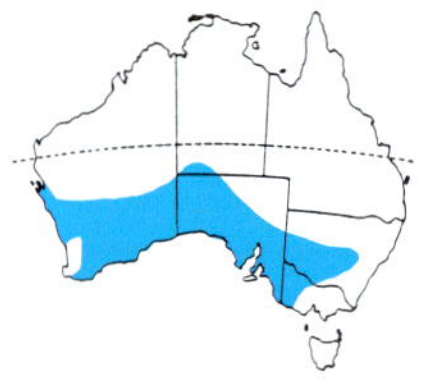

45 CALIFORNIA QUAIL

Lophortyx californicus 24 cm

A large quail species with a waving crest plume. Introduced to Australia from North America around 1860, it failed then to survive, but a further batch introduced to King Island in the 1930's, there protected from hunters, has become established. Coveys inhabit wooded country and scrub with open spaces, moving into pasture to feed on plants and insects. Roosts off the ground. Breeds usually September–January. Nests in depression under cover. **Female**: *lacks black throat, duller, crest shorter.* **Immatures**: *like females.* **Voice**: *many calls including triple notes, the middle one highest and loudest.* **Male**

46 COMMON PHEASANT

Phasianus colchicus ♂ 85 cm ♀ 60 cm

Legend has it that the original stock of this pheasant came from the river Phasis (now Rioni) in the ancient province of Colchis on the Black Sea, and that it was brought from there by the Argonauts on returning from their quest for the Golden Fleece. The scientific name perpetuates this belief. There are many hybrid forms from continuous interbreeding of this famous game bird. Only small feral populations have survived from introductions in Australia – in the south-east and south-west of the continent, in Tasmania and on some of the offshore islands. **Female**: *mottled brown.* **Voice**: *explosive 'karrk-kock'.*

Male

47 PEAFOWL

Pavo cristatus ♂ 2.4 m ♀ 90 cm

The fabled Peafowl is most often seen in zoos, parks and large estates. A native of India and Sri Lanka, it has been famous throughout history. The Old Testament tells us that Solomon (ninth century BC) imported peacocks with other riches to adorn his temple (I Kings X:22 and 2 Chronicles IX:21). Introduced to Australia, it exists in the wild only in a very few areas, such as Rottnest Island where releases were made in 1912 and 1917. The Peacock is polygamous and may breed throughout the year. Nests in a scrape in the ground. **Female:** *smaller, duller and lacks the striking train.* **Voice:** *strident screech.*

Male

37

38

39

40

41

42

43

44

45

46

47

4

48 KING QUAIL

Coturnix chinensis 14 cm

The King Quail, widespread throughout south-eastern Asia, is particularly shy and elusive. In Australia, it keeps mainly to rank vegetation, swamps, heath and crops such as lucerne and feeds on seeds and insects. If disturbed, it squats but, if forced to fly, it does so feebly and briefly, quickly dropping back to cover. Usually in pairs or coveys of 5 or 6, sometimes larger. Numbers fluctuate according to conditions. Breeds most months. Builds nest in ground hollow. **Immatures**: *like females.* **Voice**: *plaintive 3-note whistle; also calls at dusk and after dark.*

Male left, female right

49 (A) BROWN QUAIL

Coturnix australis

(B) SWAMP (Tasmanian) QUAIL

Coturnix ypsilophora 17 cm

The Brown Quail, also found in Tasmania, lives in grassland, marshes and scrub. The larger and paler Swamp Quail (illustrated), confined to Tasmania and the south-east of the mainland, lives almost exclusively in the wet places. When flushed they characteristically scatter in a whirr of wings. Breeds from spring in south and January–May in tropical north. Nests on ground under dense cover. **Sexes:** *female larger and more boldly marked.* **Voice:** *(A) loud rising whistling. (B) 'be-quick, be-quick'.*

50 STUBBLE QUAIL

Coturnix pectoralis 18 cm

The widely spread Stubble Quail lives in all types of country, commonly occurring in grassland, swamp margins and crop stubble. it has multiplied in some areas following agricultural development. Partial to seeds of weeds, plants and fallen grain, also takes insects. Companies fluctuate in size and roam widely; sometimes covering 1000 km or more. When flushed, runs rather than flies; when pressed, breaks cover with a whirr of wings in low, short and rapid flight. Breeds mainly October–February. Nests in ground scrape. Endemic. **Immatures**: *like females.* **Voice**: *'two-to-weep'.*

Female left, male right

51 RED-CHESTED BUTTON-QUAIL

Turnix pyrrhothorax ♂ 13 cm ♀ 15 cm

Distribution of this quail is patchy and thin across the north of its range. Some seasonal movement and can turn up anywhere. Singly, in pairs and small variable coveys, it frequents grasslands and open timbered country with good cover. Feeds on seeds and insects. If flushed, it flies low and straight over short distance. Breeds October–March. Nests on ground under cover. Endemic. **Recognition**: *easily confused with Little Button-quail but darker above.* **Female**: *brighter; no black markings on neck or breast.* **Immatures**: *like adult male, darker eye.* **Voice**: *repeated 'ooms'; gurgles.*

Male

52 PAINTED BUTTON-QUAIL

Turnix varia ♂ 17 cm ♀ 20 cm

This bright-plumaged quail occurs in several isolated populations and certain offshore islands. Its varied habitat includes grassy woodlands and rocky hillsides with open timber or scrub. Birds of the Abrolhos Islands are smaller and paler. Pairs and family parties forage in leaf litter for insects, plant food and seeds, leaving small circular scrapes. Erratic nomadic movements. Breeds September-March south; variably north. Nests in ground depression. Endemic. **Male**: *smaller, duller.* **Immatures**: *duller, eye brown.* **Voice**: *low repeated ventriloquial 'oom'.*

53 (A) CHESTNUT-BACKED BUTTON-QUAIL

Turnix castanota ♂ 15 cm ♀ 18 cm

(B) BUFF-BREASTED BUTTON-QUAIL

Turnix olivei ♂ 18 cm ♀ 20 cm

T. castanota (illustrated) lives in poorly grassed areas in dry savannah woodland, especially on rocky sandy ridges. The similar *T. olivei* has plain buff breast and lives in wet open areas with heath-like vegetation fringing rainforest. Both breed December-March, and nest on ground under shelter. Endemic. **Sexes:** *similar; female brighter.* **Immatures:** *like adults, darker eyes.* **Voice:** *repeated low moaning calls.*

54 BLACK-BREASTED BUTTON-QUAIL

Turnix melanogaster ♂ 16 cm ♀ 18 cm

This rare quail lives in the thick tangled lantana and vine undergrowth of dry sub-tropical and tropical rainforest. Solitary, or in small coveys, it moves slowly fossicking in leaf litter for seeds and insects, leaving scratched out bare circular depressions along its trail. A reluctant and clumsy flier, it prefers running. Probably sedentary. Breeds variably, but not during colder periods. Nests on ground in depression. Endemic. **Immatures**: *like adult males, eye dark, legs grey.* **Voice**: *female, repeated low booming; male, clucks.*

Female left, male right

55 RED-BACKED BUTTON-QUAIL

Turnix maculosa ♂ 13 cm ♀ 15 cm

So named from the rufous upper back of the female, the species, which also occurs in Indonesia, New Guinea and various tropical islands to the north, is found in Australia on the margins of wet and swampy areas, rainforest and cleared country. Some plumage variation across its range. In pairs and small coveys, it feeds on the ground on seeds and insects. Shy and elusive and not often seen. Breeds October–July. Nests under cover in ground depression. **Male**: *smaller, lacks rufous on upper back.* **Immatures**: *like adult males; dark eye.* **Voice**: *rapidly-repeated 'oom'.*

Female

56 LITTLE BUTTON-QUAIL

Turnix velox ♂ 14 cm ♀ 15 cm

Widespread and more abundant than other quails, it is also the least coloured. It lives in a wide variety of habitat ranging from open woodlands and grassy plains to spinifex in the arid inland. It feeds mainly on seeds and is highly nomadic, moving to areas where rain has provided good growth. Usually seen in small flocks, but these can be in thousands in good seasons. Breeds most months, according to rain. Nests on ground in shelter of grass tussocks. Endemic. **Sexes**: *similar except for size.* **Immatures**: *like adult male, eye dark.* **Voice**: *moaning 'oop', squeaky alarm call.*

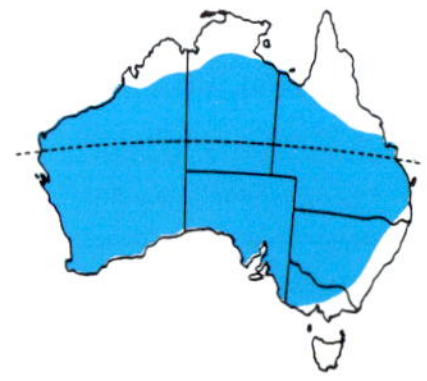

57, 58 PLAINS-WANDERER

Pedionomus torquatus ♂ 16 cm ♀ 18 cm

The female of this rare strange quail-like bird is larger and brighter than the male and is polyandrous (it mates with more than one male). To this extent it is like the button-quails but, unlike them, it has the hind toe of the typical quails. While its instinct is to freeze with the approach of danger and its crouching head-down run is quail-like, it is unlike quails in its quaint habit of standing on tip-toe to survey the land and in its weak uncertain flight. Also unlike quails, it is said to be reluctant to take flight from danger and can even be caught by hand in broad daylight. Because of these unique characteristics, taxonomists have thought it best to leave it on a twig of its own on the evolutionary tree. The spread of agriculture into its natural habitat of lightly grassed flat open plains has led to its alarming decline, although there is some evidence of it adapting to cereal crops and ungrazed paddocks. Sightings in recent years have mainly been confined to a small area of western NSW into Victoria and SA. Usually solitary, also in pairs and small parties, it forages for seeds and insects. Breeds June–January. Nests in scrape under cover. Endemic. **Immatures**: *like adults, more heavily scalloped on breast.* **Voice**: *repeated mournful 'moo'.*

Male left, female right

59 AUSTRALIAN BUSTARD

Ardeotis australis ♂ 1 300 cm ♀ 800 cm

Now common only in remote unsettled areas, this elegant and stately bird is a resident of open country with low bushes and scattered trees. In large flocks at times, but more usually seen singly or in pairs. It feeds on insects, small animals and vegetable matter. When disturbed, it freezes then walks casually away with head upstretched, taking to flight only if pressed. Probably nomadic in response to rain. Breeds variably. Nests on ground under tussock or bush. Moves to and from New Guinea. **Sexes** and **Immatures**: *alike but female smaller.* **Voice**: *low booming, but mainly silent.*

48

49

50

51

52

53

54

55

56

57

58

59

The Noisy Scrub-bird — its discovery, rediscovery and rehabilitation

Atrichornis clamosus

DISCOVERY

In 1697, the Dutch explorer, Willem de Vlamingh, noted in his journal that he had heard a 'nachtegaal' when exploring the Swan River, on which Perth now stands. As nightingales do not live in Australia, Dr. D.L. Serventy, the distinguished ornithologist, believes that it was most probably the remarkably loud melodious notes of the Noisy Scrub-bird that attracted Vlamingh's attention.

Another 145 years were to pass before the bird itself was seen when John Gilbert, John Gould's able assistant, recorded it for the first time, on 3 November 1842 at Drakesbrook 100km south of Perth near Waroona in the Darling Range, during a two-month trek to Augusta. A letter which has recently come to light from Gilbert to Gould dated 2 January 1843 reports the results of the Augusta trip. *Here reproduced is an extract relating to the discovery of the Noisy Scrub-bird, which makes most interesting reading.

'The other bird, is another allied form of Dasyornis [bristlebird] but having a totally different form of bill and quite devoid of bristles, it is about the size of the rough sketch I sent you, the whole of the upper side, dark reddish brown, each feather from the nape downwards, wings and tail beautifully barred with zigzag bands of black, throat and sides of neck white,—black chest, while breast gradually becoming buff on the vent, the tail feathers are very broad, the legs are very robust, in fact the whole body of the bird, has the osteological characters carried out to extreme strength; this is without exception the loudest of the Western Australian songsters, I have yet heard. It inhabits the densest and rankest vegetation, on the sides of hills and thick swamp grass on the banks of small running streams or swamps, and is of all birds, the most difficult I have yet had to procure, it is in fact but a mere chance the seeing the bird at all, it runs with the utmost rapidity on the ground perfectly sheltered from view by the overhanging vegetation, which renders it almost impossible to get a shot at it, for it is seldom indeed that it trusts to its wings for progression, when I first heard its extraordinary loud melodious note, I was perfectly convinced it proceeded from a bird never before seen by me, and I was tantalised still more, by watching and waiting in rain for hours together, in creeping quietly through the bush, I have frequently been as I imagined within a yard of the bird, while its loud but pleasing note fairly made my ears ring, and yet I could not see the creature, no bird in fact that I have yet met with has cost me so much time and patience, the result is, I have only obtained four specimens, two of which are only tolerable specimens, however they will be sufficiently perfect to enable you to figure, in each case I was only a few yards from the bird, and almost blew them to pieces, there is not the chance of measuring my distance, as it is impossible to see it beyond a yard or two; I have little doubt you will be well pleased with this second addition to this curious form, that this species as well as the other one, is in particular districts abundant as the common Dasyornis, I am convinced, for when at Cape Lewin, I frequently heard all three birds singing at one and the same time within twenty yards of me yet could not obtain a sight of either.'

The discovery is marked by a memorial plaque at Drakesbrook (photograph opposite) erected in 1949 when ornithologists had given up hope of its survival. It is believed to be the only memorial to a bird in Australia, and is a touching tribute to the young Gilbert who was then only twenty-nine. Gilbert again recorded the Noisy Scrub-bird towards the end of 1843 at Mount Barker and Albany during his last trip in Western Australia — a fact only revealed in 1973 when Ederic Slater, then of the CSIRO, chanced upon some previously unknown notes by Gilbert among Gould material in the British museum (Natural History). Between 1842 and 1889 there were sightings by others in the south-west corner of Western Australia, but they were few and far between and ceased in 1889.

REDISCOVERY

Despite numerous searches, 72 years were to pass when, on 17 December 1961, Harley Webster, headmaster of Albany Senior Primary School and a keen observer and photographer of birds, identified a Noisy Scrub-bird at Two Peoples Bay, 32km east of Albany on Western Australia's south coast. His dramatic discovery first broke in the *West Australian* on Christmas Day 1961 in which Vincent Serventy, the dedicated champion of conservation causes, was quoted as describing the discovery as 'the most exciting news of the century in Australian ornithology', and reports of the event rippled around the world. A month later, Harley Webster, in company with four ornithologist friends, began further searches for the Noisy Scrub-bird southwards onto the Mt Gardner promontory, and two more birds were quickly discovered, the first by Dr. G.F. Mees in a high valley on the flanks of the mountain. Within a few months, Harley Webster became acquainted with 'upwards of thirty' birds calling in almost all the mountain gullies in the vicinity of Mt Gardner. On a trip down the Mt Gardner coast with two visitors on 8 January 1963, the first nest, albeit unfinished, was found by Lindsay Sedgwick. On 6 October 1963, Harley Webster discovered a completed nest and on 18 June 1967, at Tick Flat, the first egg—125 years after Gilbert's discovery of the bird, and never previously seen or described. Young at the nest found in 1963 were seen on 2 July 1967. Since then, a great deal has been learned about the biology of the species as a result of further field studies by Webster, in which he was joined by F.N. Robinson in 1962 and later by G.T. Smith, both of CSIRO, and from successful experiments on captive breeding instituted by CSIRO in 1975.

REHABILITATION

Writing in the *Western Australian Naturalist* in May 1962, Harley Webster warned that the continued existence of the newly-found population was 'horrifyingly precarious' because of squatters; plans already surveyed to build a new township in the area to be called Casuarina; and the ever-present danger of bushfires. A highly-charged clash of interests ensued between those who wanted the town and people around the world who became concerned at the looming tragedy. HRH The Duke of Edinburgh, who was visiting Perth for the 1962 Commonwealth Games at the height of the controversy, lent his weight to the campaign to save the bird by expressing his concern to the Premier of Western Australia. Eventually, on 22 April 1966, to their lasting credit, the Government announced the cancellation of the town project and gazetted 4 639 hectares at Two Peoples Bay as a nature reserve. It is now manned by a ranger under the Western Australian Department of Conservation and Land Management; there is a good system of fire protection with 60km of firebreaks; and a large buffer zone separates the scrub-bird's territory from the surrounding farmland and forest.

During the first few years after the Noisy Scrub-bird was rediscovered, the population was estimated at from 40 to 45 pairs (based on the calls of singing males). Since then, thanks to the protective measures taken, the number has increased more than three-fold to about 130 pairs, a population believed to be the maximum that the Two Peoples Bay/Mt Gardner area can sustain.

A translocation programme to establish a second colony at Mt Many Peaks, about 12km northwest of the base area, was therefore initiated in 1983. This was undertaken jointly by Graeme Folley, Reserve Officer, Two Peoples Bay Reserve, and Don Merton, a world-renowned expert in translocating endangered birds, loaned by the New Zealand Wildlife Service. Ten males and four females were released between May and June of that year and two more females in November — sixteen birds in all. Territories were quickly established and four singing males were recorded during the winter of 1984. A second batch of eight males and eight females, were taken to Mt Many Peaks during July and August 1985 — bringing the new colony to thirty-two birds. Thirteen males have since been heard calling, which indicates that the birds are settling in well in their new territory, and have become viable in their own right. So far so good, but it is believed that it may take from two to three years before the birds mate and multiply in their new surroundings. The results achieved, however, justify optimism that the Noisy Scrub-bird is well on the way back from near extinction.

The combined action of so many people and organisations to save the bird is a landmark in the history of Australian conservation and promises to be one of the world's great conservation success stories.

*This letter is included in a file of correspondence deposited in Liverpool City Library, England by the present Earl of Derby. The full text is reproduced in the *Australian Zoologist* Vol.22 No.1 September 1985 published by the Royal Zoological Society of New South Wales.

60

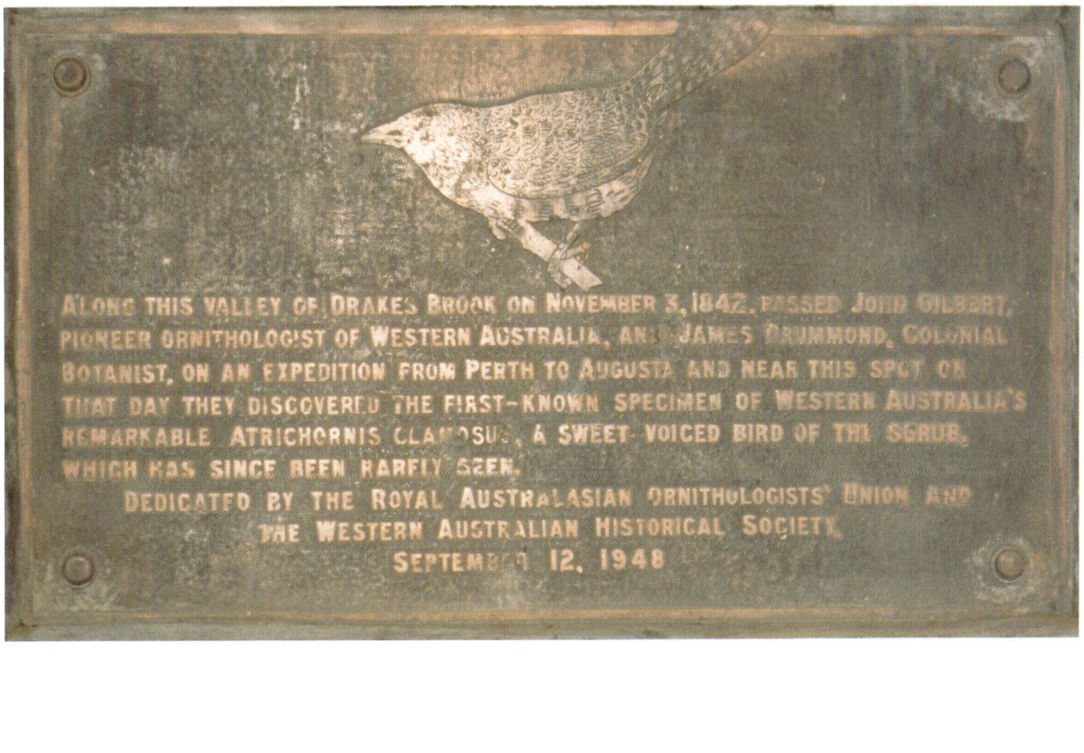
ALONG THIS VALLEY OF DRAKES BROOK ON NOVEMBER 3, 1842, PASSED JOHN GILBERT,
PIONEER ORNITHOLOGIST OF WESTERN AUSTRALIA, AND JAMES DRUMMOND, COLONIAL
BOTANIST, ON AN EXPEDITION FROM PERTH TO AUGUSTA AND NEAR THIS SPOT ON
THAT DAY THEY DISCOVERED THE FIRST-KNOWN SPECIMEN OF WESTERN AUSTRALIA'S
REMARKABLE ATRICHORNIS CLAMOSUS, A SWEET-VOICED BIRD OF THE SCRUB,
WHICH HAS SINCE BEEN RARELY SEEN.
DEDICATED BY THE ROYAL AUSTRALASIAN ORNITHOLOGISTS' UNION AND
THE WESTERN AUSTRALIAN HISTORICAL SOCIETY.
SEPTEMBER 12, 1948

5

Lyrebirds

Lyrebird
Albert's 2, 4
Superb 1, 3

First to discover the remarkable lyrebird was John Wilson, a young ex-convict who had been living with the aborigines for some five years and who, on returning to Sydney in 1797, told of seeing 'a bird of the pheasant species' near the head of the Hawkesbury River, now known as the Superb Lyrebird. The smaller Albert's Lyrebird was discovered in 1849 by Frederick Strange, an English naturalist, in the Richmond River area of north-eastern New South Wales.

The spectacular display ritual of the male is surpassed by few other species. The Superb Lyrebird begins with a marvellous expansion and quivering of the sixteen 40 cm tail feathers into a fan. The two large outer plumes, shaped like the ancient musical instrument that gives the bird its name, extend laterally – not vertically as depicted in early illustrations, including a 1/- Australian stamp. The two long central wires and the twelve egret-like feathers are thrown forward in a shimmering cascade to almost obscure the bird. The Albert's Lyrebird exhibits its tail in somewhat similar fashion, but it is little more than half the length and lacks the side lyrate feathers. A stomping dance routine and ringing outpourings of glorious song complete the act. The female remains in admiring attendance until mating, when she retires to a nest remote from the scene to incubate the single egg and rear the chick on her own.

Most famous of the lyrebird's attributes is their extraordinary gift of mimicry, possessed by both male and female, and unmatched by any other bird. In addition to their own distinctive calls, they are credited with imitating the calls and songs of many other birds, the cries of other living creatures, mechanical sounds, and even human voices. Stories of these vocal feats are legendary and, while some may be fanciful, there are records which cannot be doubted – such as those of Tom Tregellas, who lived in a large hollow log in Sherbrooke Forest for seventeen years from 1914 to 1931 and whose diary lists imitations of the calls of over forty species of birds. It also notes that the sounds of 'sawing of wood, blows of an axe, the clinking of wedges are faithfully reproduced'. But others who have spent long periods with lyrebirds filming them and recording their calls have had no similar experience. It would therefore appear that, while lyrebirds can reproduce almost any sound heard by day or by night, they adopt sounds other than the voices of birds and mammals only in special circumstances – such as being close to human habitation.

The song of the lyrebird reaches its peak during the mating period, but it is also heard at other times of the year when, it has been suggested, it continues singing for the sheer joy of it. Its gifts of extemporisation and composition are of a high musical order; to the extent that K.C. Halafoff, a music scholar, found that the Superb Lyrebird's clever blending of the calls of other birds in its repertoire 'came close to the idea of interplay of different instruments' used in the structure of contemporary orchestral compositions by such composers as Stravinsky and Prokofieff. He also discovered that all the birds studied sang in the key of C major!

Lyrebirds are vulnerable to land clearance, predation from foxes, dogs, cats and bush fires.

PHOTOGRAPHERS:
H. & J. Beste 3;
R. & D. Keller 1;
W.G. Lawler 4;
E.C.J. Smith 2.

1, 3 SUPERB LYREBIRD

Menura novaehollandiae ♂ up to 100 cm ♀ 86 cm

The elusive Superb Lyrebird lives in dark gullies in wet eucaplypt forests. Here it grubs among fallen litter for beetle, worms and crustaceans; it also probes for insects under bark of rotting logs. The birds are most in evidence during the colder months when mating takes place and when their displays and wonderful voices are at their peak. The famous display rituals of the polygamous male are usually performed on a special platform, several of which may be prepared in readiness for the mating season. These average about a metre in diameter and are about 15 cm high. A series of platforms may be used in a day. After the chick has fledged, the male moults its tail and falls silent for up to six weeks. New tails become progressively longer for up to ten years. The song is constantly improved through practice and the young birds are schooled by their elders. Roosts are often chosen in tree tops, perhaps as high as 40 metres. Breeds April–September. **Immatures**: *like female.* **Voice**: *'ca-luck, ca-luck', 'pellick, pellick', integrated with mimicry of other birds.*

Below: female Opposite top: male

2, 4 ALBERT'S LYREBIRD

Menura alberti ♂ 90 cm ♀ 65 cm

The Albert's Lyrebird is smaller and more rufous than the Superb Lyrebird. It is restricted to the dense rainforest of the rugged McPherson Range on the Queensland/New South Wales border, and lives in an almost impenetrable habitat of vines and other sub-tropical ground cover where it is exceedingly difficult to find. It is even shier than the Superb Lyrebird, and this is illustrated by the experience of Wayne Lawler the photographer of the male bird opposite, who took three years to work out a strategy to get his shots, which required digging a deep pit in the forest floor in which to obscure himself and camera. The Albert's Lyrebird behaves much like its southern relative, except for its display platform which is simply a flat trampled arena of vines and leaves. It is partial to the very large snails that abound in its area. Breeds June–July. Nests on rocky ledge, tree stump or buttress, and sometimes on a high branch. **Immatures**: *like female.* **Voice**: *perhaps less variable than that of the Superb Lyrebird and a much different territorial call.*

Below: female Opposite below: male

1

2

3

4

6

Pigeons and Doves, Cuckoos

Opposite: White-quilled Rock Pigeon

The pigeon family Columbidae belongs to the order Columbiformes, one of the main orders of birds which includes the extinct Dodo. This ungainly bird is familiar to most people from its illustration in *Alice In Wonderland* based on sketches by Dutch artists of live Dodos brought to Europe from its home in Mauritius. The Dodo disappeared around 1680 due to the numbers killed for food by visiting sailors and the depredations of pigs and monkeys introduced onto the island by the Portuguese in the 16th century.

The use of the terms 'pigeon' or 'dove' has no zoological significance, but the larger and plumper species tend to be called 'pigeons' and the smaller more graceful species with pointed tails 'doves'. Pigeons range in size from Australia's sparrow-sized Diamond Dove to New Guinea's giant Goura Pigeons. Pigeons have distinctive characteristics which separate them from most other birds: they drink by continuous sucking instead of tipping their heads back to swallow after each sip; their feathers drop out easily, perhaps as protection against some predators; and they feed their young with a regurgitated cheese-like secretion known as 'pigeon's milk'. Most pigeons lay glossy white eggs, usually a pair. The family Columbidae has about 285 species distributed throughout the world in temperate and tropical regions. Of these, twenty-five occur in Australia, including three introduced species. Thirteen are endemic, six extend to New Guinea and adjacent islands, and three further afield. There are four subfamilies, two of which are represented in Australia – Columbinae and Treroninae.

COLUMBINAE (Subfamily)

Most famous — and most tragic — member of the subfamily is the Passenger Pigeon of North America which at one time was perhaps the most abundant bird in the world. An ornithologist in 1808 estimated one flock in Kentucky at over two thousand million birds, and in 1813 John James Audubon witnessed a flock that passed over for three days and darkened the sun. A century later, as a result of slaughter on a gigantic scale, the species had been wiped out — the last surviving dying in Cincinnatti Zoo in 1914. The Australian species of this large and diverse subfamily fall into seven easily identifiable groups.

Typical Pigeons, genus *Columba.* This group of fifty-one species worldwide is represented in Australia only by the endemic White-headed Pigeon and the introduced Domestic Pigeon. Typical of the group is the Rock Dove or Rock Pigeon of temperate Europe and western Asia which is the ancestor of our familiar street pigeons and the many varieties bred from them. It has a history of domestication going back to ancient civilisations and has been used by the Romans and in the two world wars for carrying messages. 9,11

Turtle Doves, genus *Streptopelia* Widespread in Eurasia and India, but especially in Africa where ten of the fifteen species occur. Two species have been introduced into Australia – the Spotted Turtle Dove and the Laughing Turtle Dove. 3, 4

Australian Doves, genus *Geopelia* The three species of these ground-feeding doves are Australian in origin, but only the tiny Diamond Dove is endemic. 1, 2, 5

Green-winged Pigeons, genus *Chalcophaps* Also known as 'emerald doves', there are two species of these ground-living birds in the Indo-Australian region, but only the Green-winged Pigeon extends to Australia. 26

Bronzewing Pigeons The bronzewings are so named from the irridescent colour in their wings, particularly evident in the Common and Brush Bronzewings. The nine species, all endemic, occupy all of the continent, and are the most successful group of pigeons in Australia. The Common Bronzewing is the most widespread of all the Australian pigeons. The group divides into four fairly distinctive genera:
Phaps – The genus comprises the Common Bronzewing, Brush Bronzewing and the wholly ground-dwelling Flock Pigeon. 13, 23, 24
Geophaps – The two forms of the Partidge Pigeon, the closely related Squatter Pigeon and the two forms of the wholly ground-dwelling Spinifex Pigeons make up this group. 14–16, 21, 22
Ocyphaps – The ubiquitous Crested Pigeon is the sole member of this group. 20
Petrophassa – The genus comprises the rock and cliff-dwelling White-quilled and Chestnut-quilled Rock-Pigeons of the Kimberley region and Arnhem Land respectively. 18, 19

Wonga Pigeon, genus *Leucosarcia* A large pigeon comprising a one-species genus restricted to Australia with no close relatives. 25

Cuckoo Doves, genus *Macropygia* A group of eight large rainforest species of the Indo-Australian region distinguished by their brown plumage and long graduated tails. The Australian species is known simply as the Brown Pigeon. 27

TRERONINAE (Subfamily)

The Treroninae divide into nine genera worldwide, of which three are represented in Australia.

Fruit Pigeons, genus *Ptilinopus* A group of some forty-seven species, widely distributed from Australia to Sumatra, the Philippines and the Marquesas, which are completely arboreal and live mainly on a diet of fruits and berries. Four species reach Australia – the Red-crowned and Purple-crowned Pigeons, the Banded Pigeon and the Wompoo Pigeon. 6–8, 10

Topknot Pigeon, genus *Lopholaimus* The endemic Topknot Pigeon is the sole representative of this genus and seems to have no close relative among the world's pigeons. 17

Imperial Pigeons, genus *Ducula* There are thirty-six species of this well-marked group of pigeons, also called 'nutmeg' pigeons because of their liking for nutmegs. They have snake-like expandable jaws which enable them to swallow these and other fruits that are sometimes larger than their heads. They range widely beyond Austra-

lia with the New Guinea region as their stronghold. Only one, the elegant and prolific Torres Strait Pigeon regularly reaches Australia. 12

DISCOVERY OF THE TORRES STRAIT PIGEON

The Torres Strait Pigeon was the first bird occurring in Australia to have been authentically recorded by a European. The story of its discovery begins on 21 December 1605, when a squadron of two Spanish galleons and a smaller vessel sailed from Callao, Peru, under the command of the Portuguese navigator and visionary Captain Pedro Fernandez de Queiros on a mission to discover the Great South Land. On 1 May 1606 a large land mass was encountered (the New Hebrides) and, believing it to be the Great South Land, Queiros named it Austrialia del Espiritu Santo in honour of the Austrian King Philip III of Spain and formed a settlement which he called New Jerusalem. After only six weeks there, Queiros' crew seems to have mutinied and, without warning, he deserted the two other ships and sailed back to Mexico.

Thus abandoned, Captain Luis Vaez De Torres, skipper of the *San Pedroco* set off on a south-westerly course on 28 June 1606 in company with the smaller *Tres Reyes*. After about two weeks, when in latitude 21°, he gave up hope of finding land ahead and turned north — not knowing that the Great Barrier Reef then lay only 190 miles away. Landfall was made at the eastern tip of New Guinea on 14 July 1606, and Torres proceeded to work his way westwards to become the first navigator to penetrate the strait that bears his named from the east. Original documents on the voyage include a letter from Torres to King Philip written from Manila and an extensive narrative written by Torres' companion Captain Don Diego de Prado y Tovar, lost until 1920 and first published in 1930. From these we learn that thirty-four days were spent seeking a way through the barrier of the shallow waters, islands, and dangerous reefs of the continental shelf that joins New Guinea to Cape York, and that an exit west was negotiated at a latitude of 11°. Among the landings made during this leg of the voyage was one described by Prado as follows (as translated by Barwick in Stevens, 1930):

> [25 September 1606] 'we reached a flat island . . . where we anchored and found plenty of very large pigeons all white and trees of plums they call Nicaragua, they have big stones and little flesh. The whole island was full of pumice stone and seems to have had a volcano in former times; we gave it the name of Isla de vulcan quemado [Island of Extinct Volcano]; it would be a league and a half in circumference'.

Brett Hilder in his *The Voyage of Torres* (1980), identifies the island as Long Island which lies about 80 km NNE of Cape York. He reports that the plums still grow there and recognises them as the wild yellow Nicaraguan plum *Ximenia americana* which has very little flesh and ripens in September. The reference to pumice is explained as coming from the many dormant volcanoes in the South Pacific. That Long Island is the island referred to by Prado is further corroborated by its being directly under the migratory route of the Torres Strait Pigeon to its breeding islands on the Great Barrier Reef. Hilder also concludes that Torres found his escape through Endeavour Strait. The final irony was that he was unaware that the land on his port bow, and in full view for three days, was the very land his expedition had set out discover.

CUCKOOS

The cuckoo family Cuculidae consists of 127 species, well distributed in the temperate and tropical regions of the world, with fourteen species in Australia. These represent the two main subfamilies — the typical parasitic Cuculinae and the long-tailed non-parasitic Centropodinae. Cuckoos everywhere owe their name to the distinctive two-note call of the male Common Cuckoo *Cuculus canorus* which heralds spring in many countries.

THE TYPICAL PARASITIC CUCKOOS Subfamily Cuculinae

There are nearly fifty species of typical cuckoos spread throughout the world. In Australia they are represented by three main groups: genus *Cuculus* comprising five medium-sized cuckoos, of which the Oriental Cuckoo is a non-breeding visitor; genus *Chrysococcyx* comprising five small Bronze-Cuckoos, of which the Shining Bronze Cuckoo of New Zealand is a non-breeding visitor and the Black-eared Cuckoo is an endemic breeding species; and a group of two large cuckoos — the Koel and Channel-billed Cuckoo. Most cuckoos occuring in Australia are migratory, some covering great distances: the Shining Bronze-Cuckoo flies over 3000 km to overwinter in the Solomon Islands. 28-41

Typical cuckoos appear to build no nests of their own or play any part in incubating and rearing their young which are left to fend entirely for themselves on leaving the nest. Remarkably, this includes finding their way unaccompanied by parents on the long migratory routes. While certain other birds overseas also behave like cuckoos, none display the same degree of ingenuity and cunning in duping hosts into their service, especially in the choice of nests of species whose eggs are similar in size and colour to their own. The cuckoo's egg is deposited sometimes only after a fight with the hosts maybe over a few days, and it is either laid by sitting or by ejection backwards. Shining Bronze-Cuckoos have been known to force their way right through the domed nests of wrens and warblers with head poking out of the other side to lay their eggs. Cuckoos sometimes remove an egg of the rightful owner when replacing it with their own. The nests of over 100 species of Australian passerine birds are known to have been parasitised by cuckoos.

Soon after hatching, the naked and blind fledgling cuckoo starts shouldering any eggs or other nestlings out of the nest. Sometimes the young cuckoo can grow to several times the size of its foster parent (see for example the photograph opposite of a Grey Fantail feeding a young Pallid Cuckoo).

THE NON-PARASITIC COUCALS Subfamily Centropodinae

The twenty-seven species of this pheasant-like bird range from Africa through Australia to the Solomons. A single species, the Pheasant Coucal, reaches Australia and is also found in New Guinea and Timor. 42

PHOTOGRAPHERS: H. & J. Beste 2, 7, 8, 11, 19, 22, 41; N. Chaffer (NPI) p. 43; G. Chapman p. 41; 4, 13, 15, 18, 29, 30, 32, 35; A.T. Foster 17, 28; T. & P. Gardner 14, 16, 25, 33; R.A. Graves 27, 40; P.N. Green 12, 42; A.L. Hertog 6; R. & D. Keller 23; P. Klapste 24; C. Webster 5, 26; B.A. & A.G. Wells 21; E.E. Zillmann 10, 38; Authors 1, 3, 9, 20, 31, 34, 36, 37, 39.

Grey Fantail feeding a young Pallid Cuckoo

6

1 DIAMOND DOVE

Geopelia cuneata 22 cm

Australia's smallest pigeon is a bird mostly of the inland, and is nomadic as the climatic conditions dictate. It is highly gregarious, congregating in large flocks at water, flying and roosting in flocks, always at least in pairs and groups. It feeds entirely on the ground on seeds, taking some buds and insects, with an amusing toddling run. Flight is direct and swooping. Breeds all the year round in response to rain. Nests low in trees or bushes or on ground. Endemic. **Sexes**: *alike but female has duller and browner foreparts.* **Immatures**: *black-grey striping on upper parts.* **Voice**: *soft 'poor-pa-poor-papa'.*

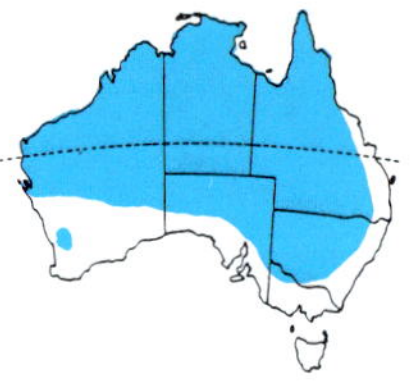

2 PEACEFUL DOVE

Geopelia placida 23 cm

This dainty native pigeon frequents open forest and scrubby woodland near water, roadsides, parks and gardens. It feeds terrestrially on a wide variety of seeds, in pairs and groups, or in loose feeding and drinking flocks. At other times it roosts quietly in trees; in the breeding season it reacts aggressively to other birds. The species is still reasonably common but may lose out in competition with introduced doves. Sedentary. Probably breeds year round. Nests in trees 6–20 m high. Also in islands to the north and north-west. **Sexes**: *alike.* **Immatures**: *paler and more brown.* **Voice**: *'doodle-do' and low churring.*

3 SPOTTED (Indian) TURTLE DOVE

Streptopelia chinensis 31 cm

A native of India and a number of other countries of South-East Asia, it was introduced to Melbourne in the 1860s and has since spread much further afield. It is very common in urban areas, especially around silos; also in developed rural areas and scrubby thickets. In pairs and groups it feeds on bread, grain and seed. In some areas it seems to be replacing native doves. Relatively tame, it flushes easily with a noisy burst. Breeds all year round. Nests in trees, ledges, or on the ground. **Sexes**: *alike.* **Immatures**: *no chequered patch.* **Voice**: *persistent gentle cooing.*

4 LAUGHING (Senegal) TURTLE DOVE

Streptopelia senegalensis 25 cm

Native of Africa, countries of the Middle East and India, the species was released in Western Australia nearly a hundred years ago. It frequents city and surburban areas, farms and silos and is seen in pairs, or small to large flocks where food is plentiful, feeding on the ground or taking seeds and buds from low bushes. It copes well with dry conditions. Probably breeds the year round. Nests in dense foliage. **Sexes**: *similar but female duller.* **Immatures**: *no speckling, no blue-grey on wing.* **Voice**: *long drawn-out cooing with a laughing quality.*

5 BAR-SHOULDERED DOVE

Geopelia humeralis 30 cm

The Bar-shouldered Dove frequents habitat near water including woodland, scrub and mangrove. In recent years there has been a gradual extension of range southwards; on the other hand it may have given ground to introduced species where the latter have established themselves. In pairs and groups, sometimes large loose flocks, it feeds on the ground close to cover, on seeds, preferring those growing in swampy areas. Sedentary. Breeds most months of the year. Nests low to 6 m high. **Sexes**: *alike.* **Immatures**: *hind neck and upper mantle brown.* **Voice**: *rollicking 'cook-a-wook'.*

6 BLACK-BANDED PIGEON [Banded Fruit-Dove]

Ptilinopus cinctus 36 cm

Living in a very restricted range of pockets of forest amongst sandstone cliffs and gorges in west Arnhem Land, this pigeon's habitat may be endangered by uranium mining. It feeds quietly and inconspicuously on native fruits, and gathers in the tree canopy in pairs or very small groups, or singly, flying from tree to tree and perching with a forward flip of the tail. Breeds in the dry season, May–November. Nests 2–5 m high. Also in Indonesia. **Sexes**: *similar.* **Immatures**: *grey replaces the white parts of adults.* **Voice**: *deep booming, soft 'coo'.*

7 PURPLE-CROWNED PIGEON [Superb Fruit-Dove]

Ptilinopus superbus 24 cm

This resplendent fruit-eating pigeon, widespread in territories to the north, occurs in Australia from Cape York southward in rainforest and mangroves. Difficult to detect in the canopy, it can betray its presence by falling fruit pieces when feeding. Nomadic. Breeds June–February. Nests in trees to 10 m high. **Recognition**: *white underside and flanks of the male.* **Female**: *small crown patch, blue instead of purple, lacks dark breastband.* **Immatures**: *like females, breast finely scalloped yellow.* **Voice**: *low 'oom' exploding into 'woops'.*

Male

8 RED-CROWNED PIGEON [Rose-crowned Fruit-Dove]

Ptilinopus regina 23 cm

Somewhat similar in distribution and habitat to its close relative the Purple-crowned Pigeon, but found also in northern Australia. It seems to be totally arboreal, even taking water from foliage, and is seen in pairs or small parties. Nomadic or partially migratory. Breeds variably October–March. Nests low in trees to 30 m high. **Recognition**: *lacks the white underparts of the Purple-crowned Pigeon.* **Sexes**: *alike, but female duller.* **Immatures**: *mostly green above.* **Voice**: *loud cooing that makes the forest 'throb'.*

9 DOMESTIC [Feral] PIGEON

Columba livia 33 cm

The familiar domesticated pigeon is descended from the wild Rock Dove of Europe, and some are still found in their original state north of the British Isles and a few other parts of the world. Feral populations of domesticated birds have spread throughout the world, including in Australia from introductions. They are now found wherever there is a city, town or village, in many agricultural areas, and even at isolated homesteads in the interior. Breeds year around. Feral birds nest almost anywhere including crevices in cliffs and on the ground. **Sexes** and **Immatures**: *alike.* **Voice**: *cooing.*

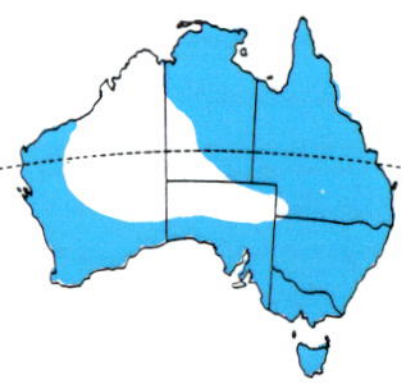

10 WOMPOO PIGEON [Wompoo Fruit-Dove]

Ptilinopus magnificus 56 cm

In tropical and sub-tropical rainforest this large fruit pigeon feeds agilely, hanging upside down from frail branches to reach fruit and berries. It spends most of its time in the tree canopy and brilliantly-plumaged as it is, it is difficult to see unless the yellow undertail catches the eye. Southern populations are threatened by rainforest clearance. Locally nomadic depending on the fruit supply. Breeds June–January. Nests high. Also in New Guinea. **Sexes**: *alike.* **Immatures**: *green patches on purple breast.* **Voice**: *loud and deep bubbling 'whompoo'.*

11 WHITE-HEADED (Baldy) PIGEON

Columba leucomela 41 cm

A large mainly arboreal fruit pigeon which is seen in tropical and subtropical rainforests and scrubs. Its numbers have declined as a result of white settlement but in some areas it has adapted to thrive on camphor laurel. Wary and elusive, it lives silently in lower tree branches, occasionally feeding on the ground in pairs and small grounds. It bursts noisily out of cover when a flock suddenly departs. Nomadic. Breeds most months. Nests in branch or vine tangle up to 18 m high. Endemic. **Sexes**: *similar but female duller.* **Immatures**: *greyer.* **Voice**: *low ventriloquial 'oom' or 'oom-coo'.*

1

2

3

4

5

6

7

8

9

10

11

6

12 TORRES STRAIT (Nutmeg) PIGEON

Ducula spilorrhoa 44 cm

The Torres Strait Pigeon, also known as the Torres Strait Imperial-Pigeon, a breeding migrant from New Guinea, arrives in the Northern Territory and Queensland in great numbers from August and departs during March and April. The Queensland birds disperse to the offshore islands of the Great Barrier Reef where they set up breeding colonies sometimes consisting of several thousand birds. They eat all sorts of large native fruits, including that of the nutmeg tree, for which they have to cross to the mainland daily, sometimes involving a round trip of sixty or so kilometres. In former times, these daily crossings were of staggering proportions. E.J. Banfield (1908) estimated that 'fully 100 000 come and go evening and morning' over Dunk Island in processions two miles wide, and that this company was 'small compared with the myriads' of the past. Even now, H.J. Frith (1982) describes these crossings as 'one of the great ornithological experiences of the tropics', but it is now necessary to go north of Cairns to witness this phenomenon, though on a lesser scale. Nests low in mangroves. **Sexes**: *alike*. **Immatures**: *grey wash over white parts*. **Voice**: *loud 'coo-woo'*.

13 FLOCK PIGEON [Bronzewing]

Phaps histrionica 30 cm

Now abundant on the vast open grass plains of the north. Competition for its food source from grazing cattle from the 1870s progressively reduced its numbers and it nearly died out following the disastrous drought of 1893–1902. It is highly nomadic throughout its range with its present stronghold centered in the Northern Territory. Breeds variably. Endemic. **Female** and **Immatures**: *head markings are brown instead of black*. **Voice**: *relatively silent, single-note murmuring when feeding, throaty 'coo' during breeding display*.

Male

14 SQUATTER PIGEON

Geopelia scripta 32 cm

The skin around the eye of the Squatter Pigeon of Cape York Peninsula is bright red, while it is blue further south. As a result of pastoral development it has become very rare in the southern part of its range, though it has fared rather better in cattle country of the north. Singles or small parties frequent grassy plains or open woodland to forage for seeds of grasses, herbs, trees and shrubs. When flushed, they either explode out of cover or squat to avoid detection. Locally nomadic. Breeds year round. Nests in scrape on ground. Endemic. **Sexes**: *alike*. **Immatures**: *markings less distinct*. **Voice**: *low 'coo-cwoop'*.

15, 16 PARTRIDGE PIGEON

Geopelia smithii 28 cm

(A) Race *blaauwi* (Yellow-eyed form); (B) Race *smithii* (Red-eyed form). The yellow-eyed form has chrome-yellow bare skin around its eyes and is found only in the Kimberley region of Western Australia where it lives in open woodland in gorges and among rocks in sandstone country. The red-eyed form has bright red bare skin around its eyes and is particularly associated with the escarpment country of Arnhem Land. The red turns to the same yellow as its relative in museum specimens and the two become indistinguishable. The species is similar in habits and habitat to the Squatter Pigeon, and also freezes or squats in cover. In pairs or coveys of from 15 to 20 birds, sometimes larger, it forages for grass seed in open tropical woodland. Flight is in partridge-like bursts close to the ground. Breeds March–October. Nests in a scrape on the ground. Endemic. **Sexes**: *alike*. **Immatures**: *upper parts flecked chestnut*. **Voice**: *'coo'*.

17 TOPKNOT PIGEON

Lopholaimus antarcticus 46 cm

During the early days of white settlement, immense flocks offered a ready supply of fresh food, but today the species is fully protected. However, the decline in size of contemporary flocks is probably due mostly to the clearance of their rainforest and adjacent woodland habitat. It is completely arboreal and even gets most of its moisture from foliage dew and raindrops, as well as its diet of native fruits. Nomadic, following fruiting trees and palms. Breeds June–December. Nests high. Endemic. **Sexes**: *alike*. **Immatures**: *like adults, bill red-brown*. **Voice**: *single low 'coo'*.

18 WHITE-QUILLED ROCK PIGEON

Petrophassa albipennis 30 cm

Though closely related to the Chestnut-quilled Rock Pigeon it lives quite separately in an area extending into the Kimberley region, where it is found on the sandstone escarpments among spinifex and stunted trees. It is sedentary and is seen singly or in small flocks foraging for grass seed and herbaceous plants. Probably breeds April–November. Nests in a loose arrangement of sticks on a rocky ledge. Endemic. **Sexes**: *alike*. **Immatures**: *less white on throat*. **Voice**: *various cooing notes*.

19 CHESTNUT-QUILLED ROCK PIGEON

Petrophassa rufipennis 32 cm

This species occurs in the rocky fastnesses of the Arnhem Land plateau and escarpments, and frequents sandstone, rainforest, woodland or spinifex. Commonly seen in small groups. In the wet season, parties move out from the protection of rocks into nearby woodland, while in the heat of summer, they shelter in cool places under rocks and in caves. The species feeds on the ground and its diet includes wattle, herb and grass seeds. Breeds perhaps February–November. Nests in rock crevices. Endemic. **Sexes**: *alike*. **Immatures**: *duller*. **Voice**: *call of 'kukarook'*.

20 CRESTED PIGEON

Ocyphaps lophotes 35 cm

Land clearance and the provision of drinking facilities for stock has actually helped this adaptable native pigeon to extend its range and, except for the sandy deserts, there are few places where this bird is absent. It feeds entirely on the ground mainly on seed, especially Paterson's Curse, but also eats vegetable plant material and insects. It takes off with whistling wings and tilts its tail on landing. Sedentary. Breeds variably. Nests in cover in bush or tree at varying heights up to 18 m. Endemic. **Sexes**: *alike*. **Immatures**: *duller*. **Voice**: *single and double 'coo' or rising 'coo-oo'*.

21, 22 SPINIFEX PIGEON

Geopelia plumifera 23 cm

(A) Race *ferruginea* (Red-bellied form), popularly known as the Red-plumed Pigeon, was scientifically named by Gould from a specimen collected by the explorer Frank Gregory on the Gascoyne River in May 1858, on the same expedition on which he discovered and named Mount Augustus, Australia's largest rock. (B) Race *plumifera* (White-bellied form), popularly known as the Plumed Pigeon, was scientifically named by Gould from a specimen collected on the Victoria River in the Northern Territory in 1839. The southern birds of (B) have white lower breasts and abdomens and are sometimes recognised as a separate race. These endemic quaint birds of erect soldierly bearing live in small flocks or coveys of 50 or more birds. They frequent arid flat open areas with seed-producing grasses, or rocky spinifex hillsides, always near water, and are seed-eaters. They are sedentary with only minor local movement even in drought seasons. They breed most months, especially after rain and nest in a scrape in the ground. Endemic. **Sexes**: *alike*. **Immatures**: *lack black and white on face and throat*. **Voice**: *repeated 'cu-woo'*.

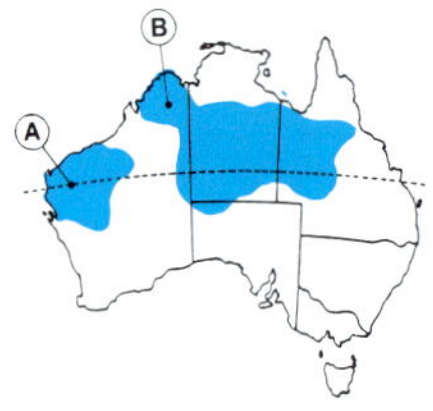

(A) race *ferruginea* left
(B) race *plumifera* right

12

13

14

15

16

17

18

19

20

21

22

6

23 COMMON BRONZEWING

Phaps chalcoptera 35 cm

The abundance of this bird provided the early settlers all over the continent with a welcome change of diet. It is the most widespread of Australia's pigeons and is found in forests, woodlands and scrubs, in all but the wettest of areas. Singly or in pairs it feeds on the ground mostly on seeds, grain and native plants. Large flocks gather at watering places, especially at dusk. Extremely wary and alert. Sedentary or locally nomadic. Breeds variably. Nests on the ground or in trees, sometimes high. Endemic. **Sexes**: *similar but female duller.* **Immatures**: *much duller.* **Voice**: *low resonant mournful ventriloquial 'com'.*

24 BRUSH BRONZEWING

Phaps elegans 30 cm

The Brush Bronzewing prefers denser cover than the Common Bronzewing and is even more secretive and guarded. Its feeding habits are similar. It is seen singly or in pairs and usually runs swiftly rather than taking to flight. Numbers have declined from interference with habitat and from the depredations of foxes and feral cats. Locally nomadic. Breeds throughout the year. Nests on the ground or low. Endemic. **Recognition**: *smaller and redder than the Common Bronzewing.* **Sexes**: *similar but female less chestnut.* **Immatures**: *like females.* **Voice**: *repeated low cooing.*

25 WONGA PIGEON

Leucosarcia melanoleuca 43 cm

This large ground-frequenting pigeon inhabits mostly coastal rainforest and scrub, sometimes on the fringes of settled areas. Solitary, in pairs only during the breeding season, it forages sedately for seeds and berries on low bushes and fallen fruit. Exceptionally alert, if disturbed it freezes or flies up with noisy wing flapping to become near-invisible in foliage. Sedentary. Breeds October–January. Nests to 10 m high. **Sexes**: *alike.* **Immatures**: *brownish rather than grey.* **Voice**: *loud continuous and monotonous cooing 'wonk' which can be heard from afar.*

26 GREEN-WINGED PIGEON [Emerald Dove]

Chalcophaps indica 25 cm

This small but colourful bird is often encountered at close range in rainforest quietly picking its way through the ground litter of forests, mangroves or lantana thickets. Unobtrusive and mostly solitary, it may also be seen sometimes in pairs and parties, feeding on seeds, fruits and berries in clearings. If disturbed it rises silently before going off to cover, from which it watches alertly. Breeds most months. Nests 4–11 m high. Widespread in southern Asia. **Female**: *white shoulder patch of male replaced with grey.* **Immatures**: *less green.* **Voice**: *extended cooing.*

Male

27 BROWN PIGEON

Macropygia phasianella 40 cm

With a range extending to Sumatra and the Philippines, the Brown Pigeon is found in Australia in the rainforests of the eastern states. Sometimes referred to as a cuckoo-dove or the Pheasant Pigeon because of its long graceful tail. It is seen singly or in pairs and is reasonably common. Tamer than most rainforest pigeons. Breeds most months. Nests at varying heights in vine tangle or tree. **Sexes**: *similar but female less pink and duller.* **Immatures**: *black-barred upper breast.* **Voice**: *repeated 'cuckoo-whup', soft clucking 'coo'.*

28 SHINING BRONZE-CUCKOO

Chrysococcyx lucidus 16 cm
Race: *lucidus*

The New Zealand breeding race of a species widespread in the Australasian region. It visits the coastal belt of eastern Australia in autumn and occasionally in spring as a passage migrant to and from the Solomon Islands. **Recognition**: *barring bronze-green; greenish sheen on crown and back of neck; male has white line on forehead.* **Sexes**: *similar but female lacks white line on forehead.* **Immatures**: *upper parts bright green.* **Voice**: *said to be indistinguishable from the Golden Bronze-Cuckoo.*

29 SHINING BRONZE-CUCKOO

Chrysococcyx lucidus 16 cm
Race: *plagosus* (Golden Bronze-Cuckoo)

The Australian breeding race of a species widespread in the Australasian region. It migrates northwards to winter, some birds also reaching the Solomon Islands. Frequents forests, woodlands and suburban areas. Breeds August–January and especially favours fairy warblers and fantails among its 60 hosts. The pale green-bronze pigmentation of the egg rubs off. **Recognition**: *barring bronze-brown.* **Sexes**: *alike.* **Immatures**: *grey-brown washed bronze-green.* **Voice**: *repeated high pitched ascending 'pee' likened to a person whistling up a dog, fading slightly.*

30 LITTLE BRONZE-CUCKOO

Chrysococcyx malayanus 15 cm

This small cuckoo is an Australian breeding race of a species widespread in areas extending to South-East Asia including New Guinea. It migrates northwards to winter. Solitary, it frequents tropical forests, woodlands and mangroves. It breeds October–January and lays its greenish-olive egg mainly in nests of warblers. **Recognition**: *no rufous, dark blue-tinged bronze-green crown and back of head, contrasting with black.* **Sexes**: *eye and eye ring – male bright red, female brown.* **Immatures**: *upperparts dull green.* **Voice**: *faint piping, a tremulous sigh and a four-note call.*

Male

31 RUFOUS-BREASTED [Gould's] BRONZE-CUCKOO

Chrysococcyx russatus 15 cm

The range of this small cuckoo overlaps that of the Cape York part of the Little Bronze-Cuckoo and their habits have much in common. It is a race of a species shared with New Guinea which extends to Borneo and Malaysia. It breeds September–January and lays its dark brown freckled egg in nests of warblers and sun-birds. **Recognition**: *upper breast and tail feathers rufous; upper parts rufous bronze-green.* **Sexes**: *eye and eye-ring – male bright red, female brown.* **Immatures**: *no rufous on breast.* **Voice**: *grasshopper-like repeated trill.*

Female

32 HORSFIELD'S BRONZE-CUCKOO

Chrysococcyx basalis 17 cm

A widespread endemic breeding species which mainly frequents timber, scrub and open country. Sometimes seen in built-up areas singing on overhead wires. Some birds move inland to winter and others migrate northwards as far afield as Malaysia and Borneo. Breeds June–December. Its whitish speckled egg is laid in the nests of over 60 host species including fairy wrens. **Recognition**: *streaked throat (not barred); broad whitish streak over the eye.* **Sexes**: *alike.* **Immatures**: *duller.* **Voice**: *a descending repeated 'see-you' whistled sorrowfully; also a sparrow-like call.*

23

24

25

26

27

28

29

30

31

32

6

33 BLACK-EARED CUCKOO

Chrysococcyx osculans 20 cm

An endemic breeding species, though with some migration to the Aru Islands and the Moluccas to winter. It occurs mostly west of the Great Dividing Range in the drier kinds of habitat. Breeds August–January and chooses the nests of birds which lay brownish eggs similar to its own – the Speckled Warbler and the Redthroat being favoured hosts. The pigmentation of its egg rubs off. **Recognition**: *no barring on underparts*. **Sexes**: *similar*. **Immatures**: *brownish-green above, pale eyebrow*. **Voice**: *repeated plaintive descending whistle; alternatively it calls 'peer-o-wit-pee-o-weer'*.

34 PALLID CUCKOO

Cuculus pallidus 33 cm

A widespread elegant Australian-breeding cuckoo found in all timbered areas. Singly or in pairs, it feeds on large insects including caterpillars and grasshoppers. It moves seasonally in complex migratory patterns, occasionally as far as the Moluccas, and populations fluctuate greatly. Breeds September–January. It has parasitised the nests of over 80 species, including many honeyeaters. **Adults**: *alike, lack mottling of the immatures*. **Voice**: *sometimes called the 'Brain-fever Bird' because of the male's frantic courting calls. Also repeated 8 or more rising notes, and at night*.

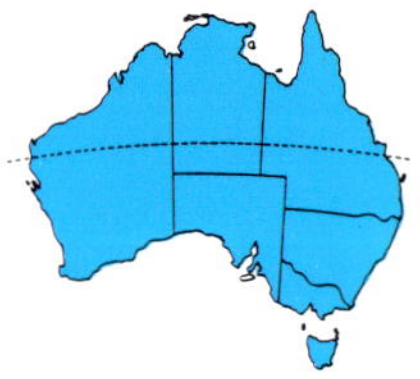

Immature

35 BRUSH CUCKOO

Cuculus variolosus 24 cm

A race of a species which extends to Malaysia and the Solomon Islands, and a winter migrant northwards, including to New Guinea. Frequenting forests and mangroves, it perches unobtrusively often alone, feeding on insects and caterpillars. It favours open cup-shaped nests to lay its faintly marked whitish egg, and some 35 hosts have been recorded, including robins, wrens and wood-swallows. **Recognition**: *square tail*. **Sexes**: *alike*. **Immatures**: *brown mottling on upperparts*. **Voice**: *6 or 7 mournful descending notes increasing in volume; also feverish accelerating crescendo of 3 ascending notes continually repeated*.

36 FAN-TAILED CUCKOO

Cuculus pyrrhophanus 27 cm

A race extending to the Aru Islands of a species ranging to New Guinea and New Caledonia. It frequents forest and woodlands, but is absent from the drier inland areas. While mainly arboreal, it often feeds on the ground and is partial to caterpillars. Characteristically, it fans and tilts its tail when alighting. Breeds August–September. Chooses domed nests, such as thornbills or occasionally the open nests of honeyeaters, to place its egg. Nomadic or migratory. **Sexes**: *alike*. **Immatures**: *upperparts and head brown*. **Voice**: *a mournful trembling cascade of falling notes; also swelling two-note upwards syllables*.

37 CHESTNUT-BREASTED CUCKOO

Cuculus castaneiventris 23 cm

This attractively-plumaged cuckoo, abundant in New Guinea, is found only in the remote rainforests of Cape York Peninsula as a small breeding population – though some migration to overwinter in New Guinea is probable. From a low perch, it watches for caterpillars, insects and their larvae, and darts to seize them. **Sexes**: *alike*. **Immatures**: *upperparts cinammon brown*. **Voice**: *descending trembling and other notes similar to Fan-tailed Cuckoo*.

38 ORIENTAL CUCKOO

Cuculus saturatus 33 cm

This bird, rare in Australia, is a non-breeding migrant from Asia, and is seen from November to April in the forests and woodlands of northern and eastern Australia. Singly, in pairs or small groups, it perches quietly on dead tree limbs or hidden in dense foliage and feeds on insects and their larvae taken from foliage or on the ground. It has a swift falcon-like flight. **Sexes**: *alike*. **Immatures**: *upperparts red-brown rather than grey*. **Voice**: *'kak-ak-ak-aak', like the Dollarbird, also a low subdued trill*.

39, 40 [Common] KOEL

Eudynamics scolopacea 46 cm

The Koel, a race of a species ranging to India and China, is a regular breeding migrant from New Guinea, where it is widespread. Birds begin their southward movement in August and most have moved north by April. It is seen singly, in pairs or small groups, at times with other birds in fruiting trees. The female is quieter and less obtrusive than the male. Breeds November–January. The female lays her marbled salmon-pink egg or eggs in the nest of one of a number of hosts such as Friarbirds or other large honeyeaters, often choosing a foster parent whose eggs resemble her own. **Female**: *mottled and barred like immature illustrated, but has broad cream streak under eye to side of neck: eye red*. **Voice**: *monotonous endlessly-repeated 'koo-eel' rising in a feverish accelerating crescendo. Also at night*.

Male left, immature right

41 CHANNEL-BILLED CUCKOO

Scythrops novaehollandiae 65 cm

A very large cuckoo which migrates to Australia to breed from territories ranging from New Guinea to Sulawesi. It begins arriving in September with the onset of the wet season in the north and it is usually absent by March. It is sometimes called the Stormbird or Rainbird because its call often presages wet weather. It favours tall timber and feeds on figs and native fruits. It is sometimes seen in flocks of up to fifty birds. Breeds October–December. Blotched off-white eggs (sometimes 2 or more) are placed in the nests of crows, currawongs and other large birds. **Sexes**: *alike*. **Immatures**: *buff head and tips to feathers*. **Voice**: *very loud raucous screaming 'quork' repeated 6 or more times, including at night and in flight*.

42 PHEASANT COUCAL

Centropus phasianinus 66 cm

The Pheasant Coucal is a sedentary species which extends to New Guinea. Its long broad tail makes up more than half its length and seems to encumber its laboured ungainly flight. The complexly-patterned plumage, predominantly black during breeding, becomes red-brown at other times. It lives in rank vegetation in damp locations, feeding on small animals and insects. Solitary. Breeds August–February sometimes twice in quick succession. Chicks are truly 'ugly ducklings' when hatched, with black bodies covered with developing feathers that look like white threads. The only Australian cuckoo not to parasitise the nests of other birds. It builds its own nest with open canopy on ground or up to 2 m in sugar-cane, pandanus etc. **Sexes**: *alike*. **Immatures**: *mottled*. **Voice**: *a series of 'woof' or 'hoop' booming notes with a short pause after the first two, then gradually quickening and descending in volume and tone*.

33

34

35

36

37

38

39

40

41

42

7

Parrots

Opposite: Galah

Worldwide, parrots are distributed throughout the tropical belt and southwards into the temperate latitudes – to include tropical Africa, southern Asia, the entire Australian Region and South America. They are absent in Europe, northern Asia and North America, whose only native parrot, the Carolina Parakeet, tragically became extinct in 1920. Of a world population of 332 species, fifty-five occur in Australia. All are confined to the Australian Faunal Region. Forty-eight species of great variety and beauty live only in Australia and are a rich endowment. They occupy every niche from rainforest to desert and many species are commonly seen in gardens, parks, golf courses and even city streets. (The Little Corellas on pages 2 and 3 were photographed roosting in a baobab (or boab) tree in the main street of Derby in Western Australia.)

Despite the great diversity of voice, diet and size, from the metre-long Macaws of South America to the 9 cm Pigmy Parrots of New Guinea (the size of a Weebill), all parrots are at once recognisable through their remarkably uniform diagnostic characteristics. These include short hooked bills, hinged upper mandibles as well as the lower ones, prominent nostrils set in fleshy ceres, and two toes forward and two backward (zygodactylous). And whether known as cockatoos, lorikeets, parrots, parakeets, lovebirds, or simply as a budgerigar, they all belong to the order Psittaciformes. Concepts on the classification of the numerous and diverse genera within the order vary widely. Here it suffices to state that the arrangement used by Joseph Forshaw in the 1981 revision of his fine *Australian Parrots*, illustrated by William Cooper, provides the framework for the descriptive paragraphs which follow. A single family Psittacidae is recognised and this is divided into three subfamilies: Cacatuinae (Cockatoos), Loriinae (Lorikeets) and Psittacinae (Typical Parrots). The Cacatuinae and Psittacinae are further divided into tribes.

COCKATOOS Subfamily Cacatuinae 1–12

Australian cockatoos, long popular throughout the world as cage birds, are highly intelligent and some can perform tricks as well as talk. Some live to a great age in captivity – one such example being the famous Sulphur-crested Cockatoo called 'Cocky Bennett' which died in Sydney in 1916 at the alleged age of 120. In his latter years, he was practically naked and used to scream out 'One b. . . feather more and I'll fly!' A Red-tailed Black Cockatoo, still living at Taronga Zoo Sydney, is approaching a similar age (see Species Note).

In July 1606, Torres (see page 40) spent three days at Milne in New Guinea before sailing westwards through the straits that bear his name. In describing the birds of the area, his companion Prado wrote: 'We saw numbers of parrots, some very white with a crest of yellow feathers and the beak and feet black . . . . .' This is a clear reference to the Sulphur-crested Cockatoo, and pre-dates Prado's description of the Torres Strait Pigeon later in the voyage which is often cited as the first authentic record of an Australian bird.

There are two tribes: Cacatuini comprising the large cockatoos, Calopsittacini with the cockatiel as its sole member. Of the subfamily's nineteen known species, twelve occur in Australia and nine are endemic. Most are characterised by a conspicuous movable crest, which is raised immediately after alighting and when the bird is alarmed or excited.

Large Cockatoos Tribe Cacatuini

Black Cockatoos The massive Palm cockatoo and the three endemic long-tailed black cockatoos are the world's largest parrots after the Macaws of South America. The spectacular crest and scarlet cheeks of the Palm Cockatoo, and the colour bands of the long-tailed cockatoos are their distinguishing features.

White Cockatoos The four predominantly white cockatoos are colour-washed under their wings, and spectacular movable coloured crests distinguish the Sulphur-crested and the Major Mitchell's Cockatoo. The latter is endemic, also the corellas except for a small colony of Little Corellas in New Guinea. Little Corellas sometimes gather in enormous flocks; one seen in recent years in the north-west was estimated at 60 000 to 70 000 birds.

Gang Gang Cockatoo and Galah These two engaging cockatoos are strikingly different from each other and from all other cockatoos.

Cockatiel Tribe Calopsittacini

A small cockatoo with a long gradated tail. 13

LORIKEETS Subfamily Loriinae 18–24

Lorikeets and Lories These spectacular parrots range from Bali in the west to the Marquesas in the east. Of about fifty-five species, only seven are represented in Australia, five of them being endemic. They wear tight glossy plumage mostly brilliantly coloured green, red and blue. Known as 'brush-tongues', they feed on nectar, pollen and fruit; and live mainly in the treetops. They are highly sociable and gather in large noisy flocks. They fly at great speed, screeching as they go. The subfamily comprises the genus *Trichoglossus*, a group of three small to medium-sized parrots with narrow, pointed tail feathers, typified by the ubiquitous Rainbow Lorikeet; the genus *Glossopsitta*, a group of three small parrots with short, wedge-shaped tails, typified by the Musk Lorikeet; and the Varied Lorikeet.

TYPICAL PARROTS Subfamiy Psittacinae

The species embraced by this large and varied subfamily are known as 'typical parrots' and most of the world's parrots belong to it. There are eight tribes, three of which are represented in Australia.

Fig Parrots Tribe Cyclopsittacini 17

The small fruit-eating fig parrots comprise this tribe, in which there is but one species divided into eight subspecies. Three of these occur in Australia; the rest are distibuted in New Guinea and its adjacent islands. The strange specific name *diophthalma*, meaning double-eyed, was given to the fig parrots because of a spot close to the eye in some varieties in New Guinea which gives the impression of a second eye.

7

Fruit Eating and Long-tailed Parrots Tribe Psittaculini 14–16, 25–31

This tribe consists of the fruit-eating Red-cheeked and Eclectus Parrots, and five long-tailed medium-sized parrots. The tribes' distinctive characteristics include a smooth-surfaced, usually pale-coloured bill and a contrastingly coloured rump or underwing coverts.

Broad-tailed Parrots Tribe Platycecini 32–60 and pages 68, 69

Most of the Platycercini are characterized by a broad long tapered tail and a slight blunt bill and are among the best known of Australia's parrots. They include two of the world's three ground-dwelling parrots. The tribe extends to Fiji and New Zealand, but is absent from New Guinea. Nine genera occur in Australia and these are grouped for convenience as follows:

Ringneck Parrots (including the Red-capped Parrot) There are two species of ringnecks, each with two subspecies. The Red-capped Parrot, though related, is distinct and is one of the few birds confined to the south-western corner of the continent to have a genus of its own.

Rosellas There are eight forms of rosellas widely spread over the continent. Three have white cheeks (the Northern, Pale-headed and Eastern Rosellas), four have blue cheeks (the Crimson, Adelaide, Yellow and Green Rosellas) and one, the Western Rosella, is smaller and sexually dimorphic. Rosellas at times hybridise where populations meet. The Eastern Rosella was first noted by early settlers at Rose Hill who called it the 'Rosehiller'. (The town of Rose Hill was given the aboriginal name Parramatta in 1791.)

'Psephotus' group (including the Blue Bonnet) The six species of this group are medium-sized, slender, long-tailed grass parrots. They include the 'anthill parrots' which excavate their nests in termite mounds, and these are typified by the endangered Golden-shouldered Parrot. (see article on the Paradise Parrot, page 70.)

'Neophema' group There are seven members of this genus of small, graceful grass parrots. They are mainly ground-frequenting and largely southern in distribution. Some, such as the Turquoise and Scarlet-chested Parrots, are among the most beautiful of birds. None are common, whilst a few, such as the Orange-bellied Parrot, still require careful biological study and management if they are to survive.

Swift Parrot This species has a brush tongue like the lorikeets and certain features like the rosellas.

Budgerigar Happily, the small world-renowned Budgerigar, native to Australia, is still widespread in the wild.

Ground Parrot Originally distributed around the coastal belt from Queensland southwards and across to south-western Australia and used to be fairly common. From first settlement it rated as a game bird, like quail and snipe, and its strong scent enabled dogs to point to it. The Ground Parrot is now regarded as an endangered species and much effort is being made to preserve it.

Night Parrot Discovered in 1845 by the explorer John McDouall Stuart at Lake Goyder in the north-eastern corner of South Australia. The little that is known of this elusive bird derives mostly from the sparse records inherited from 1875 to 1885 when it was comparatively abundant, and when most of the twenty-two specimens in the world's museums were collected. Records of confirmed sightings petered out in 1912, but during the last two decades, at least four further sightings have found acceptance. Brian Powell, a station manager at Partacoona, twenty miles north of Quorn in South Australia, reported seeing a Night Parrot on 15 August 1969 in saltbush country, devoid of spinifex, and watched it for fifteen minutes as close as 6 ft. He recalled having seen Night Parrots in the area five times in six years. Lou Ward of Glenayle Station northeast of Wiluna in Western Australia reported a night sighting in May 1979 on a largely unvegetated stony flat 'so close he could have caught it with a net'. (The sighting was authenticated by Simon Bennett who subsequently visited the area and interviewed him.) R.J. Rowlands and his wife saw a Night Parrot in the Simpson Desert in October 1980 between Birdsville and Lake Muncoonie, at about fifty metres. Camel-mounted daylight sweeps in June 1979 organised by Rex Ellis of Transcontinental Safaris with Shane Parker, Curator of Birds at the South Australian Museum, succeeded in flushing Night Parrots twice on Coopers Creek due east of Lake Perigundi. John McKean, of the Conservation Commission of the Northern Territory, reported in 1985 that several sightings of the Night Parrot had been made since 1975, in the general Keep River National Park area, north-western Northern Territory including one by himself in April 1982. Most recent of the many abortive attempts made to find the bird have been a series of beats by large parties of enthusiasts in 1984 and 1985 on the Coopers Creek flood plains, not far from the original point of Stuart's discovery, during expeditions mounted by Bert Bolton's Outback Tours in co-operation with the members of the New South Wales Field Ornithologists Club. Although no longer seen in some of its former areas, these recent widely-separated sightings suggest that the Night Parrot is far from extinct. Because it appears to be highly nomadic and able to adjust to seasonal fluctuations of food and water, and enjoys the protection of nocturnal and secretive habits in a remote habitat, there is hope for its survival.

Opposite: Rainbow Lorikeet

PHOTOGRAPHERS: H. & J. Beste 8, 17, 21, 47, 48; C.A.C. Cameron 24; G. Chapman 5, 10, 58, 59; T. & P. Gardner 34; C.H.H. Jerrard (NPI) p. 71; R.D. Mackay 16; S. Parish 11; T. Pescott 56; K. & B. Richards 19, 51; L. Robinson p. 68 (lower). 42, 45, 52, 53; V. Serventy p. 69; T.A. Waite 36; B.A. & A.G. Wells 49; Authors pp. 53, 55, 68 (top). 1, 2, 3, 4, 6, 7, 9, 12, 13, 14, 15, 18, 20, 22, 23, 25, 26, 27, 28, 29, 30, 31, 32, 33, 35, 37, 38, 39, 40, 41, 43, 44, 46, 50, 54, 55, 57.

ARTIST: T.R. Lindsey 60.

ARTICLES: The Paradise Parrot and John Gilbert p. 70; Australia's First Ornithological Record? p. 72.

7

1 PALM COCKATOO

Probosciger aterrimus 60 cm

This majestic cockatoo, widespread in New Guinea, is found in Australia only in the tropical rainforests of the north on Cape York. It has a beak strong enough to bend fencing wire and to crack open the hard pandanus nuts which it favours. Under stress, its bare facial patches flush crimson. They roost at night in the topmost branches of tall trees and gather by day in small playful parties in a favoured big tree in open country. Sedentary. Breeds August-February. Nests in trunk hollows 8–12 m high. **Female***: small face patch.* **Immatures***: yellow markings.* **Voice***: very loud whistles, wails and screeches.*

2 GLOSSY BLACK COCKATOO

Calyptorhynchus lathami 50 cm

Distinguished from most other cockatoos by the absence of a crest, this small cockatoo has duller plumage than its name implies. Rather uncommon, it is seen in pairs and small parties in woodland, especially casuarina stands in the south-east, and on Kangaroo Island. It feeds mainly on casuarina seeds and on larvae. It is tame and approachable and its flight has a floating quality. Nomadic. Breeds March–August. Endemic. **Sexes***: similar but female has yellow patches on head and neck.* **Immatures***: like females but wings spotted with yellow.* **Voice***: resembles a filing or sawing sound; also a lugubrious single note, softer than Red-tailed Black Cockatoo.*

Female left Male right

3, 4 RED-TAILED BLACK COCKATOO

Calyptorhynchus banksii 61 cm (north) 52 cm (south)

When Captain Cook's *Endeavour* was beached for repairs at the mouth of the Endeavour River in 1770, Sydney Parkinson, draughtsman to Joseph Banks, made a pencil drawing of a female of this species. His sketch, now in the British Museum (Natural History) is the first known illustration of a land bird from the east coast of Australia. Banks took the specimen back to England and the often-used name of Banksian Cockatoo marks his association with the species. The birds illustrated were photographed at Taronga Zoo, Sydney and are believed to be centenarians. The Head Bird Keeper, Les Clayton, reports that they are in excellent feather and condition and that the female is still laying eggs!

This handsome bird occurs in flocks of up to 200 in the north but in the south-east it is less numerous. In the wheat belt of SW Australia huge flocks of 1 000 or more gather to feed on 'double-gee' and introduced burr. Birds are smaller in the south-west and in the interior. Ranging through forest and woodland, it is partial to nuts, fruit, casuarina and acacia seeds, and to the destructive longicorn beetle. Wary and difficult to approach. Nomadic. Breeds April–October. Nests in large tree hole. Endemic. **Immatures***: like females.* **Voice***: harsh grating shrieks and loud mournful single notes.*

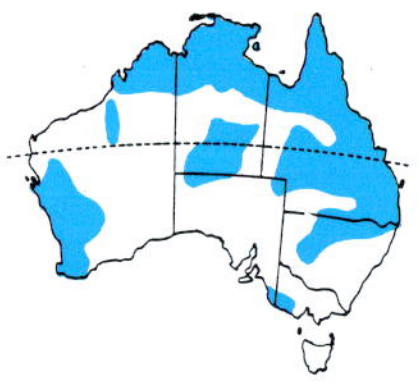

Male left, female right

5 THE WHITE-TAILED BLACK COCKATOOS

(A) SHORT-BILLED FORM

Calyptorhynchus funereus latirostris 55-60 cm

Frequents the drier inland areas of WA where it feeds on banksias, grevilleas etc. It breeds in the wandoo and salmon gum forests in the east of its range and returns to the higher rainfall areas towards the coast after breeding.

(B) LONG-BILLED FORM *Calyptorhynchus baudinii* 55–60 cm

Confined to the heavy timber of the south-westcorner, it feeds mainly on gumnuts of the tall marri eucalypts which it is able to extract with its long specially-adapted bill. In the non-breeding season, it may be seen in company with its short-billed relative. The two forms do not however interbreed. Breeds August–December and nests in large tree hollows. Endemic.

Male*: cheek patch smaller and dull white, bill grey-black.* **Immatures***: similar to adult female.* **Voice**: High-pitched drawn-out whistle 'whee-la'.

Female (A)

6 YELLOW-TAILED BLACK COCKATOO

Calyptorhynchus funereus 69 cm

This large cockatoo is characterised by its long tail—about half its total length. Usually seen in small flocks, but these can sometimes be quite large, perhaps of 200 birds or more. It is nomadic and seen in most types of timbered country. It feeds mostly on the seeds of forest trees and also uses its strong bill to strip bark and probe deeply into the timber for the larvae of wood borers. Breeds March–April (north) July–January (south), and is dependent on finding old trees with large hollows for nesting sites. Endemic. **Sexes***: Males: face patches smaller and less pronounced; bills dark grey; eyes dark brown; eye rings pink.* **Females***: bills bone; eye rings dark grey.* **Immatures***: like females.* **Voice***: noisy and their wailing 'wee-yu' or 'wy-la' is often heard before birds come into view.*

Female

1

2

3

4

5

6

7

7 SULPHUR-CRESTED (White) COCKATOO

Cacatua galerita 49 cm

The intelligent and popular white cocky is a familiar sight in the Australian bush in varied habitats of forest, woodland and open country. Often in large flocks, it forages on the ground for seed, grain, roots and nuts while lookouts in the trees stand on sentry duty. It is extremely wary and raises its spectacular sulphur-yellow crest when alarmed or angry. Clamorous and conspicuous, especially in the roosting trees at dawn and dusk. Breeds May–August (north), August–January (south). Also in New Guinea and Indonesia. Introduced in WA. **Sexes**: *male – dark brown eye, female – dark red eye.* **Immatures**: *faint grey markings.* **Voice**: *raucous screeches.*

8 MAJOR MITCHELL'S (Pink) COCKATOO

Cacatua leadbeateri 36 cm

In 1838 the explorer Sir Thomas Mitchell wrote in praise of this 'beautiful species', rhapsodising over its 'pink-coloured wings and glowing crest' and it is named after him. Wary. Usually seen in pairs or small flocks, and uncommon in most areas, it inhabits the arid scrublands of the interior where it feeds on seeds, roots, berries, nuts and wood boring grubs. Breeds August–October. Nests in tree hollows 5–20 m high. Endemic. **Male**: *eye – dark brown, abdomen – salmon pink.* **Female**: *eye – red, abdomen – white, crest – yellow band may be broader.* **Immatures**: *paler than adults and eye light brown in females.* **Voice**: *crotchety screams.*

Female

9 LITTLE (Bare-eyed) CORELLA

Cacatua sanguinea 42 cm

The English buccaneer William Dampier was the first to record the Little Corella during his 1699 voyage to New Holland in HMS *Roebuck*. On 21 August his men landed on Rosemary Island off Australia's west coast where they described it as 'a Sort of white Parrots which flew a great many together', and they are still there. The Little Corella sometimes gathers in enormous flocks – as many as 60 000–70 000 have been seen at a time in the north-west – and these can envelope whole trees to create a picture of a cherry orchard in full bloom. It frequents a variety of habitats, including open woodland, scrub, grasslands and suburban trees and feeds mainly on seeds, also on fruit shoots and insects. It invariably roosts near water. Nomadic. Breeds June–October or variably. Nests in tree hollows or occasionally cliff holes and anthills. A small population extends to New Guinea. In the extreme south-west there is a larger population with decidedly longer upper mandibles, and whose snowy plumage often becomes heavily soiled with earth as a result of digging for bulbs and roots.
Sexes and **immatures**: *similar.* **Voice**: *a range of raucous calls and screeches.*

10 LONG-BILLED CORELLA

Cacatua tenuirostris 38 cm

Although now uncommon to rare in many of its former haunts, the Long-billed Corella is still occasionally seen in very large flocks, frequenting woodland and adjacent grasslands, seldom far from water. Sentinels are posted when flocks are feeding or drinking. Immatures are nomadic. Breeds August–December. Nests in tree hollows. Endemic. **Sexes**: *similar.* **Immatures**: *shorter mandibles.* **Voice**: *a range of raucous calls and screeches.*

11 GALAH

Cacatua roseicapilla 36 cm

One of the most beautiful sights Australia has to offer is that of a huge flock of Galahs, presenting glorious sweeps of warm pink or silvery grey against the sky as they wheel in the air, settle like massed blooms in trees, or forage on the dark soil of wheatfields. It is one of the most numerous and ubiquitous of the parrots, found in all types of open country, feeding on seeds, grains, fruits, blossoms and shoots as well as insects and their larvae. Adults sedentary, immatures nomadic. Breeds July–December. Nests 4–20 m high. Endemic. **Sexes**: *similar but male's eye is dark brown, female's pink.* **Immatures**: *duller.* **Voice**: *extremely varied, loud whistles, strident shrieks and screams.*

Male

12 GANG-GANG COCKATOO

Callocephalon fimbriatum 36 cm

This tame, droll bird with quizzical eye and creaky voice often enters city parks and suburban gardens in search of seeds and berries when in summer it descends from mountain forests to lower valleys and coastal regions. It is almost entirely arboreal and becomes so immersed in feeding as to be indifferent to the approach of humans. Breeds October–January. Nests very high in tree hole. Endemic. **Male**: *crest and head bright red.* **Female**: *crest and head grey.* **Immatures**: *like females, with underside of tail barred grey.* **Voice**: *two or three grinding notes which can almost be reproduced by rubbing the thumb over an inflated balloon.*

Young male

7

8

9

10

11

12

7

13 COCKATIEL (Quarrion)

Nymphicus hollandicus 33 cm

This attractive small crested parrot with its pretty peach cheek patches is seen abundantly in pairs and small flocks in open or lightly timbered country near water. It forages mainly on the ground for grass seed, grain, fruits and berries. Sometimes it effaces itself by perching lengthwise along the branch of a dead tree. Flight is straight and swift. Highly nomadic in the north, seasonally migratory in the south. Breeds April–August (north), August–December (south). Nests in tree hollow. Endemic. **Voice**: *soft but far-carrying warbling 'queel-queel'.*

Female, male, junvenile

14, 15 ECLECTUS PARROT

Eclectus roratus 43 cm

The remarkable feature of this species is that the female sports the more brilliantly-coloured plumage. As this is the reverse of normal in the bird world, it was for many years believed that the male and female were different species. The parrot is widely distributed in New Guinea and other islands to the north but, in Australia, it is confined to the rainforests of Cape York Peninsula, where it was discovered in 1913. It is usually seen in pairs and parties foraging in the tree canopy for berries, fruit, nuts and seeds and its strident cries are a characteristic of the forest. Family groups of up to eight birds are supportive in rearing the nestlings. At night they gather to roost together in screeching commotion. Sedentary. Breeds July–February. Nests in a high tree hollow near water. **Immatures**: *similar to adults of each sex.* **Voice**: *male utters note resembling 'quork'; female has a screeching whistle. Also resonant clanging, wailing.*

Male left, female right

16 RED-CHEEKED PARROT

Geoffroyus geoffroyi 25 cm

The dense tropical jungle of the Iron Range area on Cape York Peninsula is the only haunt in Australia of the Red-cheeked Parrot, where it was discovered in 1913, although it had been widely known in New Guinea and Indonesia. Plentiful within its restricted range, pairs or family parties gather in favoured trees to feed on seeds and fruit. Not easily seen amidst dense foliage, its flight between trees is distinctive, with quick shallow wingbeats and no undulating or gliding. Sedentary. Breeds August–November. Nests high. **Immatures**: *like females.* **Voice**: *loud screeching, metallic chattering.*

Female left, male right

17 DOUBLE EYED FIG-PARROT

Cyclopsitta diophthalma 16 cm

(A) Race *marshalli* (Marshall's)
(B) Race *macleayana* (Red-browed)
(C) Race *coxeni* (Blue-browed)

Widespread in New Guinea, this, the smallest parrot in Australia, occurs in the rainforests of eastern Australia as three isolated races, of which the race *coxeni* is now exceedingly rare. It feeds silently in a tree, mainly on native figs, betraying its presence by dropping a steady stream of fruit particles from above. Breeds August–October. Nests in decaying tree branches. **Female**: *less red on face.* **Immatures**: *like females.* **Voice**: *high-pitched 'tseet'.*

(B) Male

18 MUSK LORIKEET

Glossopsitta concinna 23 cm

Acquiring its name from a musky smell said to be associated with it, this is an extremely noisy bird which betrays its presence by highpitched chattering and screeching when it is heedlessly engrossed in feeding high among flowering eucalypts and other blossom or fruit trees. It also eats insects and their larvae. It lives in open timbered country, farmland or gardens, forming strongly bonded pairs. Sometimes it joins flocks of mixed lorikeets. Seasonally nomadic. Breeds August–January. Nests high in tree hollows. Endemic. **Sexes**: *alike.* **Immatures**: *duller; eye brown, bill dark brown.* **Voice**: *brassy screeches, chattering.*

19 VARIED LORIKEET

Psitteuteles versicolor 20 cm

The Varied Lorikeet is usually seen in large flocks feasting voraciously and messily on the nectar and pollen of flowering gums and paperbarks. Berries, seeds and insects are also taken. Noisy, numerous and truculent, it may chase off other species competing for food, although it is often in the company of Rainbow Lorikeets. Nomadic. Breeds mainly April–August. Nests high in a tree hollow. Endemic. **Sexes**: *alike.* **Immatures**: *crown dull green showing some red.* **Voice**: *very shrill screeching and chattering, softer twittering.*

20 LITTLE LORIKEET

Glossopsitta pusilla 16 cm

Similar in habits to the Musk Lorikeet, the noisy Little Lorikeet is fearless and approachable when it is absorbed in feeding. It is seen in flocks which frequent flowering eucalypts and other blossom and fruitbearing trees, often in company with other lori-keets. Nomadic. Breeds August–January. Nests in tree holes above 6 m, often near water. Endemic. **Recognition:** *differentiated in flight from the Purple-crowned Lorikeet by yellow-green underwings instead of red.* **Sexes:** *alike.* **Immatures:** *face duller, bill and eye brown.* **Voice:** *shrill 'sssit', twittering.*

21 PURPLE-CROWNED LORIKEET

Glossopsitta porphyrocephala 17 cm

Widely distributed along the southern fringes of the continent, it is the south-west's only endemic lorikeet. Its chief haunts are mallee and drier forests, where it is abundant, but it also lives close to the coast. In smaller parties or large flocks it feeds in flowering eucalypts and visits gardens and orchards in search of pollen, nectar, fruit, insects and their larvae. Nomadic, sometimes ranging far inland. Breeds September–December. Nests in hollow trees near water. Endemic. **Sexes**: *alike.* **Immatures**: *sometimes no purple on crown.* **Voice**: *loud shrill buzzing.*

22 RAINBOW LORIKEET

Trichoglossus haematodus 32 cm

Large flamboyant mobs of the lovely Rainbow Lorikeet congregate noisily in flowering gums to take nectar and pollen. They also come readily as tame visitors to gardens and tourist parks such as that at Currumbin on the Queensland Gold Coast where thousands of birds descend daily on the honey trays provided, even clambering over spectators in their eagerness. They occur in all types of timbered country. Introduced in WA. Breeds August–January, or variably in the north. Nests in tree hollows. Also in New Guinea and South Pacific islands. **Sexes**: *alike.* **Immatures**: *duller.* **Voice**: *metallic screeching.*

23 RED-COLLARED LORIKEET

Trichoglossus rubritorquis 31 cm

Its plumage is similar to that of the Rainbow Lorikeet except for the difference in nape and belly colour. The two are sometimes regarded as races of the same species and their habits are similar. The Red-collared is more abundantly found in the wetter more northerly parts of its tropical habitat. Mostly sedentary. Nests in tree hole near water. Endemic. **Sexes**: *alike.* **Immatures**: *duller, bill and eye brown.* **Voice**: *like Rainbow Lorikeet.*

24 SCALY-BREASTED LORIKEET

Trichoglossus chlorolepidotus 24 cm

This predominantly green parrot is usually seen in flocks, often in company with Rainbow Lorikeets. Its favourite haunts are open forests, where it feeds on the nectar and pollen of eucalypts and melaleucas. Fruits, seeds and cultivated grains also form part of its diet. In its swift flight, the orange underwings are distinctive. Nomadic to follow flowering trees. Breeds May–January. Nests in tree hollows. Endemic. **Sexes**: *alike.* **Immatures**: *shorter tail, eye and bill brownish.* **Voice**: *more shrill than the Rainbow Lorikeet.*

13

14

15

16

17

18

19

20

21

22

23

24

7

25, 26 RED-WINGED PARROT

Aprosmictus erythropterus 33 cm

The Red-winged Parrot stands out as among the most spectacularly handsome of Australia's parrots. In varied habitats from the tropics to the south-eastern inland, rarely far from water, it may be seen in all types of timbered scrub country, open sandy plains, or in the mangroves of the north. It spends its day in the treetops, foraging acrobatically for nectar, blossoms, berries and insects. It comes to ground only to drink, or occasionally to investigate fallen fruit or nuts. Generally common, it appears in pairs, family parties or small flocks. The young male takes at least two years to acquire adult plumage and to begin breeding, remaining with the family group until that time. Breeds from May (north), August–February (south). Nests deep in a hollow tree with a high entrance. Also in New Guinea. **Immatures**: *like adult females*. **Voice**: *flight call 'crillik-crillik,' soft feeding chatter, alarm screeching*.

Male left, female right

27 KING PARROT

Alisterus scapularis 44 cm

This regal parrot mainly frequents mountain forests, but often visits gardens and parks. It feeds on native fruits and seeds in flowering trees, and also raids orchards and cereal crops. During winter it migrates from the highlands to coastal plains, then returns to breed, October–January. Nests deep in hollow tree with entrance sometimes over 10 m high. Endemic. **Female**: *mostly green, lower breast and belly red*. **Immatures**: *like adult females, eye brown*. **Voice**: *three or four penetrating pure musical notes*.

Male

28, 29 SUPERB PARROT

Polytelis swainsonii 42 cm

This aptly named Superb Parrot has one of the most restricted ranges of any Australian parrot and is found usually between the Namoi–Castlereagh River system of New South Wales and the Murray–Lachlan system of southern new South Wales and northern Victoria. Intensive clearing of its principal habitat for agricultural development and illegal trapping for the cage-bird market have reduced its number, and some fear that the species is endangered. It is usually seen in well-bonded pairs or small flocks and is seldom found far from water. It spends most of its time on the ground foraging for seeds, sometimes along roadsides. It is partial to cultivated grain crops and spilt grain, and will also take nectar from flowering eucalypts. It is gracefully streamlined both in repose and in effortless flight. It is nomadic throughout its range, and the entire population migrates to New South Wales to breed in September–December. During the breeding season, flocks seem to be composed entirely of males. It nests very high in tree holes. Endemic. **Immatures:** *face blue-green, eye brown*. **Voice:** *deep warble, twittering*.

Male left, female right

30 PRINCESS (Alexandra's) PARROT

Polytelis alexandrae 46 cm

John Gould named this parrot in honour of Princess Alexandra. It is an uncommon bird of the dry interior. In pairs or small flocks, it feeds on or near the ground searching for seeds, especially those of spinifex. It is also partial to acacia blossoms and mistletoe berries. Highly nomadic. Breeds September–January. Usually nests in a hollow in a eucalypt near water, often in colonies. Endemic. **Female**: *duller, shorter, tail shorter*. **Immatures**: *like adult females*. **Voice**: *mainly silent; rolling unmusical note often in flight, softer notes and twittering*.

Male

31 REGENT (Smoker) PARROT

Polytelis anthopeplus 41 cm

The Regent Parrot is notable for its striking plumage, flocks resembling shafts of sunlight as they fly swiftly among the treetops. Uncommon in the east, abundant in the south-west where the birds are greener and duller and may appear in large flocks. Near water, in woodland, scrub or mallee it forages mainly on the ground for seeds, also in flowering gums. Seasonally migratory in WA. Breeds August–January. Nests in tree hollows. Endemic. **Female**: *duller, pink markings on undertail*. **Immatures**: *like adult females*. **Voice**: *rolling warble rather like the carol of the Australian Magpie*.

Male

32 YELLOW ROSELLA

Platycercus flaveolus 34 cm

This 'blue-cheeked' rosella is usually seen in pairs and parties along inland rivers, mainly the Murray and Murrumbidgee, amongst red river gums or in mallee, grassland or orchards associated with the river systems. It has a diet of seeds, berries, blossoms, insects and their larvae. Relatively timid. Sedentary. Breeds September–January. Nests in tree hollows at varying heights. Endemic. **Sexes**: *alike*. **Immatures**: *back and wings olive-green, underparts yellow-green*. **Voice**: *like Crimson Rosella*.

33 ADELAIDE ROSELLA

Platycercus adelaidae 36 cm

Inhabiting the Mount Lofty Ranges and adjacent areas including suburban Adelaide, it appears in pairs and parties amongst eucalypts and is also at home in cultivated environments. It is similar in its behaviour to the other 'blue-cheeked' rosellas. It often feeds on spilt grain around farms and drinks from stock troughs. Plumage colour may vary: breast and back markings are usually red, but orange-yellow may predominate in some birds. Sedentary. Breeds September–February. Nests high in tree hollows. Endemic. **Sexes**: *alike*. **Immatures**: *dark green with red markings*. **Voice**: *like Crimson Rosella*.

34, 35 CRIMSON ROSELLA

Platycercus elegans 37 cm

Often called the Mountain Lowry, the Crimson Rosella is the most widely distributed of the 'blue-cheeked' rosellas. It is represented by two populations, birds in the north being smaller and darker. Its natural habitat is rainforest and eucalypt woodland, both heavily timbered and more open, but it commonly frequents suburban gardens and parks. Often seen on the ground, it feeds on seeds, nectar, fruit and scale insects, and is usually in pairs or family parties, with flocks of largely immature birds being formed at times. It is a confident and confiding bird, allowing a close approach especially when feeding. Mostly sedentary. Breeds September–February. Nests 8–30 m high. Endemic. **Sexes**: *alike*. **Immatures**: *may be green up to 18 months*. **Voice**: *shrill calls, piping and whistling, soft feeding chatter*.

Adult left, immature right

36 GREEN ROSELLA

Platycercus caledonicus 36 cm

This 'blue-cheeked' rosella is confined to Tasmania and the larger islands of Bass Strait. It is abundant in Tasmania in all wooded habitats and is a familiar sight on the fringes of urban areas. Usually in pairs or parties, flocks may gather in winter. It feeds on the ground or in the trees and is partial to hawthorn berries. It also raids orchards and grain crops. Sedentary. Breeds September–January. Nests high in tree hollows. Endemic. **Sexes:** *alike*. **Immatures:** *head and underparts yellow-green*. **Voice:** *very loud, piercing 'cussick, cussick' flight call; double bell-like whistle, fluting and piping*.

25

26

27

28

29

30

31

32

33

34

35

36

7

37 EASTERN ROSELLA

Platycercus eximius 34 cm

Prolific and familiar, this vibrant parrot is one of the 'white-cheeked' group of rosellas. Usually in pairs or family parties, though larger flocks may congregate during winter. It frequents woodlands and grasslands and is often seen in the suburbs and around farms. It forages quietly on the ground or in trees for seeds, nectar, fruit and insects. Sedentary. Breeds August–January. Nests in hollow. Endemic. **Female**: *duller on head and breast.* **Immatures**: *crown and nape patchy green.* **Voice**: *pleasant whistling, metallic, 'kwink-kwink-kwink'.*

Male

38 PALE-HEADED ROSELLA

Platycercus adscitus 32 cm

Another of the 'white-cheeked' rosellas, its plumage varies considerably. North of Cairns, birds have a patch of pale violet on the lower face instead of white. There are hybrids between the Pale-headed and Eastern Rosellas where their ranges overlap. The preferred habitat is open forest, clearings, or edges of grassland; it also visits orchards grainfields and patches of weed such as thistle and burr. Sedentary. Breeds September–December (south), February–June (north). Endemic. **Sexes**: *alike.* **Immatures**: *duller, may show grey or red head markings.* **Voice**: *similar to Eastern Rosella, perhaps harsher.*

39 NORTHERN ROSELLA

Platycercus venustus 30 cm

The remote distribution of this rosella of the 'white-cheeked' group separates it from the other rosellas. It is a quiet bird, not easy to locate, approachable and only takes refuge in nearby trees when disturbed. It prefers open forest near water, or mangroves, but visits cropfields and gardens in search of seeds. It is usually seen in pairs or family groups and is not given to flocking. Mainly sedentary. Breeds June–August. Nests in tree or post hollows. Endemic. **Sexes**: *alike.* **Immatures**: *duller.* **Voice**: *short succession of two-syllable metallic notes.*

40 WESTERN ROSELLA

Platycercus icterotis 26 cm

This, the smallest and quietest of the rosellas, which has yellow cheeks, is confined to the south-west corner of Western Australia. It frequents open country and sparse woodland and is usually seen on the ground feeding on grass seeds, often at roadsides. Mostly sedentary. Breeds August–December. Nests in hollow tree limbs. Endemic. **Female**: *less brilliantly coloured.* **Immatures**: *lack yellow cheeks, less red on underparts.* **Voice**: *soft musical repeated whistles.*

Male

41 PORT LINCOLN PARROT

Barnardius zonarius 37 cm

(A) Race *zonarius* (Port Lincoln)
(B) Race *semitorquatus* (Twenty-eight)

The species occurs in two forms: the yellow-breasted Port Lincoln Parrot, found mainly in the inland areas, and the green-breasted Twenty-eight Parrot, found only in the wetter heavily-timbered south-west corner of Western Australia. The two populations interbreed where they meet. Breeds August–February. Endemic. **Sexes**: *similar.* **Immatures**: *duller.* **Voices**: (A) – *high-pitched whistles resembling 'kwink';* (B) – *double-noted 'twenty-eight' call from which it is named.*

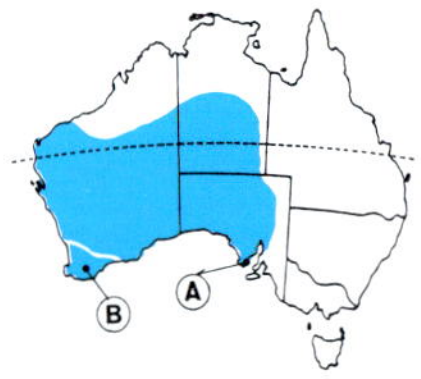

(A) Illustrated

42 MALLEE RINGNECK

Barnardius barnardi 35 cm

(A) Race *barnardi* (Mallee Ringneck)
(B) Race *macgillivrayi* (Cloncurry Parrot)

This species includes the smaller isolated Cloncurry Parrot of north-west Queensland, which has pale yellow underparts and lacks a red frontal band. Southern birds favour mallee scrub; northern birds eucalypt watercourse. Pairs and family parties forage for seed, blossoms, insects and their larvae. Sedentary. Breeds August–January. Nests in tree holes. Endemic. (A) **Male**: *back and mantle deep blue-black.* (B) **Sexes**: *similar.* **Immatures**: *like adult females.* **Voice**: *ringing 'twink-wink-wink'.*

(A) Female

43 RED-CAPPED PARROT

Purpureicephalus spurius 37 cm

Called the King Parrot in Western Australia, it occurs only in a relatively small area in the south-west corner. In pairs and small groups, it feeds quietly in the treetops, favouring one particular eucalypt, the marri, both for its seeds and for nesting. It also feeds on grass seed and cultivated fruit. Sedentary. Breeds August–December. Nests high. Endemic. **Female**: *duller.* **Immatures**: *duller, with crown green.* **Voice**: *grating metallic call, whistling in flight, alarm shrieks.*

Male

44 RED-RUMPED PARROT

Psephotus haematonotus 28 cm

In pairs or small parties, it mostly frequents lightly-timbered grasslands, cultivated paddocks and trees bordering watercourses. It congregates in flocks in winter. It spends most of its time on the ground foraging for grass seeds or spilt grain and is partial to the leaves and blossom of certain thistles. Mainly sedentary. Breeds August–December. Nests in hollow stumps and limbs. Endemic. **Female**: *much duller and more grey-green.* **Immatures**: *paler.* **Voice**: *shrill two-syllable whistle and pleasant warbling.*

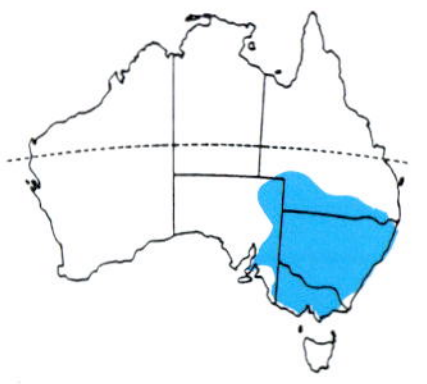

Male

45 MULGA PARROT

Psephotus varius 30 cm

Widespread, but nowhere abundant, this beautiful bird of the interior lives quietly and inconspicuously in lightly timbered country and scrub. Solitary, in pairs or small parties it is often seen on the ground feeding on grasses, herbaceous plants and shrubs. Sedentary or nomadic during drought. Breeds July–December or after rains. Nests in tree holes. Endemic. **Female**: *head, neck and upper breast green-brown, dull red on nape and shoulder, rump green.* **Immatures**: *duller.* **Voice**: *repeated whistling 'jeep'.*

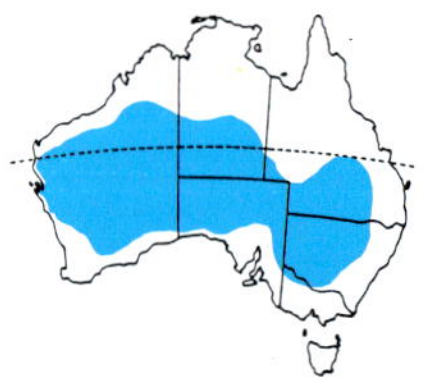

Male

46 PARADISE PARROT

Psephotus pulcherrimus 30 cm

Acclaimed as Australia's most beautiful parrot, it is now feared to be extinct. It used to frequent the sparsely timbered grasslands of south-east Queensland and northern NSW in pairs and small flocks and appeared to be wholly seed-eating. It bred August–December, sometimes later, and nested in cavities excavated in termite mounds, river banks and occasionally low in a tree stump. It laid from three to five eggs. **Female**: *duller.* **Voice**: *'sweet, plaintive piping, not unlike that of a young bird'* (Chisholm). (See also pages 170 and 71). The mounted specimen of a male bird photographed opposite was donated to the National Photographic Index of Australian Wildlife in 1970 by the late Sir Edward Hallstrom.

47, 48 GOLDEN-SHOULDERED PARROT

Psephotus chrysopterygius 27 cm

The survival of this rare and beautiful little parrot, found only in a relatively small area of Cape York, is threatened by the depredations of illegal bird trappers. The species is noted for its skill in excavating nesting chambers in the slab-sided termite mounds which abound in the area. Remarkably, all of these point north/south so that a minimum area is exposed to the heat of the mid-day tropical sun. A tunnel about 40 cm long is enlarged at the end into a chamber in which the eggs are laid, then incubated by the female. When the young hatch they remain with the family group, the males not attaining the full plumage glory until their second year. Seen in pairs and flocks, it feeds on grass seeds and herbaceous plants. Sedentary. Breeds April–September. Endemic. **Immatures**: *like adult females; young males have brighter colour.* **Voice**: *soft whistled 'fweep'.*

Female left, male right

37

38

39

40

41

42

43

44

45

46

47

48

7

49 HOODED PARROT

Psephotus dissimilis 27 cm

The sooty black forehead of this Northern Territory bird gives it its name and distinguishes it from its close relative, the Golden-shouldered Parrot of Cape York. The Hooded Parrot is perhaps more often seen in flocks, otherwise their feeding and nesting habits are identical and the two birds are sometimes regarded as races of a single species. Sedentary, local movement. Breeds April–July. Nests in termite mound. Endemic. **Female**: *lacks black crown, in general light olive-green.* **Immatures**: *duller.* **Voice**: *similar to Golden-shouldered Parrot.*

Male

50 BLUE BONNET

Northiella haematogaster 35 cm

(A) Race *haematogaster* (Yellow-vented)
(B) Race *haematorrhous* (Red-vented)
(C) Race *pallescens* (Lake Eyre)
(D) Race *narethae* (Nullarbor)

Races (A) and (B) intergrade where they meet; (C) is pallid; (D) is isolated and distinctive. A species of semi-arid open country where trees provide nesting hollows and is a ground feeder. Breeds August–December. Nests below 10 m. Endemic. **Female**: *less blue on face, duller red or yellow on belly*. **Immatures**: *duller.* **Voice**: *whistling, harsh 'chuck'.*

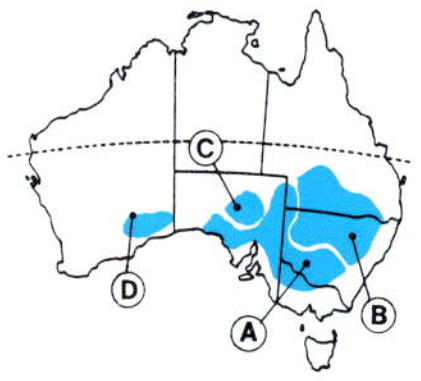

(B) Male

51 BOURKE'S PARROT

Neophema bourkii 22 cm

Crepuscular – moving after dusk and before dawn – and semi-nocturnal, it is sometimes active even in total darkness. It frequents mulga and inland scrub and spends much time on the ground feeling, merely fluttering off to cover if disturbed. Usually in pairs or small family parties, but flocks may throng at water during drought. Sedentary or nomadic. Breeds August–December or after rains. Nests in tree 1–3 m from ground. Endemic. **Male**: *similar but forehead sky blue.* **Immatures**: *like adult females, less pink on breast.* **Voice**: *soft 'chu-wee' in flight, also warbling and whistling.*

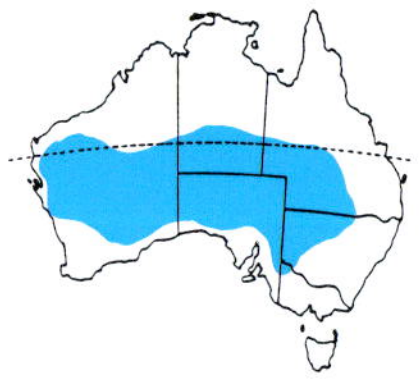

Female

52 BLUE-WINGED PARROT

Neophema chrysostoma 23 cm

The Blue-winged Parrot is a gregarious bird that has adapted to a variety of habitats from forested valleys to coastal dunes, mallee and inland saltbush plains. It feeds mainly on the ground and is approachable. If disturbed the whole flock will rise as one and fly to cover, then cautiously return in small groups. Seasonally migratory throughout its range, including to and from Tasmania. Breeds October–January. Nests in tree hollow, sometimes very high. Endemic. **Female**: *duller and more olive-green.* **Immatures**: *duller, less blue.* **Voice**: *soft twittering, melodious tinkling notes.*

53 TURQUOISE PARROT

Neophema pulchella 21 cm

Once common around Sydney, the Turquoise Parrot was portrayed by the convict artist Watling in the earliest days of colonial settlement. It is found in open forest, timbered rocky slopes and ridges, or along watercourses and has responded well to the little-disturbed country of the Warrambungle National Park. It feeds on the ground in pairs or small flocks. Nomadic. Breeds August–December. Nests in a dead tree-spout, hollow limb or fence post. Endemic. **Female**: *paler, no red on wing*. **Immatures**: *duller; young males have faint red wing patch.* **Voice**: *two-syllable whistle usually in flight.*

Male

54 ELEGANT PARROT

Neophema elegans 24 cm

While closely resembling the Blue-winged Parrot, its yellow markings are brighter and it lacks the extensive blue on the wing. Although its distribution is interrupted by the Nullarbor Plain, the two populations appear to be identical. In the east, it is sometimes seen in the company of Blue-winged Parrots where their ranges overlap; in the west, it is extending its range and numbers. A ground feeder, it is not easily flushed but when disturbed will rise high and fly off. Partly nomadic. Breeds August –November. Nests in tree holes. Endemic. **Sexes** and **immatures**: *similar.* **Voice**: *sharp 'zit-zit' in flight.*

55 SCARLET-CHESTED PARROT

Neophema splendida 21 cm

Very rare in some of its former haunts, this bird is now generally found only in the inland areas where it frequents mallee, mulga scrub, spinifex, eucalypt and casuarina woodland. Unobtrusively in pairs or small parties, it feeds on seeds on the ground. It seldom goes to water and probably gets its moisture from dew or plants. Breeds August–January or after rains. Nests in a tree hollow. It has been seen carrying aromatic leaves into the nest presumably for deodorising or humidifying purposes. Endemic. **Immatures**: *duller.* **Voice**: *feeble twittering.*

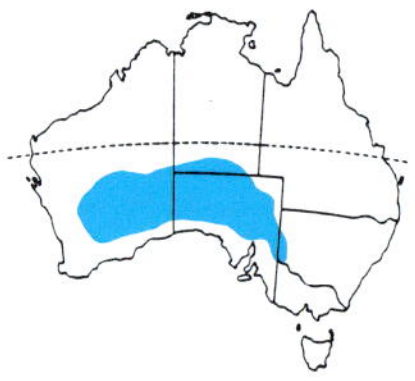

Female left, male right

56 ORANGE-BELLIED PARROT

Neophema chrysogaster 21 cm

The total population of this very rare species, found only in Tasmania and in small parts of Victoria and South Australia, is estimated to be down to only 180 birds. In recent years, an intensive survey of the species has been sponsored by the World Wildlife Fund Australia, in which Governments, scientific bodies and amateurs have co-operated, and a 'Recovery Plan' has been adopted in an effort to save it from extinction. Mainland birds migrate to Tasmania to breed during the summer. Nests in tree hollows. Endemic. **Female**: *duller.* **Immatures**: *duller than adult females.* **Voice**: *distinctive continuous buzzing.*

57 ROCK PARROT

Neophema petrophila 23 cm

The Rock Parrot frequents often bleak and exposed coastal areas and rocky offshore islands. In pairs and small flocks it feeds inconspicuously on the ground on seeds and fruits of littoral plants. Relatively tame but, when alarmed, it flits off to nearby rocks and bushes. Between coast and islands the flight is swift and high, and it circles overhead before landing. Mostly sedentary. Breeds August–October on islands. Nests in rock crevices or under overhanging ledges. Endemic. **Sexes**: *alike.* **Immatures**: *lack blue band on forehead.* **Voice**: *'tsit-tseet', twittering.*

58 SWIFT PARROT

Lathamus discolor 25 cm

Often seen together with lorikeets, this swift-flying parrot is characterised by a thin pointed red-brown tail. It has a brush tongue, like the lorikeets, though it is unrelated and its activities are similar. It clambers with agility amidst topmost foliage in search of nectar, berries, fruit and insects such as lerp-scale. When feeding it chatters busily. It breeds in Tasmania September–January and most migrate to the mainland to overwinter. Nests in tree holes 6–18 m high. Endemic. **Sexes**: *alike.* **Immatures**: *duller.* **Voice**: *torrent of notes 'clink-clink', chattering, fluting rather than screeching.*

59 GROUND (Swamp) PARROT

Pezoporus wallicus 32 cm

This ground-frequenting species makes its home in damp coastal heath and grasslands. Timid and elusive, its long legs enable it to retreat swiftly through the undergrowth when threatened. When flushed, it rises with a rush of wingbeats and flies away on a zigzag course at great speed, alternatively beating and gliding before diving back to cover. It is particularly active and vocal towards nightfall and again at sunrise. Breeds September–December and nests on the ground. Endemic. **Sexes**: *alike.* **Immatures**: *heavier black breast markings.* **Voice**: *thin high-pitched ascending notes.*

60 NIGHT PARROT

Pezoporus occidentalis 24 cm

The elusive and rarely seen Night Parrot ranges widely over the vast arid inland areas, and is probably highly nomadic. It frequents spinifex when in seed, and salt-lake and flood-plain vegetation at other times. It remains silently concealed by day and, if flushed, flies a short distance and runs off at a tangent to cover on landing. It apparently feeds, drinks and calls only after dark. Breeds – recorded in July and August. Nests found tunnelled into clumps of spinifex and 'samphire bush'. Endemic. **Sexes**: *alike.* **Immatures**: *duller.* **Voice**: *sweet, low, two-tone whistle uttered frequently. Frog-like croaks when alarmed.*

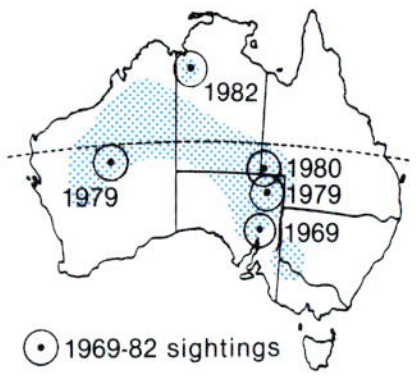

49

50

51

52

53

54

55

56

57

58

59

60

7

BUDGERIGAR

Melopsittacus undulatus 20 cm

The remarkable story of this charming little native Australian begins with the collection of a specimen at Parramatta in 1804. This in itself was unusual since the bird is not normally seen east of the Great Dividing Range except in periods of drought. The first description was based on this specimen, which was exhibited in London at Bullock's Museum as the 'Undulated Parakeet'. Gould later called it the 'Warbling Grass Parrakeet' – a name which persisted into this century until superceded in popular use by 'Shell Parrot'. The now universal name Budgerigar, which has evolved through several spellings, appears to be aboriginal in origin and means 'good bird' or 'good food'.

The Budgerigar is thought to be the most numerous of all Australian parrots and it occurs throughout the vast inland areas of the continent. It is highly nomadic and ranges over great distances in quest of water and seeding grasses.

It is a great survivor. During periods of high temperature it will remain totally inert in the shade of trees and bushes to conserve body moisture. It has a prolific capacity to breed when conditions are favourable and it can make up losses suffered through drought and heat-wave at an astonishing rate. Populations were even able to recover from the holocaust of the prolonged heat-wave of January 1932 when thermometers burst and when temperatures in parts of South Australia reached 126°F (50°C). From the many harrowing reports published in the *South Australian Ornithologist* of April 1933, it would seem that literally millions must have perished.

Budgerigars will often gather in extremely large flocks. These will sometimes settle in a dead tree with every twig covered so that it appears to be totally clothed in fresh green leaves. On occasion, flocks reach truly staggering proportions, even darkening the sky – one such case being recorded in South Australia in December 1931 when myriads of birds were seen passing overhead for three hours.

When John Gould returned to England in 1840, he took with him four aviary-bred birds. Within a few years, they were being bred in England as well as in Belgium, Holland and France. Exports from Australia also increased, with ten thousand birds being sent to Europe in one batch in 1868.

In due course, yellow and blue strains began to appear in the birds bred overseas and other mutations followed. In the wild, the Budgerigar is almost always green, the only exception known to the authors being a sighting by Keith Davey of Newcastle of a few isolated blue birds in a large flock recorded 75 km west of Poeppel's Corner during the west/east crossing of the Simpson Desert with the Leyland Brothers in 1966. Interest in selective breeding in Australia only began with the arrival of some yellow birds from England in about 1900. Thus, this amazing little bird, now world famous, returned to the land of its ancestors in new garb to trigger the breeding of every known colour variety and several new mutations in the aviaries of Australia. **Sexes**: *alike but female has brown cere when breeding*. **Immatures**: *duller, may lack black throat spots*. **Voice**: *animated conversational chatter, pretty warbling*.

7 The Paradise Parrot and John Gilbert

Mention of the Paradise Parrot leads inevitably to the name of John Gilbert, who discovered it. He was born at Clerkenwell, London on 14 March 1812 – a fact only recently established by C.T. Fisher through a renewed search of parish records. When working as a taxidermist at the Zoological Society in London, he met the great John Gould who brought him to Australia as his assistant – arriving in Hobart on 19 September 1838. He spent five and a half of the last years of his short life wandering widely in Australia collecting specimens of birds and their eggs, mammals, reptiles, insects and plants for Gould. The distinguished naturalist, Alec Chisholm, credits him with discovering, almost single-handed, more new and little-known birds than any other man before or since, and estimates that sixty to seventy of the species named by Gould were collected by Gilbert, including approximately forty new species. In 1844, Gilbert's wanderings took him to the Darling Downs, where he spent some months. In a letter dated 8 June 1844 to Gould, then in London, Gilbert reported the discovery of

> 'a totally new Parrot of the size and form of *Platycercus haematogaster* [Blue Bonnet] but without exception the most beautiful of the whole tribe I have ever yet seen in Australia; the mingling of the beautiful shades of green, is its most conspicuous and beautiful character.'

There followed a detailed description (later to be used verbatim by Gould), a list of seventy other species of birds seen in the area, and this request:

> 'If you have not already honoured my poor name in your works, I know of no species that w[oul]d delight me more to see Gilberti attached to than this beautiful Bird.'

In his reply, dated 20 December 1844, Gould wrote:

> 'I am especially delighted about the new *Platycercus*. . . . As I have lately named a *Pachycephala Gilberti* [now known as Gilbert's Whistler], I cannot with propriety add your name to this new kind. You know it is a mode of nomenclature which I very much set my face against and it is only in the most particular instances that I do it. . . .'

Gilbert did not live to receive this letter, nor was he to know that Gould was to name his new parrot *Platycercus pulcherrimus* (meaning very pretty). Sadly, the name *Pachycephala gilberti* could not be sustained (see species note at Section 10), and Gilbert's name remains unrepresented amongst scientific bird names. One can feel sympathy for the view expressed by Alec Chisholm in 1944 that 'It is an anomaly that awaits adjustment'.

On 1 October 1844, Gilbert set out from Jimbour Station, near Dalby, as a member of Ludwig Leichhardt's Port Essington Expedition. His comments on the Paradise Parrot form a valuable part of the little that is known about the species and began with this observation in the letter of 8 June 1844 to Gould

> 'It is in habits truly a grass-eating Parrot, assembling in small families, and feeding in high grass, all the specimens I have killed have had their stomach crops fully distended with grass seeds exclusively.'

During his nine months with the Leichhardt Expedition, there were nine points along the route where he made diary entries recording the presence of his 'new Platycercus'. (These points are indicated on the accompanying map). On 2 November 1844, when about 80 km north of Chinchilla, Gilbert wrote:

> 'I observed several of the new Green Parrot. Its pretty note at once distinguishes it from all other Parrots; it may be described as *tit-sweet* with a rather plaintive modulation. I have never remarked this bird uttering any harsh or disagreeable cry, as is common to most of the tribe'.

The species was recorded as 'fairly abundant' on the Robinson River, but the diary entries ceased on 24 December in the vicinity of the Comet River – until the Expedition reached the Mitchell River some 1200 km to the north. On 17 June 1845, Gilbert wrote 'I was surprised today to meet with my 'new Platycercus' of the Downs'. He reported it as 'common' on 20 June and 'in great numbers' on 24 June. On 28 June, three days after leaving the Mitchell River on a westerly route, Gilbert concluded his entry on the day's journey thus:

> 'the remaining part of the stage was through a beautiful open country thickly studded with Lotus ponds at one of which we camped'.

Shortly after turning in that night, the camp was attacked by aborigines and Gilbert was killed by one of their spears. The survivors buried him the next day and lit a large fire over his grave to hide it from the natives.

By following the remarkably detailed observations in Leichhardt's journal and his accompanying map, one can plot the route of the Expedition on a modern map with a fair degree of confidence (see map on this page), and deduce that a point where he marked 'Mr Gilbert killed' lies approximately 56 km west of Dunbar Station on a latitude of 15°15′. But the actual site of the tragedy on the ground had never been discovered – until recently.

In 1979, Brian Dalton, Professor of History at James Cook University at Townsville, was approached by Mr Jim Bessell, a former cane farmer, now living in Camoweal, Queensland, with the news that he had found the precise spot where Leichhardt had set up camp on that fateful night. He had made it a life-long quest to find Gilbert's grave, and a response to his request for help was readily given. The fascinating story of the detective work that ensued cannot be adequately told here, but happily we can look forward to a book on the subject by Professor Dalton. Suffice it to note that, thanks to the cooperation from the Army, the Airforce, a botanist, an earth scientist, who located the fire sites in the area using a caesium magnatometer, and a colleague who carbondated the charcoal, the hearth of the camp fire was found in 1982 and the location of Gilbert's grave, 100 metres away, in October 1985. (The grave remains to be excavated at the time of going to press). A memorial to Gilbert, subscribed by the colonists of New South Wales was erected in St. James' Church Sydney in 1854.

During the early days of settlement of Queensland, the Paradise Parrot was commonly seen in certain areas of the south-east corner, and in northern New South Wales. They were trapped extensively. Many were exported and the London dealers are credited with naming them the 'Paradise Parrot'.

By 1890, its numbers had been so reduced that exports ceased, but birds were still sold in Brisbane until the end of the century (at five shillings a pair!).

Forshaw believes that Gilbert's patchy sightings indicate that the species was never common and may already have been declining. It is clear, however, that its decline was hastened not only by profligate trapping and the uninhibited egg collecting then practised, but by the large-scale pastoral development in the middle of its range pursued from the 1840s which deprived it of the grasslands on which it depended. There followed, between 1893 and 1902, one of the most severe droughts ever experienced, with appalling losses of stock and wildlife, after which the Paradise Parrot was no longer seen in its old haunts.

In 1918, Alec Chisholm campaigned for information on the missing bird. Most responses were of earlier sightings, but there were two current reports: one of May 1918 of several birds at Augathella in Queensland's central west, and a surprising one from a police constable at Coen in Cape York Peninsula of sightings in the 'far north'. The constable added:

> 'I have also seen the Golden-shouldered Parrot, and it is similar in habits to the Scarlet-shouldered, but not so plentiful. Both of these birds keep moving back when settlers take up land close to their habitat'.

An interesting observation, since Gilbert's 1845 sightings on the Mitchell River have always been suspect because of the occurrence of the Golden-shouldered Parrot in the same area – a bird not described as a separate species until 1857. Subsequently, the constable's successor corroborated the sightings in the 'far north' stating that the parrots were 'moderately plentiful . . . in the neighbourhood of the Archer River'.

On 11 December 1921, C.H.H. Jerrard, a keen naturalist and photographer of Gayndah, Queensland reported seeing seven or eight adult Paradise Parrots and some young ones in the upper Burnett River valley. Photographs of a pair of birds and their nest in a termite's mound followed (three of which are reproduced opposite). Sadly, the eggs proved to be infertile. Jerrard reported other pairs in the area in 1921 and 1922, some with broods, but he saw the bird no more after 1927. Arnold McGill, who has read Jerrard's diaries, says that, of six more nests examined, the majority of the eggs were also infertile. Sporadic reports of the Paradise Parrot persist but, regrettably, there have been no reliable sightings for many years and, save for childhood memories of this beautiful bird in a cage, no-one now living is known to have seen it alive.

Alec Chisholm was one of the last to do so – in the Burnett River valley on 1 November 1922 – and in later years wrote:

> 'I am now merely toying with a recollection of the varied colours, the shapely form, and the prideful bearing of the exquisite little bird as it appeared on a summer day in a secluded part of Queensland'.

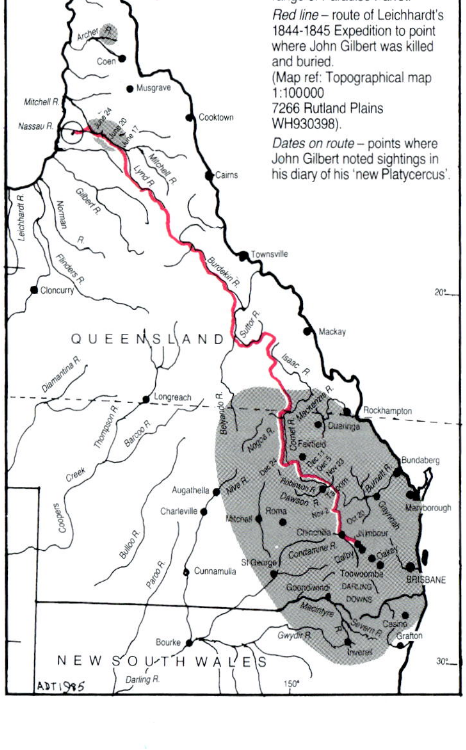

7 Australia's First Ornithological Record?

On the world map of Gerard Mercator of 1569, there is an inscription (illustrated opposite) marking a place called *Psitacorum Regio*, Land of Parrots. A free translation follows:

> Region of Parrots — so named by the Portuguese because of the unheard of size of the birds at that place. Driven there when sailing to Calicut by a *libeccio* [south-west] wind, they followed the coast of this land for 2000 miles without finding an end to it. There is no doubt therefore that they had reached the Southern Continent.

As the accompanying map shows, it lies in a latitude approximately corresponding with that of Tasmania on a vast imaginary continent which juts northwards on the eastern boundary of the Indian Ocean on a line roughly coinciding with the west coast of Australia. It is, therefore, tempting to entertain the idea that it may have been Australian parrots that impressed the early Portuguese navigators, and this led to our making extensive enquiries, including overseas, to trace any research done on the subject — but without result. Commenting on the inscription in his writings on the early history of Australian zoology, Gilbert Whitely wrote: 'If [the inscription] were anything else than mythical, it would be Australia's first ornithological record'. Here was a motive for investigating the question and we offer our findings.

THE GREAT AGE OF DISCOVERY

The Period Mercator produced his map within only a few years from the beginnings of the Great Age of Discovery by Portugal and Spain — with voyages of Columbus to America in 1492, Vasco da Gama to India in 1500, and Magellan to the Pacific in 1520. Huge areas of the world still remained to be discovered, including Australia, and the mapping of territories beyond Europe, the Middle East and northern Africa was very much in its infancy.

The Treaty of Tordesillas In 1494 this treaty divided the world into two equal spheres of interest between Spain and Portugal. The Lines of Demarcation are included in our map which shows the extent to which Australia falls within the Portuguese domain.

Navigation Mariners could determine latitude fairly correctly in the sixteenth century, but they had no means of measuring longitude until the advent of the chronometer in the eighteenth century and vast errors in navigation resulted.

The spice trade It is difficult now to appreciate either the scale or the riches involved in the spice trade, and Portugal became a very rich country as a result of establishing its sea route to India and breaking the monopoly previously enjoyed by Venice.

MERCATOR'S 1569 MAP

The land occupying the bottom of Mercator's map is part of what he conceived as a vast continent around the South Pole and, except for the northern coast of Terra del Fuego, is mostly imaginary. The inclusion of the promontory jutting above the Tropic of Capricorn, however, was influenced by the writings of Marco Polo (1254-1324?), but erroneously named Maletur for Marco Polo's Malaya. Also, the deep bite to below the sixtieth parallel in the Pacific allowed for reports from India that 'certain people [probably the Javanese] had sailed south to such a degree that they find a day of 4 hours, that is, in the 63 degree of latitude'. The open water there doubtless also took account that in 1503 Vasco da Gama was driven south in a storm to a point where, on 22 May, 'the day was of eight hours' [which works out at 55° of latitude]. Mercator's reputation was such that his imaginary Southern Continent, with the inscriptions upon it, were copied on world maps for over a century — until the shape of Australia itself began to emerge as a result of the chartings by the Dutch in the seventeenth century. *Psitacorum Regio* was shown on at least forty maps, although in abbreviated form and in various languages.

As to the sources used, all we have is the general statement by Mercator on his 1569 map that he 'compared the marine charts of the Castilians and Portuguese and read many accounts of voyages, reconciling all these carefully with each other'.

EARLIER MAPS

We wish to acknowledge the extent to which our further researches were simplified by Rodney W. Shirley's monumental work, *The Mapping of the World* (1984).

The *Psitacorum* maps The earliest *psitacorum* inscription found was recorded on a set of gores on an Anonymous Map published in Nurenberg c.1535. It was shown simply as *Psitacorum Terra* without explanatory text, and differed from later maps in being placed about 20° (1200 miles) to the east of them. Then came a gore map of a terrestrial globe by Mercator of 1541 inscribed with a legend in terms that somewhat differed from those of his 1569 map:

> [literal translation] Region of parrots discovered by the Portuguese in 1500 who sailed along 2000 miles; so called because it has parrots of unheard-of size, measuring three cubits in length.

This was followed by a map by Abraham Ortelius of Antwerp of 1564 with a somewhat different version of the inscription:

> [literal translation] Region of Parrots so called by the Portuguese, who sailed along its coast for 2000 miles, however they did not find its end, on account of the incredible abundance of birds seen there. The inhabitants are naked and are handsome, and of tall stature.

The *Papagalli* maps The enquiry took on a new dimension when we noticed a place called *Papagalli Regio* on the Anonymous Map of c.1535, as well as *Psitacorum Terra*. As both terms translate as Land of Parrots, the clear implication was that the cartographer recognised two places in the world as renowned for their parrots. Six other maps, ranging from 1516 to 1574, were identified as naming Brazil *Bresilia sive Terra Papagalli* (Brazil or Land of Parrots). Two of these, dated 1570 and 1574, also showed *Psitacorum Terra* on the Southern Continent.

We were curious to know why the cartographers had used different names for the parrots, and sought the guidance of Alan Treloar (see p. 223). He was of the view that *Papagalli* was a Latinised word from the Portuguese *papagaios*, which in itself is an adaption of the Arabic form of a west African word for parrots; whereas *psitacorum* is derived through the Greek *psitakos*, which is of Indian origin. He felt that the distinction strengthened our case.

TERRA PAPAGALLI

On 9 March 1500, a fleet of thirteen ships under the command of Pedro Alvares Cabral sailed from Lisbon for Calicut in India to extend Portugal's control over the spice trade. Veering westwards to avoid the doldrums and the south-east trades, the fleet chanced upon the coast of Brazil, where it stayed from 23 April to 2 May at a place Cabral named Porto Seguro. The very large and colourful macaws seem to have made a greater impression than anything else seen in Brazil, and the new land became popularly known as *Terra da papagaios*. A despatch on Cabral's expedition to the Doge of Venice from his diplomatic representative in Lisbon of 27 June 1501, two days after the return of the first ship, included this passage:

> Above the Cape of Good Hope towards the west they have discovered a new land. They call it *La Terra di Papaga*, because some are found there which are one and a half cubits in length, of various colours. We saw two of these. They judged that this was mainland because they ran along the coast more than 2000 miles but did not find the end. It is inhabited by nude and handsome men.

[Similarities of parts of the Mercator and Ortelius *Psitacorum* inscriptions with this report are difficult to explain].

PORTUGAL'S MARITIME OPERATIONS

The Hakluyt Society's volume on Cabral's voyage (see p.225) cites an estimate (in a work by Liuz de Figuiredo Falcao of 1859) that, from Vasco da Gama's voyage in 1497 to the annexation of Portugal by Spain in 1580, no less than 620 ships sailed to the Indies from Portugal. While they would have been engaged primarily in trade, it seems more than probable that some of them would have undertaken voyages for the purpose of discovering new lands — for as K.G. McIntyre points out in his *The Secret Discovery of Australia* (1977):

> Portugal made a fetish of the minute and thorough examination of every speck of land, every island however small, in her hemisphere; and there is no reason to believe that the Australian sector alone would be neglected.

That they may have probed directly eastwards from the Cape taking advantage of the Roaring Forties seems a real possibility. This would be consistent with the fact that Mercator placed his *Psitacorum Regio* in this latitude, also his statement in the 1569 inscription that the Portuguese were driven there by a *libeccio* [south-west] wind. This wind is well known in the Mediterranean, as with the *sirocco* [south-east], and these terms appear to have been commonly used by seamen of the period for the quarters of the wind anywhere at sea. The *Grande Dizionario della Lingua Italiana* also defines the *libeccio* as 'a bringer of rain and tempest', 'almost always squally' and 'very violent'. Another point which may not be irrelevant is that when in 1611 the Dutchman Hendrik Brouwer decided to test an easterly bearing from the Cape as the first leg of a quicker route to the Indies, he was able to draw on the experience of the many Dutchmen who had previously sailed with the Portuguese.

THE PARROTS

There are seventeen species of parrots whose ranges reach the western coastline of Australia. The largest are the White-tailed (65cm) and the Red-tailed (60cm) Black Cockatoos, both to be seen south of Perth. By comparison, the macaws of South America vary in size from 85-100cm. From the descriptions given in the Hakluyt volume, the two taken back to Lisbon in 1502 were probably Scarlet Macaws. These measure 85cm or nearly two cubits (the length of a forearm from the tip of the middle finger to the elbow) on the basis of a Roman cubit of 44.5cm. Mercator's reference in his inscription of 1541 to the *Psitacorum* birds being three cubits in length can be taken as romantic licence. Ortelius refers to 'the incredible abundance of birds seen there'. Perhaps the great flocks of Little Corellas to be seen north of Perth — referred to by the English buccaneer William Dampier in 1699 'which flew a great many together' — were those intended.

CONCLUSION

The circumstantial evidence in these notes would appear to be enough to justify a belief that if the official records of Portugal's great discoveries had not all been lost in the great earthquake that destroyed Lisbon on 1 November 1755, records of contacts with the coasts of Australia would be found with the details upon which the *Psitacorum* inscriptions were based. There would seem to be no reason therefore to dismiss Mercator's assertion that 'There is no doubt therefore that they [the Portuguese] had reached the Southern Continent'. And it would seem that it was by its parrots that this land was best remembered by its discoverers — just as it was with Brazil.

The help and advice given by those named on page 223 is gratefully acknowledged.

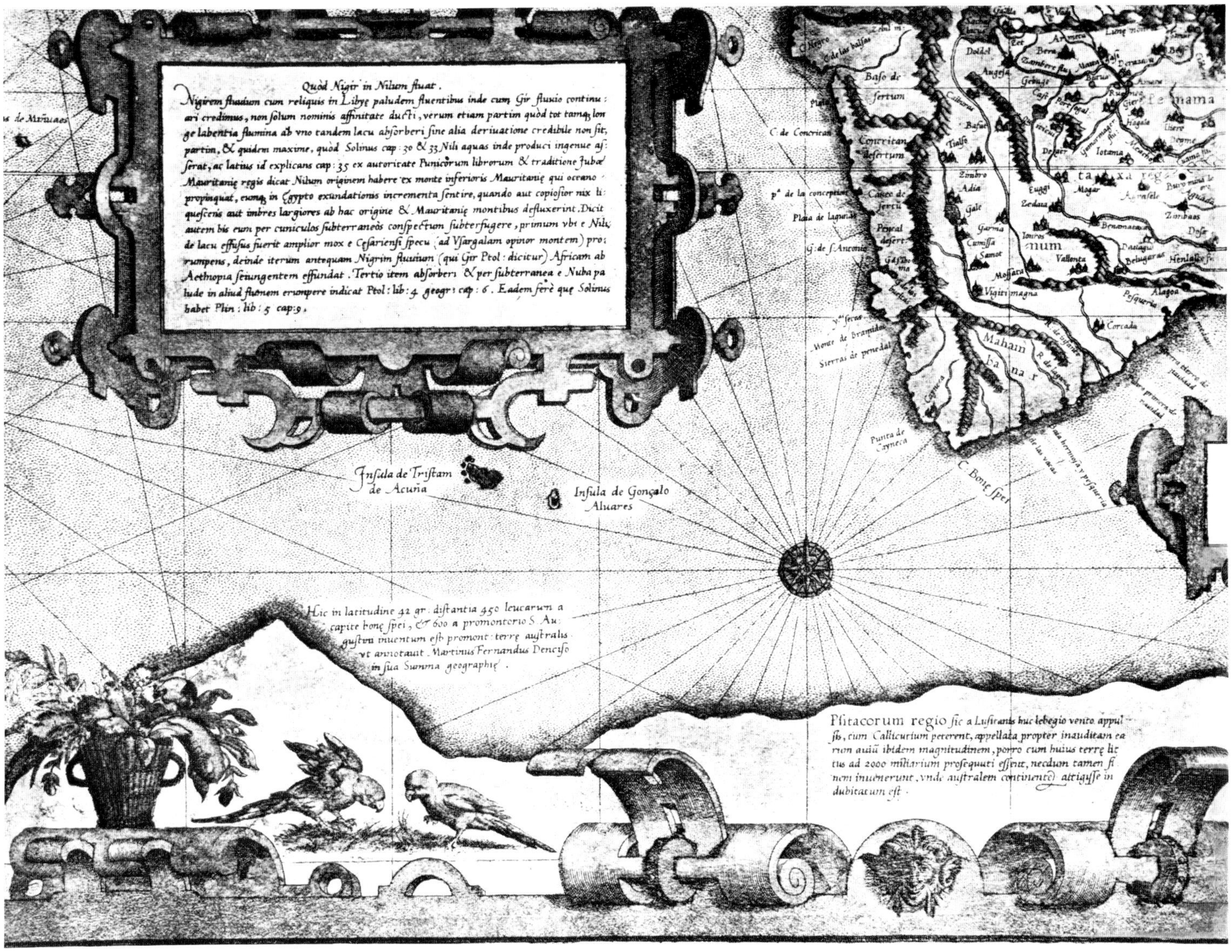

Extract from Gerard Mercator's world map of 1569 carrying the inscription *Psitacorum Regio,* reproduced from a copy in atlas form published by the International Hydrographic Bureau, Monaco in 1931

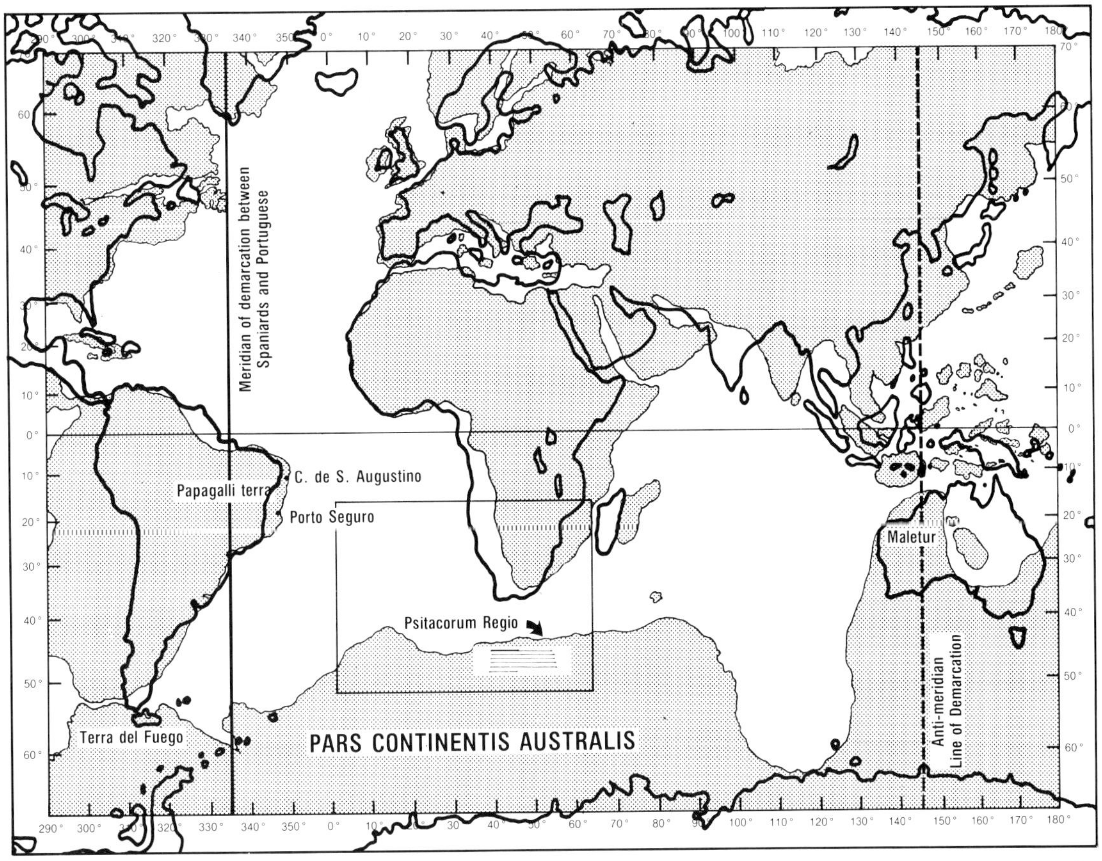

Outline of part of Gerard Mercator's world map of 1569 shows the location of the above extract. The modern outline superimposed for comparison allows for a difference of 23° between Mercator's prime meridian and Greenwich. *Papagalli Terra,* an early name for Brazil, and the Anti-meridian have been added for reference.

8

Cuckoo-shrikes and Allies, Woodswallows, Swallow-like Birds, Sittellas and Treecreepers

CUCKOO-SHRIKES AND TRILLERS Family Campephagidae

Spreading far afield through Australasia to southern Asia and Africa, this family of over seventy species is represented in Australia by seven, one of which, the Ground Cuckoo-shrike, is endemic. While the predominantly grey Cuckoo-shrikes superficially resemble some cuckoos, with their slender bodies, long tails and undulating flight, and have notched bills like the shrikes of the northern hemisphere, they are not closely rlated to either. The two trillers are accomplished songsters, and the males go into eclipse plumage after breeding. The Cicadabird – formerly known as a 'caterpillar-eater' does indeed utter sound like a cicada. The nests of the birds in this group, except for the Ground Cuckoo-shrike, are disproportionately small – some only large enough for one egg.

1–10

WOODSWALLOWS Family Artamidae

This family of ten species extends from Australasia east to Fiji and north-west as far as India. Australia is their stronghold with six breeding species, four of which are endemic. They are not, as the name suggests, swallows, although like them their wings are pointed, some have forked tails, and they hawk for insects on the wing. These attractive soft brown or sooty grey birds are highly social and may often be seen packed tightly together on a branch preening one another; when birds in the middle fly off the remainder immediately close ranks to fill the gap. All have a fairly extensive range. The White-browed, Masked and White-breasted Woodswallows are either migratory or nomadic, whereas the Little, Dusky and Black-faced Woodswallows are mainly sedentary.

11–16

SWIFTS AND SWIFTLETS Family Apodiade

The swift family (not to be confused with the swallow family which is not closely related) is poorly represented in Australia, with only one resident breeder (the Grey Swiftlet) and two regular visitors (the Spine-tailed and Fork-tailed Swifts). The other two (the Glossy and Uniform Swiftlets) are seldom observed and are believed to be accidental visitors rarely extending south of the Cape York area. Swifts spend most of their lives on the wing, even sleeping and copulating. They are differentiated from swallows by their longer and narrower crescent-shaped wings and their very fast flight, which has been estimated to exceed 150 kph in dives. They cannot perch and have to cling to vertical surfaces when grounded.

17–20

SWALLOWS AND MARTINS Family Hirundinidae

This worldwide family of some eighty species is represented in Australia by only three swallows and two martins. However, two of the swallows (the White-backed and Welcome) and one of the martins (the Fairy) are endemic. The martins are distinguishable from the swallows by their slightly smaller size and by the absence of a deeply forked tail. Though closely related, the four Australian breeding species exhibit a remarkable diversity of nesting behaviour. The Welcome Swallow and the two martins make good use of man-made structures as nesting sites, such as under the eaves of buildings. Many odd and unexpected places have been recorded – including the adoption by Welcome Swallows of a Cobb & Co coach near Melbourne in 1889 and the ferry-boat at Wiseman's Ferry near Sydney more recently.

21–25

SITTELLAS Family Neosittidae

There are two species of sittellas, one found in New Guinea and the other, the Varied Sittella, in Australia. Although superficially resembling the nuthatches of the norther hemisphere, they are believed to have evolved independently and are unrelated. They will almost always be found wherever there is woodland and dry forest. There are five quite distinctive forms and these keep to particular areas of the continent, except for fairly narrow zones of overlap where they interbreed and produce intermediate forms. They wander widely within their areas, usually travelling in small flocks. Their days are spent busily scurrying about among the branches of trees probing the crevices and prising up bark in search of insects. They are extremely acrobatic and, unlike the treecreepers, they usually work along branches from the tip and spiral down the trunk. They are communal breeders, living in family groups with several birds of a group helping to build the nest, usually in a deep fork, and feeding the young.

26–30

TREECREEPERS Family Climacteridae

Of the seven species of treecreepers comprising this family, six occur in Australia and all of these are endemic. The other species lives in New Guinea. Treecreepers feed mostly on tree trunks and spiral upwards probing under the bark for grubs and insects. On reaching the end of their climb, they dive to the base of the next tree of their choice to repeat the process. Although this feeding behaviour is the same as that of the treecreepers of the northern hemisphere, the Australian birds do not have the advantage of their stiff pointed tail feathers for support, and the two families are not closely related. Although sometimes called 'woodpeckers' locally they are not related, and Australia is one of the few countries in the world where the woodpecker family is unrepresented.

31–36

Opposite, top: Black-faced Cuckoo-shrike left; White-browed Woodswallow right
Opposite, bottom: White-backed Swallow left; White-browed Treecreeper right

PHOTOGRAPHERS: H. & J. Beste 28, 29; N. Chaffer 5, 6; G. Chapman p. 75 (top left); 7, 26, 34; R. Dencio 20; C.B. Frith 19; T. & P. Gardner 12, 18; J. Gray 27; E. & D. Hosking 22; K. Ireland 9, 16; R. & D. Keller 24; P. Klapste 17; W.J. Labbett p. 75 (bottom right) 33; J. Purnell 4, 10, 36; K. & B. Richards 31, 35; M.J. Seyfort 25; T.A. Waite 15; G. Weber p. 75 (bottom left) 21; C. Webster 1, 13; E. Whitbourn 3; E.E. Zillmann 30; Authors p. 75 (top right) 2, 8, 11, 14, 23.

8

1 BLACK-FACED CUCKOO-SHRIKE

Coracina novaehollandiae 33 cm

The Black-faced Cuckoo-Shrike is also known as the 'Blue Jay' or the 'Shufflewing', the latter because it folds and refolds its wings on alighting. It inhabits open timber, sometimes in small flocks, and comes to parks and gardens. It eats insects, berries and fruit, sometimes hovering, then diving to the ground. Partly nomadic. Breeds August–February. Nests 5–15 m high. Extralimital range: New Guinea, Indonesia and S.E. Asia. **Sexes**: *alike.* **Immatures**: *less black on face.* **Voice**: *2 or more drawn out, husky 'cheeeer' sounds.*

2 LITTLE [White-bellied] CUCKOO-SHRIKE

Coracina papuensis 28 cm

Only the northern race *hypoleuca* (illustrated), which spans the top end of the continent, has a white belly. The larger race *robusta*, which occupies eastern Australia, has a grey breast and belly. It is usually seen in pairs and small flocks in forest and woodland, where it feeds mainly in the tree canopies. Some seasonal or nomadic movement. Breeds August–March. Also in New Guinea, Moluccas and the Solomons. **Sexes**: *alike.* **Immatures**: *brown mottling.* **Voice**: *a rather reedy 'kiseek, kiseek', other calls like the Black-faced Cuckoo-Shrike.*

3 BARRED [Yellow-eyed] CUCKOO-SHRIKE

Coracina lineata 28 cm

Singly and in pairs when breeding, and in flocks at other times, this bird frequents the coastal rainforests. It is constantly on the move foraging for its diet of fruit, berries and insects, often in the company of fruit pigeons and figbirds. Non-breeding birds roost communally at night. Breeds October–January. Nests in trees up to 20 m from the ground. Extralimital range: New Guinea to the south-west Pacific. **Recognition**: *barred underparts and yellow eye.* **Sexes**: *alike.* **Immatures**: *barring on breast is less distinct.* **Voice**: *series of soft notes and pleasant chatter.*

4 GROUND CUCKOO-SHRIKE

Coracina maxima 36 cm

A communal breeder, the species is seen mainly in groups of three or four, all of whom help to feed the young. Very wary, it is hard to approach in open country. At a distance it appears black and white, like a long-tailed magpie. It feeds on the ground in sparsely timbered areas, mainly inland. It nods the head pigeon-like when walking, and on alighting flicks the wings, more so when alarmed or calling. Nomadic. Breeds August–December. Nests 3–15 m high. Endemic. **Sexes**: *alike.* **Immatures**: *faint barring.* **Voice**: *peculiar rippling call note, and a series of piercing cries.*

5, 6 VARIED TRILLER

Lalage leucomela 18 cm

In Australia, this species lives mainly in coastal tropical and sub-tropical monsoon rainforests as well as mangroves, river margins and gardens. It eats both insects and fruit, scrutinising the foliage, and is sometimes seen in mixed feeding flocks. Sedentary, moving inland during the wet season. Breeds August–April. Nests 1–2 m high. Extralimital range: New Guinea and adjacent islands. **Recognition**: *the female Cicadabird also has a barred abdomen, but is a much bigger bird.* **Immatures**: *like adult females.* **Voice**: *trilling, drawn-out 'k-a-r-r-'.*

Male left, female right

7, 8, 9 WHITE-WINGED TRILLER

Lalage sueurii 18 cm

When breeding, the male endlessly pours out beautiful musical notes, even when sitting on the nest and in flight. The female is noticeably quieter. Seen singly, in pairs or groups, it frequents lightly-timbered areas including suburban parks and gardens, foraging in foliage, amongst blossoms, or on the ground. Breeds September–February. The nest, a shallow cup, is often built with other species in the same tree, from 1–10 m high. After breeding, males moult into an eclipse plumage of dull brown but this is rarely seen in the south. Migratory. Extralimital range: Timor, Indonesia and Sulawesi (Celebes). **Sexes**: *similar in non-breeding season.* **Immatures**: *like females.* **Voice**: *persistent melodious trilling and warbling.*

Male (breeding) left, male (eclipse) centre, female right

10 CICADABIRD

Coracina tenuirostris 27 cm

The male of this small cuckoo-shrike utters a high-pitched buzzing during the breeding season resembling the stridulations of the cicada. Arboreal and inconspicuous in the high foliage searching for insects singly or in pairs. West of the Gulf of Carpentaria, the birds are smaller. Seasonally migratory. Breeds October–March and nests in trees 3–28 m above ground. Extralimital range: widespread throughout the south-western Pacific, New Guinea and adjacent islands. **Male**: *lacks the barred breast.* **Immatures**: *like adult females.* **Voice**: *male (as above); female a sharp whistling 'cheep'.*

Female

11 WHITE-BROWED WOODSWALLOW

Artamus superciliosus 19 cm

This strikingly plumaged bird is perhaps the most beautiful of the woodswallows. It is highly nomadic in search of food and wanders in large flocks, often with the Masked Woodswallow to which it is closely related, and frequents open forest, woodland, orchards and gardens. Its brush tongue enables it to imbibe nectar and it also takes insects. Breeds August–January (usually when conditions are suitable). Migratory after breeding. Endemic. **Female**: *noticeably duller.* **Immatures**: *mottled.* **Voice**: *sparrow-like chirps in flight.*

Male

12 MASKED WOODSWALLOW

Artamus personatus 19 cm

The black face and throat of this species are prominent distinguishing marks. It may be seen with its White-browed relative in flocks, sometimes comprising more than 100 birds. It chatters and hawks on the wing for insects. In spring huge flocks join with the White-browed Woodswallow over southern Australia prior to dispersing to start breeding. It breeds August–December and nests in bushes, fence posts etc. usually 1–2 m above ground. Migratory after breeding. Endemic. **Sexes**: *similar.* **Immatures**: *upper parts mottled.* **Voice**: *soft chattering.*

1

2

3

4

5

6

7

8

9

10

11

12

8

13 WHITE-BREASTED WOODSWALLOW

Artamus leucorhynchus 17 cm

Perches in groups of ten or more and together set up a constant animated chattering and chirping. It feeds on insects taken in the air, and is found in well-timbered country near water. Sedentary in the north. Nomadic and seasonally migratory further south. Breeds August–January. Nests in a variety of situations, sometimes using a Peewee's old nest. Extralimital range: species extends to Borneo, Philippines and Fiji. **Recognition**: *a white rump and an all-black tail*. **Sexes**: *alike*. **Immatures**: *mottled*. **Voice**: *harsh 'cheeek', continuous chattering*.

14 BLACK-FACED WOODSWALLOW

Artamus cinereus 18 cm

One of the commonest birds of inland Australia where it is mainly sedentary, preferring sparsely vegetated open savannah. It lives in pairs or family groups which roost in clusters at night. Occasional large winter flocks occur in arid areas. It hawks for insects on the wing and takes passing prey while perched, but also feeds on the ground. Breeds August–January. In arid regions breeding is quickly responsive to rains. Nests 1–5 m above ground in a bush or on a stump. **Sexes**: *alike*. **Immatures**: *grey-brown mottling*. **Voice**: *a short subdued rolling chatter*.

15 DUSKY WOODSWALLOW

Artamus cyanopterus 18 cm

In pairs and family groups, it frequents open forests and partly-cleared lands. The western population – separated from the eastern by some 2 500 kms – is partial to open tracts within the jarrah and karri forests. Like both the Black-faced and Little Woodswallows, it hawks for insects on the wing and roosts in close clusters. Breeds August–January. Nests within 11 m of the ground in a wide variety of situations. Eastern population nomadic after breeding. Endemic. **Recognition**: *the white wing edge distinguishes this species from the other woodswallows*. **Sexes**: *alike*. **Immatures**: *mottled*. **Voice**: *constant chattering*.

16 LITTLE WOODSWALLOW

Artamus minor 12 cm

This, the smallest of the woodswallows, is seen in pairs and family groups. The inland birds frequent rocky outcrops; the birds on the WA coast haunt the cliffs and those in Queensland inhabit the woodland fringes. The species takes insects while floating high in graceful flight and low over water, sometimes with other woodswallows. Occasionally sips nectar. They roost in clusters at night. Breeds October–January. Nests in cliffs and rock crannies in WA and in trees elsewhere. Endemic. **Recognition**: *lacks white wing edge*. **Sexes**: *alike*. **Immatures**: *mottled*. **Voice**: *softer than other woodswallows*.

17 GLOSSY SWIFTLET

Collocalia esculenta 11 cm

This non-breeding summer visitor has been seen only on a few occasions in Australia – in Cape York Peninsula and near Mackay. The smallest of the swiftlets, and distinguishable by its white belly and contrasting glossy black upper parts. It is seen in small parties or flocks and is reported to fly more slowly and lower than other small swifts. Extralimital range: Malaysian Peninsula, New Guinea, Solomon Islands, New Caledonia and New Hebrides.
Sexes: *alike*.

18 GREY [White-rumped] SWIFTLET

Collocalia spodiopygia 11 cm

Flocks of this small swiftlet circling high betray breeding sites located in caves and gorges. These have been found at Cairns, near Mackay, Tully River, Chillagoe Caves, Dunk and Bedarra Islands. It breeds September–February, and nests in dense colonies. The nest is a tiny cup of grasses cemented to the rock face with saliva, and the bird guides itself in dim cave interiors with echoes from its radar system. Caves are abandoned after nesting. Extralimital range: some islands adjacent to New Guinea and in the western Pacific. **Sexes**: *alike*.

19 FORK-TAILED SWIFT

Apus pacificus 19 cm

A regular non-breeding migrant, arriving from northern Asia in early October and leaving in mid-April. Sometimes in huge flocks. It will go anywhere where insects are plentiful and favours low pressure zones where the humidity is high. Like all swifts, it takes insects in the air, and skims over water to drink. It apparently sleeps on the wing, soaring at considerable heights. It has excellent twilight vision. **Sexes**: *alike*.

20 SPINE-TAILED SWIFT [White-throated Needletail]

Hirundapus caudacutus 20 cm

Arrives mid-October in eastern and south-eastern Australia from Asia, and departing mid-April. It is one of the world's fastest and most powerful flyers, achieving cruising speeds of 50–130 kph, and is seen soaring high in parties to very large flocks. It hawks for insects on the wing, and skims lakes and other stretches of water to drink. It apparently sleeps soaring on the wing, and has excellent twilight vision. Occasionally clings to trees or rock ledges with its sharp claws using its tail spines as a prop. **Sexes**: *alike*.

21 WHITE-BACKED SWALLOW

Cheramoeca leucosternum 15 cm

In small flocks and colonies, it frequents open country including the drier inland areas. It has a graceful, fluttering flight and takes insects on the wing. It breeds August–December and nests in burrows drilled into a sandy creek bank or road cutting and is the only Australian swallow to do so. It is highly sociable and up to twenty birds may roost at night in a nesting burrow. Clusters of birds have also been found hibernating in a burrow in a torpid state during cold and inclement weather. Endemic. **Sexes**: *alike*. **Immatures**: *duller*. **Voice**: *an attractive twittering song, with a 'check check' call in flight*.

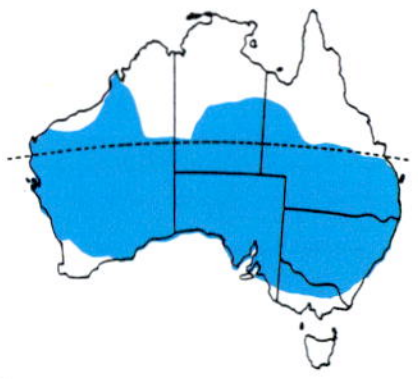

22 BARN SWALLOW

Hirundo rustica 17 cm

A non-breeding summer migrant from Asia, which is common almost throughout the northern hemisphere. It regularly occurs in the Kimberley area, but elsewhere sightings are sporadic. Most are in northern Australia but it could turn up anywhere. Like the Welcome Swallow, its flight is swift and graceful and it hawks for insects on the wing, sometimes close to water, over which it also skims to drink. **Recognition**: *thin blue-black band below throat*. **Sexes**: *alike*. **Immatures**: *duller*. **Voice**: *high-pitched 'tswit' or 'tswee' and rapid chatter*.

23 WELCOME SWALLOW

Hirundo neoxena 13 cm

Familiar and endearing, this endemic swallow is indeed welcome when, in eastern Australia, it returns with the spring from overwintering in the north to breed – many to the same beautifully-made and softly-lined mud nest year after year with the same partner. In autumn, lines of birds twittering on telegraph wires gathering to take off for their winter migration make a pretty sight. The birds in the west are mainly sedentary. **Recognition**: *the deeply forked tail and absence of a white rump distinguish this bird and the Barn Swallow from the martins*. **Sexes**: *alike*. **Immatures**: *duller*. **Voice**: *sweet twittering 'seet'*.

24 TREE MARTIN

Cecropis nigricans 13 cm

Probably the most common and widespread of the swallow-like birds in Australia. It is partial to the bigger trees which offer plenty of nesting sites, in open eucalypt woodland and likes to be near water. It feeds on the wing on insects over tree canopies or water. Breads July–January and nests in loose colonies in small holes in trees, interstices in buildings, or cliff holes. Migrates northwards to overwinter, including to New Guinea. In southern Australia, large flocks gather for the autumn migration. Range extends to Indonesia and New Caledonia. **Sexes**: *alike*. **Immatures**: *browner*. **Voice**: *high-pitched twitter*.

13

14

15

16

17

18

19

20

21

22

23

24

8

25 FAIRY MARTIN

Cecropis ariel 12 cm

Gregarious and favours open country near water. Takes insects in the air or over water. Breeds August–January in the south and May-March in the far north. The distinctive bottle-shaped nest of mud pellets, has a down-sloping spout as an entrance, and is often built in large clusters on a variety of surfaces, including under bridges and caves. Several birds may combine to build a single nest, and new birds often take over old nests or build new ones over them. Southern birds are seasonally migratory. **Recognition**: *rusty-brown head. Endemic.* **Sexes** and **Immatures**: *alike.* **Voice**: *noisy twittering especially at dawn.*

26 (A) VARIED SITTELLA (White-winged)

Daphoenositta chrysoptera leucoptera 10 cm

(B) VARIED SITTELLA (Striated)

Daphoenositta chrysoptera striata

(Also see Introduction) 10 cm

These two races have white wing patches instead of orange. (A) (illustrated) frequents woodlands and light timber. (B) *striata* (not illustrated) has the darkest head and is the most streaked of the sittellas. It frequents forest and scrub. The two races breed July–October. **Sexes** and **immatures**: *some differences.* **Voices**: *see 27.*

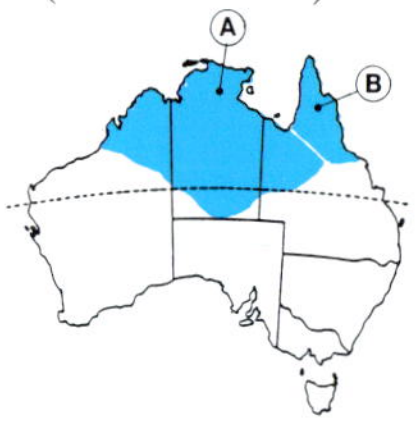

27 VARIED SITTELLA (Orange-Winged)

Daphoenositta chrysoptera chrysoptera 12 cm

(Also see Introduction)

An orange wing-patch and a dark grey-brown head characterise this the least distinctively marked nominate race. It lives in most types of forest, except rainforest, and lightly wooded country. **Sexes**: *similar, but throat of female is dusky white and speckled.* **Immatures**: *paler, underparts speckled.* **Voice**: *varied thin high-pitched incessant 'psees', 'dear-dip', 'wits' and 'chips', often in flight.*

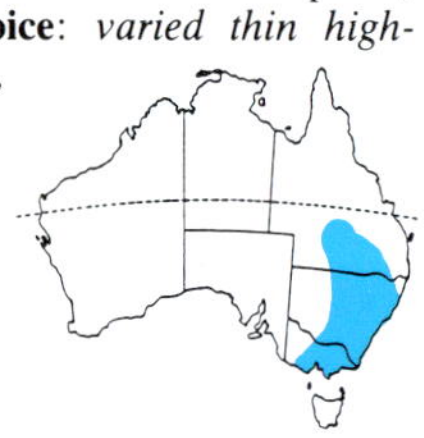

28, 29 VARIED SITTELLA (Black-capped)

Daphoenositta chrysoptera pileata 12 cm

(Also see Introduction)

The most widespread of the orange-winged sittellas of southern Australia. It is distinguished by its bold and striking markings and by its white (not mottled) underparts. It frequents forest and woodland of the coastal areas and timber near water of the arid inland. Breeds usually September–January. **Sexes**: *the black cap of the female extends over the forehead. The faces may be white or black in either sex (Boles 1980).* **Immatures**: *crown in greyish-black.* **Voice**: *see 27.*

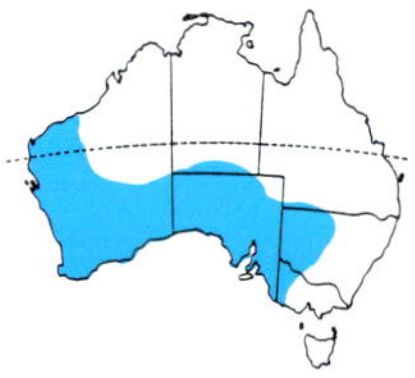

30 VARIED SITTELLA (White-headed)

Daphoenositta chrysoptera leucocephala 12 cm

(Also see Introduction)

The head and neck of this orange-winged sittella are wholly white. It lives in forest and woodland from the high country of the Great Dividing Range to the more open coastal plains. Breeds July–October and February–March. **Sexes**: *similar, but the throat of the female is dusky instead of white.* **Immatures**: *some may have grey heads.* **Voice**: *see 27.*

Male left, female right

31 WHITE-THROATED TREECREEPER

Climacteris leucophaea 17 cm

The species includes the smaller Queensland race *minor* 'Little Treecreeper', found only on the Atherton Tablelands. An agile bird, which can even hop along the underside of a branch, it zig-zags briskly up a tree probing under bark with its long beak for insects, spiders, and especially ants. It frequents a variety of wooded habitats. Breeds August–January, and nests in tree cavities from 5 m above ground. Endemic. **Male**: *lacks orange neck spot.* **Immatures**: *chestnut rump.* **Voice**: *bursts of loud piping notes in a variety of songs briskly repeated.*

Female

32 RED-BROWED TREECREEPER

Climacteris erythrops 15 cm

Less conspicuous than the White-throated Treecreeper, with which it shares its range. It frequents rainforest or dense eucalypt forest and, where the populations overlap, the Red-browed Treecreeper takes to the higher trees. It breeds August–January and nests in trees 6–30 m above ground. Endemic. **Females**: *upper breast streaked.* **Immatures**: *no eyebrow marking.* **Voice**: *similar but quieter than its noisy relative, the White-throated Tree-creeper and the song is less sustained.*

Male

33 WHITE-BROWED TREECREEPER

Climacteris affinis 14 cm

Closely resembles the Red-browed Treecreeper in plumage and voice. It is partial to a habitat of mulga, casuarina, and other sparse vegetation and often forages on the ground among fallen timber for food. It will however behave like other Tree-creepers where taller timber borders its perferred habitat. Breeds September–November and nests low, perhaps in a tree hollow or log. Endemic. **Male**: *upper breast grey (at p.75 lower right).* **Immatures**: *no marked streaking of underparts.* **Voice**: *loud strident call and weak song.*

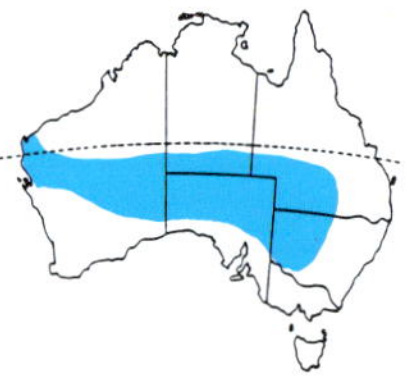

Female

34 RUFOUS TREECREEPER

Climacteris rufa 17 cm

The most colourful of the treecreepers. It is often seen in family groups of three to five and is a communal breeder. It frequents the open woodland, moist jarrah forest, and mallee of the south western parts of Australia. Breeds September–December and nests in a tree up to 8 m above the ground and occasionally in a hollow log. All members of the family may help to feed nestlings. Endemic. **Sexes**: *similar but female has breast feathers edged with red.* **Immatures**: *plainer with rufous rump.* **Voice**: *shrill repeated whistling like the Brown Tree-creeper, but higher in pitch.*

Male

35 BROWN TREECREEPER

Climacteris picumnus 17 cm

The northern race *melanota* 'Black Treecreeper' holds a special place in the annals of ornithology as it was first collected for John Gould by John Gilbert on the day in 1845 he was speared by aborigines on Leichhardt's first expedition. The species lives in groups of up to five birds in a variety of habitat and is common in eucalypt woodland. Breeds June–January (but earlier inland and later in the east) and nests in trees 1–15 m above ground, sometimes in a fence post. Endemic. **Female**: *faint rufous breast markings.* **Immatures**: *underparts faintly streaked.* **Voice**: *repeated strident notes.*

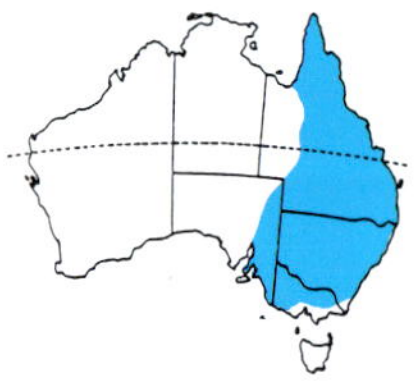

Male

36 BLACK-TAILED TREECREEPER

Climacteris melanura 20 cm

(A) race *melanura*; (B) race *wellsi* (Allied Treecreeper). The latter race is paler and smaller and apparently lives quite separately from its northern relative. The species is the largest and darkest of the treecreepers. Like the Brown Treecreeper, it usually keeps together in small groups and frequents open woodland and savannah. Breeds August–January and nests in tree hollows 4–10 m above ground. Endemic. **Female**: *smaller, white throat and upper breast streaked rufous.* **Immatures**: *faint white streaks on throat.* **Voice**: *strident sharp calls.*

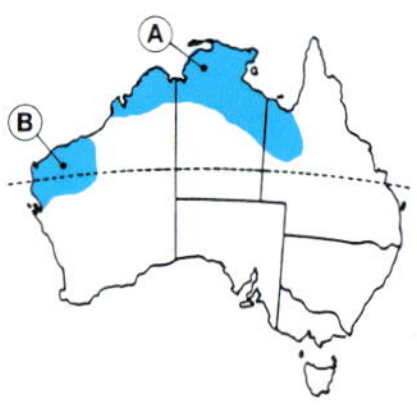

Male

25

26

27

28

29

30

31

32

33

34

35

36

9 Kingfishers

This Section embraces Australia's representatives of the kingfisher family Alcedinidae, the bee-eater family Meropidae and the roller family Coraciidae – three of the ten families that comprise the worldwide order Coraciiformes. Most of its members are brightly coloured with their three front toes partly joined, or syndactylous.

KINGFISHERS Family Alcedinidae

Of a world total of approximately 90 species of kingfishers, ten occur as breeding species in Australia. As Europe has only one species and America only three, Australia is indeed fortunate. The Laughing Kookaburra, the largest kingfisher in the world, and the Red-backed Kingfisher are endemic to Australia. The rest are confined to the Australian Region, except for the Mangrove Kingfisher which has a widespread range extending from northern Polynesia to the Red Sea. Three of the species – the Sacred, White-tailed and Mangrove Kingfishers – migrate to islands to the north to overwinter after breeding. Kingfishers divide into two subfamilies: Alcedininae which includes the typical fishing kingfishers, and Daceloninae, the so-called forest kingfishers. 1–9, 12–15

Alcedininae Fishing kingfishers Most eye-catching of the fishing kingfishers are the small vividly-coloured members of the genus *Ceyx*, represented in Australia by the Azure and Little Kingfishers, also easily recognised by their short tails. Fishing kingfishers burrow into river and creek banks to nest.

Daceloninae Forest kingfishers The forest or woodland species of kingfishers are larger and more varied, and may live far from water, like the Forest Kingfisher. Their diet mostly consists of insects, reptiles, amphibians, small mammals and even nestlings. But, given the opportunity, they are not averse to diving for a fish. The largest group of the forest kingfishers belongs to the genus *Halcyon* which ranges widely as far as Africa and much of Asia. Five of Australia's kingfishers are of this genus. The ancient Greeks believed that kingfishers nested at sea and called them *Halkyons, hals* (sea) and *kyon* (conceiving), or *Halcyon* in Latin, and that the gods calmed the waters for them at nesting time – hence 'halcyon days'. The forest kingfishers also nest in holes dug in banks, but some, like the Kookaburra, nest in trees.

RAINBOW BEE-EATER Family Meropidae 10

The graceful Rainbow Bee-eater is Australia's sole representative of this family of twenty-four species which ranges widely through the tropical and temperate areas of the Old World. The Australian species may migrate after breeding to overwinter as far as Indonesia (Sunda Islands, Borneo and Sulawesi) and the Solomon Islands – and occasionally even to Japan.

ROLLERS Family Coraciidae 11

The Dollarbird alone represents this family in Australia. Its seventeen species, like the bee-eaters, range widely throughout the tropical and temperate regions of the Old World including Japan, China and India. The Australian breeding species migrates to New Guinea and beyond to overwinter. Rollers are so named because of their spectacular rolling and tumbling aerobatics during courtship displays.

THE LAUGHING KOOKABURRA

The Laughing Kookaburra was first illustrated in 1776 in a book by a Frenchman, Pierre Sonnerat. It is believed to have been figured from a specimen given to him at the Cape of Good Hope by Joseph Banks on the *Endeavour's* return voyage to England in 1770. It has been known by various colloquial names including 'Laughing Jackass' and the 'Bushman's Clock' – the latter presumably because its laughter is heard at its loudest in the early morning and at sunset. The now universal name Kookaburra is aboriginal in origin. Legends of the bird lie deep in their folklore, and it has been known by many other names among the various tribes. Forty-five of them are given by Alec Chisholm in his book *Bird Wonders of Australia* (1934) who adds that his list is not exhaustive.

Much that is new on the social behaviour of this interesting bird has been discovered by naturalist Veronica Parry as a result of years of painstaking observations. In her book *Kookaburra* (1970) she reveals, for instance, that they live for 20 years or more; remain in one area the year round; form pairs for life; live in clans comprising several generations; and that the young adults stay with the clan as auxilliaries, sometimes for years, to help defend territory, feed new broods and care for the fledglings. 12, 14, 15

Opposite: Red-backed Kingfisher

PHOTOGRAPHERS: Fr. J. Bell (NPI) 11; G. Chapman 5; J. Carnemolla (APL) 15; T. & P. Gardner 6, 10; A.L. Hertog 3; J. Purnell 9; L. Robinson 1; C. Spiker 13; C. Webster 2; E.E. Zillmann 4; Authors p. 83; 7, 8, 12, 14.

9

1 WHITE-TAILED [Buff-Breasted Paradise-] KINGFISHER

Tanysiptera sylvia 36 cm

Unlike any other of Australia's kingfishers, this small jewel of the rainforests of Cape York Peninsula is graced with two long (18 cm) stiff white tail feathers. In November, the birds arrive from New Guinea in a burst to breed, but once the bustle of territory-claiming and nest-building is over, they become quieter and retiring. Nesting takes place in a low termite mound and three or four weeks are spent tunnelling into it and excavating the nesting chamber (about 15 cm in diameter). By the time the young have been reared in the confines of such a small space, the parents' plumage is stained and worn. The adults begin returning to New Guinea in March and most have left by the end of April. **Sexes**: *alike.* **Immatures**: *lack long tail streamers and have fine scallopings of light grey on breast.* **Voice**: *series of 'chop- chop' notes, loud scream, persistent trilling.*

2 AZURE KINGFISHER

Ceyx azureus 19 cm

A true kingfisher which dives to seize fish and other aquatic life in either salt or fresh water. It perches for long periods on a low branch overhanging water, quiet, solitary and intent, waiting to dart for its prey. Sedentary. The northern birds are smaller and bluer on the flanks. Its range extends through New Guinea to Indonesia. Breeds September–January (south), October–April (north). Nests in a channel dug in a stream bank, with a terminal chamber lined with fishbones and shells. **Sexes**: *alike.* **Immatures**: *less glossy, paler underparts.* **Voice**: *shrill whistling in flight.*

3 LITTLE KINGFISHER

Ceyx pusillus 13 cm

The Little Kingfisher lives in New Guinea and adjacent islands, and in two separate populations in Australia – in Cape York Peninsula and the Northern Territory. Birds west of Normanton have bluer backs and larger wing patches. A true kingfisher, it frequents mangroves, estuaries and rainforest creeks where it dives for small fish. It has been known to catch as many as eight a day. The smallest of the kingfishers in Australia, it is solitary and seldom seen. Breeds October–March. Nests in creek banks or low in termite mounds. **Sexes**: *alike.* **Immatures**: *areas of turquoise-grey.* **Voice**: *very high shrill whistle, usually in flight.*

4 FOREST KINGFISHER

Halcyon macleayii 23 cm

Extending to New Guinea and adjacent islands, the two Australian populations are separated by the Gulf of Carpentaria. Singly or in family parties, it frequents forest, swamps, mangroves and woodland, but may be found far from water. Frogs, worms and insects are its diet. Breeds August–December, sometimes twice. Nests in termite mounds and in river banks. Eastern birds migrate to New Guinea. **Recognition**: *violet-blue crown and white underparts. Western birds – back bluer and white wing patch, seen in flight, larger.* **Males**: *back of neck white.* **Immatures**: *duller.* **Voice**: *trills and whistles.*

Female

5 RED-BACKED KINGFISHER

Halcyon pyrrhopygia 24 cm

Widespread, and adapted to live in the arid and semi-arid areas, often far from water – presumably obtaining enough moisture from its food of insects, frogs, reptiles and even young birds. Singly, in breeding pairs or in very small parties, it frequents a variety of habitat, often in timber along dry rivers and creeks. Highly nomadic. Southern birds migrate northwards to winter. Breeds August–February. Nests in a burrow in a creek bank or in a tree hollow. Endemic. **Recognition**: *orange-tan rump, streaked crown.* **Sexes**: *similar but female duller.* **Immatures**: *white-flecked wings.* **Voice**: *mournful descending whistle.*

6 SACRED KINGFISHER

Halcyon sancta 23 cm

The Sacred Kingfisher, whose range extends from Australasia to Indonesia, is a familiar sight throughout all but the arid centre of the Australian continent. It perches high, is silent and often solitary and feeds on aquatic insects or small animals. After breeding, the deep buff of the breast is often worn and discoloured to dull ivory. Many birds migrate northwards to winter. Breeds September–March and nests in a creek bank, termite mound or tree hollow. **Sexes**: *similar but female duller.* **Immatures**: *like females.* **Voice**: *repeated 'kee-kee-kee' and a series of churring scolding notes.*

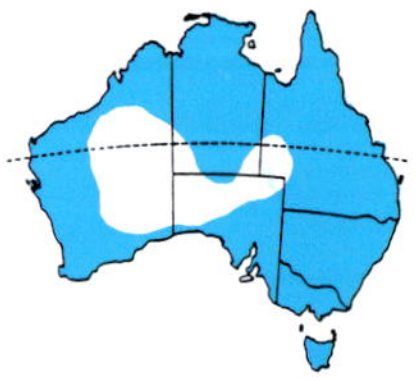

7, 8 YELLOW-BILLED KINGFISHER

Halcyon torotoro 21 cm

In Australia, the small and inconspicuous Yellow-billed Kingfisher, widespread in New Guinea, occurs in the remote tropical rainforests of Cape York Peninsula. Usually solitary, it lives in relative obscurity, but during and after breeding it is seen in pairs and sometimes small family parties. It perches on a branch to scout with concentrated intensity for insects, lizards and worms. Some migration to overwinter in New Guinea suspected. Breeds November–January and nests in termite mounds or hollow tree limbs. **Sexes**: *similar but male has black neck patch, female has black head patch.* **Immatures**: *flecked with black.* **Voice**: *mournful ascending trilling call.*

Male left, female right

9 MANGROVE [Collared] KINGFISHER

Halcyon chloris 29 cm

In Australia this large kingfisher, which extends throughout the south-west Pacific to the Red Sea, is predominantly a mangrove dweller, whereas elsewhere its habitat is varied. It can be seen perching on the bare branch tops and it feeds on insects, small animals and nestlings. Southern birds move northwards to winter. Breeds September–March and nests in tree hollows, termite mounds or earth banks. **Recognition**: *like Sacred but larger and whiter underparts.* **Female**: *duller.* **Immatures**: *flecked white.* **Voice**: *some notes like Sacred Kingfisher, some like Kookaburra.*

10 RAINBOW BEE-EATER

Merops ornatus 28 cm

The Rainbow Bee-eater is aptly named for it displays every hue of the rainbow when pirouetting on the wing, and does eat bees which, with other insects, form a substantial part of its diet. The sting is rendered harmless and the bee killed before it is swallowed. It occurs in mallee and all types of open country, especially where there are sandy river banks. Breeds November–January and nests at end of a tunnel in a bank or on the ground. Migrates northwards to overwinter. **Sexes**: *similar but central tail feathers shorter in female.* **Immatures**: *lack black throat bar.* **Voice**: *soft musical trilling.*

Male

11 DOLLARBIRD

Eurystomus orientalis 30 cm

The Australian Dollarbird is so named for the pale blue coin-spot visible on its wing during flight. It arrives from its overwintering territories to the north to breed in October and departs in February. It perches strategically to dart out and seize large flying insects, returning to its perch to batter the prey before swallowing it. It is usually seen singly or in pairs and frequents woodlands and timbered environs of settled areas. It nests in cavities of high trees on the edges of forests or in open woodland. **Female**: *paler.* **Immatures**: *duller.* **Voice**: *harsh squawk followed by calls of 'kak-kak-kak'.*

1

2

3

4

5

6

7

8

9

10

11

9

12

12, 14, 15 LAUGHING KOOKABURRA

Dacelo novaeguineae 47 cm

Few native species have adapted to city and suburban life as successfully as the endemic Laughing Kookaburra, and it is as familiar in gardens and backyards as it is in its natural state in large-timbered country and woodlands. The northern birds are smaller; those in Western Australia, Tasmania and Kangaroo Island are descended from releases from eastern Australia. The Kookaburra sits patiently for long periods scanning for prey. It eats snakes, lizards, rodents, small birds and insects, and bashes its victim to pulp before swallowing it. Breeds September–January. Nests in tree holes. **Sexes**: *alike, but males often have blue patch on rump.* **Immatures**: *shorter tail, bill all-black.* **Voice**: *the famous call is usually joined by several birds in chorus which can rise in a riotous and deafening crescendo, particulary at dawn and dusk.*

Above, and below opposite (also see page 82)

13 BLUE-WINGED KOOKABURRA

Dacelo leachii 44 cm

The range of this smaller blue-winged relative of the Laughing Kookaburra extends right across the tropical and sub-tropical top end of the continent and northwards into New Guinea. There is some variation throughout its range, the western population generally being paler and the Cape York population smaller. In Queensland where the territories of the Blue-winged and Laughing Kookaburras overlap, the former will be found more often in the wetter tropical coastal fringes. It rests quietly during the heat of the day, and is more retiring than its relative. Otherwise, the habits and diet of the two species are similar. Sedentary. Breeds September—January. Nests in a termite mound or tree hollow. **Female:** *rufous tail barred blue-black.* **Immatures:** *brown on underparts, head paler.* **Voice:** *does not laugh but utters very loud fiendish howls especially at dawn, usually with several birds joining in.*

Above opposite

13

14

15

10

Robins, Flycatchers and Thrush-like Birds

This Section includes the fifty-four Australian representatives of the huge world-wide assemblage of some 645 species of insect-eating songbirds which, in the past, have collectively been known as Muscicapidae (meaning fly-catchers). In one way or another all the Muscicapidae have something in common, but some are sufficiently different to enable them to be separated into smaller groups — although the lines between them may not be sharp. Two family divisions are adopted here for the Australian species: the Pachycephalidae (comprising the flycatchers and their allies) and the Turdidae (comprising the true thrushes). Also included is the introduced Bulbul of the family Pycnonotidae.

FLYCATCHERS (INCLUDING ROBINS, WHISTLERS AND FANTAILS) Family Pachycephalidae

The small to medium members of this diverse family range widely from the Australian Region to Africa and southern Asia. While they superficially resemble the Old World flycatchers (for which the name Muscicapidae is retained), they differ from them in certain anatomical and behavioural respects. Of a total of 225 species of Pachycephalidae, fifty-one occur in Australia. They fall into the following seven groups.

Robins There are seventeen species of robins in Australia, of which thirteen are endemic. Although called robins, they are in fact very specialised and colourful flycatchers. The breasts of the males of five of the species are red and this, together with the gentle and engaging ways of these birds, inevitably led to their being likened to the Robin Redbreast *Erithacus rubecula* of Europe and collectively called robins from the earliest days of settlement. The European Robin, however, is a member of the family of true thrushes and only in the case of the two ground-frequenting scrub-robins, which are more closely allied to the true thrushes, is there perhaps a family relationship. 1-24

Flycatchers Flycatchers feed entirely on insects and spiders, and characteristically sally forth from a perch to seize their prey in the air or on the ground. Their broad flattened bills are specially adapted for catching prey on the wing, with mouths edged with stiff bristles to extend the catching area. They separate into two categories:

— **The 'Microeca' Flycatchers** The members of this genus, which includes the Jacky Winter, closely resemble the robins. Their nests are the smallest of any Australian bird, hence the generic name *Microeca* (Greek meaning 'little house'). These are seven species spread between Australia and New Guinea, of which four occur in Australia.

— **The Monarch Flycatchers** This large group of ninety-one species extends from the Australian Region, including the Pacific islands, to Africa south of the Sahara and to south Asia, and is sometimes treated as a family on its own (Monarchidae). Twelve species of monarch flycatchers occur in Australia, including the Boat-billed Flycatcher. They are easily separated from the *Microeca* flycatchers by their strongly contrasting plumage patterns, and most of them are larger. Some are also sexually dimorphic (their plumages differ). 25-43

Fantails Unafraid and inquisitive, the fantails are conspicuous for their fanned tails and gyrating movements whilst catching insects in the air, or perhaps just flitting in playful abandon. Of about thirty-eight species, four (including the Willie Wagtail) occur in Australia. 44-47

Crested Shrike-tit Despite a much heavier bill and a pronounced crest, this colourful endemic species has affinities with both the whistlers and the thrushes. There are three widely separated populations, each differing somewhat in plumage. 49, 52-53

Crested Bellbird In some ways, this gifted ventriloquial songster with erectile crest resembles the whistlers. It is unrelated to the Bell Miner or Bellbird (which is a honeyeater) and its voice is quite different. 50-51

Whistlers Known as whistlers from their melodious whistling calls, they are also noted for their large rounded heads — from which they get the generic name *Pachycephala* (meaning thick-head). They are widespread in the Australian Region to Malaysia and the Philippines. There are eight members of the genus in Australia, six of them endemic. 54-69

Shrike-thrushes The four species in this mostly Australian genus have a close affinity with the whistlers. They share some of the characteristics of the family of true thrushes (family Turdidae) including the gift of glorious melodic song. The shrike part of the name is presumed to have derived from certain similarities in appearance with the grey shrikes of the northern hemisphere. 70-74

TRUE THRUSHES Family Turdidae

These, the true thrushes, are small to medium-sized birds, and are among the world's finest songsters. Included in the family are the European Robin, the American Robin and the celebrated Nightingale. There are about 310 species worldwide, of which only one occurs naturally in Australia, the Song Thrush and Blackbird being introduced. 75-78

BULBULS Family Turdidae

The pert, sweet-voiced introduced Red-whiskered Bulbul is the only representative in Australia of an Old World family of about 110 species. It has spread only slowly, but is one of Sydney's most familiar birds. 45

Opposite: Red-capped Robin

PHOTOGRAPHERS: H. & J. Beste 35; C.A.C. Cameron p. 89; N. Chaffer 48; G. Chapman 14, 15, 16, 20, 23, 26, 27, 28, 36, 38, 41, 49, 50, 55, 56, 58, 60, 63, 67, 68, 73; B. Chudleigh (ANT) 76; F.M. Dowling 10; J. Frazier 11; T. & P. Gardner 7, 18, 21, 29, 33, 37; J. Gray 39, 42, 51; P.N. Green 46, 62; P. Jordan 44; R. & D. Keller 32; W. J. Labbett 54, 64; R.B. Legge 4; J.R. Napier 3,6; F. & B.J. Park (ANT) 78; J. Purnell 1, 2, 5, 40, 53, 70, 74, 75; K. & B. Richards 65; L. Robinson 8, 22, 57; L.F. Schick 77; E.C. Slater 9; T.A. Waite 13, 66; G. Weber 43; B.A. & A. G. Wells 61; E.E. Zillmann 52; Authors 12, 17, 19, 24, 25, 30, 34, 45, 47, 59, 69, 71, 72. **ARTIST:** T.R. Lindsey 31 (from a photograph by R.D. Mackay)

10

1, 4 FLAME ROBIN

Petroica phoenicea 14 cm

In the Australian Alps, the Flame Robin, like the Robin Redbreast of Europe, is sometimes seen in snow-clad country where its brilliant lambent orange breast stands out in glorious contrast. It favours the dry forests and woodlands of the high country while breeding and until the onset of winter, when it descends to lower altitudes and spreads widely over open country to the coast and along the Murray River and its tributaries. Some birds may also migrate from Tasmania to the mainland at this time. Normally in pairs, the Flame Robin also congregates in flocks and is the only red-breast to do so. It will sit patiently on a post waiting to dive to the ground to snatch an insect or a worm. Sometimes it flicks its wings and tail while perched. The male may breed still in his immature brown plumage. Breeds September–January. Nests low in tree cavities and rock fissures. Endemic. **Immatures**: *like adult females but with buff markings.* **Voice**: *soft musical 'you-may-come, if-you-will, to-the-sea'.*

Male above, female below

2, 5 ROSE ROBIN

Petroica rosea 12 cm

The dainty Rose Robin, of delicate hue, is Australia's smallest robin and perhaps the most captivating. Singly and in pairs, it frequents the rainforests, dense gullies and creeks of the south-east during the summer, some dispersing to the north of its range to overwinter, where it is seen more often in drier country. It feeds mainly on insects taken from foliage, sometimes on the ground or on the wing – where it displays much of the agility of a flycatcher. It flicks its tail when perched and flits quietly about with tail fanned. Inquisitive and trusting, it will sometimes approach to inspect a person within a few metres. Breeds September–January. Nests 1–20 m high, often fosters cuckoo eggs. Endemic. **Immatures**: *like adult females.* **Voice**: *sweet trilling song and 'tick-tick-tick' calls.*

Male above, female below

3, 6 PINK ROBIN

Petroica rodinogaster 14 cm

The Pink Robin is more often seen in Tasmania and its offshore islands than anywhere else in its somewhat restricted southerly range. In quiet and unobstrusive pairs or family groups, this trusting bird mostly lives and breeds in the fern gullies of dense wet forest. Males tend to remain in their breeding areas for most of the year, but wandering birds are sometimes seen in dry forest country and woody suburbia during the non-breeding season. The species is heard calling from cover more often than it is seen. It feeds on flying and ground insects. Breeds September–December. Nests in a leafy tree up to 20 m high. Endemic. **Recognition**: *the only 'red' robin with no white in tail. Its coloration is also variable.* **Immatures**: *like adult female.* **Voice**: *slight 'tick', simple but pleasant warble.*

Male above, female below

7, 10 SCARLET ROBIN

Petroica multicolor 14 cm

The handsome Scarlet Robin with its white cap and glossy black head and throat lives in two separate populations in the south-eastern and south-western corners of the continent. It also occurs in islands of the South Pacific. It breeds in high forest country, then moves to lower open areas, sometimes to cultivated farmland or suburbia. Like other robins, the Scarlet is an insect-eater, feeding on the ground. Breeds July–January and nests in a hollow stump or tree fork up to 19 m from the ground. The tiny nest is favoured by cuckoos. **Immatures**: *like females, but duller, no red at first.* **Voice**: *a succession of sweet trilling notes.*

Male above, female below

8, 11 RED-CAPPED ROBIN

Petroica goodenovii 12 cm

Mainly a bird of the drier inland, this pretty robin is seen also in the edges of rainforest as well as open forest and scrub. It is absent from most of the tropic north and areas of high rainfall in the south, including Tasmania. It flicks its wings and tail when perched, and takes insects on the ground, in foliage, and sometimes on the wing. Sedentary, but nomadic in drought conditions. Some seasonal movement to the coast in winter. Adept at a feigned injury display when its nest is threatened. Breeds July–December. Nests in a tree up to 10 m above ground. Endemic. **Immatures**: *like adult females, but have pale red wash.* **Voice**: *call note somewhat like the ticking of a clock, also feeble insect-like song.*

Male above, female below

9, 12 HOODED ROBIN

Melanodryas cucullata 17 cm

This smartly turned-out pied robin is wary and unobtrusive. Usually in pairs, it is seen in open drier forest or mulga scrub, particularly where there is fallen timber on which it perches. It may be nomadic to follow food supplies, mainly insects either gathered on the ground or taken on the wing. Its feigned injury display when its nest is threatened is remarkable and includes tumbling over and over. Breeds July–December (south), March–April (north). Nests in a tree fork or hollow 5–6 m from ground. Endemic. **Immatures**: *like adult females.* **Voice**: *yapping notes by night and day, especially prominent in the dawn chorus.*

Male above, female below

1

2

3

4

5

6

7

8

9

10

11

12

10

13 DUSKY ROBIN
Melanodryas vittata 16 cm

Confined to Tasmania and Bass Strait islands, where it is moderately common, the species is sombre and sedate compared with the other robins. It prefers sparse forest and woodland with fallen timber and debris. It perches low and motionless waiting to dart to the ground or in the air to catch its prey. Feeds on insects, spiders and snails, and seeds in winter. Sedentary. Breeds July–December. Nests below 6 m. Endemic. **Sexes**: *alike.* **Immatures**: *less pronounced white markings.* **Voice**: *low 'choo-wee-choo-wee-er' monotonously repeated.*

14 MANGROVE ROBIN
Eopsaltria pulverulenta 16 cm

As its name suggests, this is a denizen of mangroves. Its range is limited to the northern coasts. Quiet and seldom breaking cover, it is difficult to spot though it is not shy. It sits on a perch, darting to seize insects and small crustaceans in foliage or mud, and is seen singly, in pairs and family parties, depending on the season. Sedentary. Breeds mainly in the winter dry period, April–September. Nests in shallow cup to 5 m over mud. Also in New Guinea and adjacent islands. **Sexes**: *alike.* **Immatures**: *paler.* **Voice**: *two drawn-out plaintive notes.*

15 WHITE-BREASTED ROBIN
Eopsaltria georgiana 14 cm

Where the plumage of the Eastern and Western Yellow Robins is yellow, that of the White-breasted Robin is white. Otherwise, the birds are much alike. It lives in dense moist vegetation along creeks, in gullies, or the understorey of karri, occasionally visiting country gardens and orchards. Like the Yellow Robins, it will cling sideways to a tree trunk, low down, to watch for insects and their larvae. Some seasonal movement. Breeds July–January. Nests in tree or shrub near water to 6 m high. Endemic. **Sexes**: *alike.* **Immatures**: *like adults but may be mottled brown.* **Voice**: *somewhat harsh 'chit-chit'.*

16 WESTERN YELLOW ROBIN
Eopsaltria griseogularis 16 cm

Like the Eastern Yellow Robin, except for a grey chest and upper breast instead of yellow, and similar in habits. It is partial to the drier habitats of the south-west and is found in open forest, woodland, scrub and mallee. Some winter movement from highland to lowland. Breeds July–December. Builds its nest in a tree fork or branch, up to 8 m high, and decorates it with strips of bark or lichen. Endemic. **Sexes**: *alike.* **Immatures**: *duller.* **Voice**: *repeated piping, harsh 'chit-chit'.*

17 EASTERN YELLOW ROBIN
Eopsaltria australis 17 cm

One of the first birds to call at dawn and amongst the last to fall quiet at nightfall. It favours dry forest and woodland and frequently appears in picnic grounds and suburban gardens. Friendly and inquisitive, it feeds on the ground on insects, and characteristically clings sideways on a tree trunk to watch for prey. Some migration to the coast in winter. Breeds July–December. Nests usually below 7 m but occasionally up to 25 m. Endemic. **Sexes**: *alike.* **Immatures**: *duller.* **Voice**: *'tchew-tchew'; soft piping.*

18 PALE-YELLOW ROBIN
Tregellasia capito 13 cm

The two widely-separated populations of the Pale-yellow Robin live in coastal and upland rainforest. It is the most common of the birds on the Atherton Tablelands. It perches in middle level foliage or clings to a tree trunk ready to sally forth to catch flies and other insects. It moves singly, in pairs or family parties, but seldom leaves dense cover. It is unafraid and inquiring. Sedentary. Breeds July–January. Nests in neat cup, mostly amongst lawyer vine creepers or low in a tree. Endemic. **Sexes**: *alike.* **Immatures**: *like adults but more brown and rufous.* **Voice**: *series of mournful whistles.*

19 WHITE-FACED ROBIN
Tregellasia leucops 12 cm

Widespread in New Guinea, but in Australia the range of this tame, cute-faced little bird is restricted to a small area of tropical rainforest of Cape York Peninsula. Its pattern of behaviour is similar to that of the Pale Yellow Robin. It frequents middle levels of foliage, and is found in pairs and sometimes largish feeding parties foraging for insects and small crustaceans. Sedentary. Breeds July–January. Nests in lawyer vine, or at a junction of tree fork and vine, low down. **Sexes**: *alike.* **Immatures**: *like adults but duller.* **Voice**: *occasional short harsh 'chee-chee'.*

20, 21 WHITE-BROWED ROBIN
Poecilodryas superciliosa 17 cm
(A) race *superciliosa*
(B) race *cerviniventris* (Buff-sided Robin)

This brisk and lively endemic robin divides into two races at the Gulf of Carpentaria. In the east it lives in rainforest bordering streams and in woodland near the coast; in the west it frequents mangroves. Usually in pairs. It is largely a ground feeder, and flits and hops from foliage to ground and back again, dropping its wings and flicking its tail as it forages for caterpillars, green ants, beetles and other insects, or crabs, molluscs and other small invertebrates in mangroves. It also clings to tree trunks probing crevices for prey. Although shy, it can be observed at close range if approached quietly. Sedentary. Breeds July–March. Nests in tree fork up to 9 m high. **Recognition**: *race (B) has orange-buff flanks and under tail coverts, and sooty-black crown and face patch.* **Sexes**: *alike.* **Immatures**: *like adults.* **Voice**: *loud repeated piping whistle.*

Race *cerviniventris* left, *superciliosa* right

22 GREY-HEADED ROBIN
Heteromyias albispecularis 17 cm

The Grey-headed Robin is a sturdy bird which is commonly seen and heard in the rainforests of the Atherton Tablelands and the coastal highlands of northern Queensland. It often appears on the edges of clearings and tracks and clinging to tree trunks. Solitary or in pairs, it spends much time on the ground foraging. Sedentary. Breeds August–January. Nests in a tangle of vine, shrub or tree 1–10 m high. Endemic. **Recognition**: *characteristic upright posture.* **Sexes**: *alike.* **Immatures**: *patchy brown and white underparts.* **Voice**: *gentle whistling.*

23 SOUTHERN SCRUB-ROBIN
Drymodes brunneopygia 23 cm

This scrub-robin lives in mulga and mallee scrublands, especially where there is broombush. Usually seen only singly or in pairs, it hops bouncily along the ground or flits to low branches, flicking wings and tail. Feeds on insects. Sedentary. Breeds July–January. Nests in shallow lined ground depression, only one egg to a clutch. Endemic. **Sexes**: *alike.* **Immatures**: *light brown and white.* **Voice**: *animated and far-carrying calls of 'chip-pip-er-ee'; will answer imitations.*

24 NORTHERN SCRUB-ROBIN
Drymodes superciliaris 21 cm

The remote habitat on Cape York Peninsula of this ground-frequenting bird makes it one of Australia's least-known species. It was recorded in the Roper River area of the Northern Territory in 1910, but there have been no recent sightings there. Solitary or in pairs, it frequents rainforest floor or the edges of clearings, hopping from place to place and turning over litter for insects. Sedentary. Breeds October–January. Nests in depression in ground. Widespread in New Guinea. **Sexes**: *alike.* **Immatures**: *lighter brown head and neck.* **Voice**: *a loud shrill whistle.*

13

14

15

16

17

18

19

20

21

22

23

24

25 LITTLE YELLOW [Yellow-legged] FLYCATCHER

Microeca griseoceps 12 cm

An unobtrusive bird with distinctive yellow legs, sometimes mistaken for a Lemon-breasted Flycatcher or a Pale Yellow Robin. It lives mostly in New Guinea and was discovered in Australia only in 1913. Here it is confined to a small area of tropical rainforest and bordering woodland in Cape York Peninsula. An Australian nest was not found until 1977. Singly or in pairs, it is active amongst foliage, stirring up insects and often rising to catch them on the wing. Breeds November–January. Small cup-shaped nest. **Sexes**: *alike.* **Immatures**: *duller.* **Voice**: *zzt-zzt.*

26, 27 LEMON-BREASTED [Lemon-bellied] FLYCATCHER

Microeca flavigaster 14 cm

The depth and hue of yellow of the underparts of the Lemon-breasted Flycatcher varies from east to west – from bright yellow south of Cape York to white in the Kimberley. (The latter population, known as the Brown-tailed or Kimberley Flycatcher, was formerly regarded as a separate species.) Singly and in pairs, the Lemon-breasted Flycatcher lives in eucalypt forest, woodland and rainforest clearings in Queensland; in dense vegetation on the Gulf of Carpentaria; in rainforest and sometimes mangroves in the Northern Territory; and in dense mangroves in the Kimberley. The eastern birds feed on a variety of insects, including ants, and caterpillars; the Kimberley birds on flies and invertebrates of the mangroves, including crabs. Eastern and northern populations breed August–January. Tiny cup-shaped nests are built in a tree 10–12 m high. Also in New Guinea. **Recognition**: *the Kimberley birds, which resemble the Jacky Winter, lack the white-edged tail.* **Sexes**: *alike.* **Immatures**: *paler.* **Voice**: *varied sweet and musical phrases.*

Kimberley form left, Cape York form right

28 JACKY WINTER

Microeca leucophaea 14 cm

Well known for its clear sweet song, this affectionately-regarded brown flycatcher is found over most of the continent in well wooded areas. Unusually amongst perching birds, it is able to hover in the air for some seconds, with the white-sided tail often fanned in flight. It is an assiduous feeder, even after dusk has fallen, adroitly taking all kinds of spiders, moths, flies and butterflies. Sedentary. Breeds July–February. Nests in tiny frail saucer to 20 m high. May be parasitised by cuckoos. Also in New Guinea. **Sexes**: *alike.* **Immatures**: *mottled.* **Voice**: *varied and profuse phrases: 'jacky, jacky' or 'peter, peter'.*

29 SPECTACLED MONARCH (Flycatcher)

Monarcha trivirgatus 16 cm

The Spectacled Monarch's fussy activity of jerking and tumbling seems designed to flush out prey of flies, moths, spiders and snails from the foliage of middle and upper layers of forest or woodland canopy. It seldom takes its food on the ground or on the wing. Some southern birds migrate northwards in winter. Breeds October–February. Nests in tree fork or vine tangle 1–7 m high. Also in New Guinea and islands to the north. **Sexes**: *alike.* **Immatures**: *less black on the head.* **Voice**: *repeated whirring 'pree-pree-pree'; whistling.*

30 BLACK-FACED MONARCH (Flycatcher)

Monarcha melanopsis 19 cm

In tropical or temperate rainforest or moist gullies, this handsome bird sedately forages for insects amongst middle level foliage, more rarely taking them in flight, hovering briefly. It may be seen in small groups during migration. The migratory pattern is complex: southern birds move northwards in winter, northern ones may be sedentary or move to bolster resident populations in New Guinea. Breeds October–January. Nests in tree fork 1–12 m high. **Sexes**: *alike.* **Immatures**: *grey where adults are black.* **Voice**: *clear whistling 'why-yew-which-you'; chattering.*

31 BLACK-WINGED MONARCH (Pearly Flycatcher)

Monarcha frater 15 cm

A breeding migrant from New Guinea, it arrives in Australia in October and leaves in March, although some may remain over the winter. It was considered a race of the Black-faced Monarch, but is now retained as a separate species because of differences in plumage, notably pearl-grey upperparts which give it the alternate common name. It sits quietly on a branch with tail quivering, on the watch for insects. Nest is goblet-shaped, in a tree fork to 10 m high. **Sexes**: *alike.* **Immatures**: *grey and buff instead of black markings.* **Voice**: *like Black-faced Monarch.*

32 BOAT-BILLED FLYCATCHER [Yellow-breasted Boatbill]

Machaerirhynchus flaviventer 12 cm

The Boat-billed Flycatcher is a lively little bird which works its way at great speed through the foliage of rainforest, creek vegetation or eucalyptus, snapping up insects; or it may rise into the air from its perch to take flying insects on the wing. Sometimes it cocks its tail like a wren. It is solitary and sedentary. Breeds September–March. Nests in forked branch 2–25 m high. Also in New Guinea and adjacent islands. **Females**: *paler.* **Immatures**: *like adult females.* **Voice**: *pretty trilling song and twittering.*

Male

33 WHITE-EARED MONARCH (Flycatcher)

Monarcha leucotis 14 cm

Hovering and fluttering around the treetops in wet eucalypts or rainforest to snap up insects flushed from foliage or bark crevices, this flycatcher may appear in small groups. Possibly some seasonal migration southwards to breed, the breeding season being September–January. In the northern part of its range it nests so high in the tree crown that they were undescribed for 70 years; in the south, nests are lower. Endemic. **Sexes**: *alike.* **Immatures**: *brown-and-cream where adults are black-and-white.* **Voice**: *three-syllable song phrase with last one descending; whistling.*

34 FRILLED MONARCH (Flycatcher)

Arses telescophthalmus 16 cm

Named from its habit of frequently raising a frilly white collar, especially in courting displays. In rainforest and adjacent eucalypts it flits from branch to branch to pick insects from the foliage; occasionally it spirals up the trunk like a treecreeper. A vivid blue eye ring and the absence of a black chest band separates it from the Pied Monarch. Sedentary. Breeds September–January. Nests in a cup slung between two vines 1–10 m high. Also in New Guinea. **Male**: *lacks grey lores.* **Immatures**: *black areas tinged brown.* **Voice**: *soft churring, whistles and creaks.*

35 PIED MONARCH (Flycatcher)

Arses kaupi 16 cm

The Pied is a more southerly relative of the Frilled Monarch. Like the latter, it raises a frilly white collar, often as it briskly and boldly flits and hops in spirals up a tree trunk in the rainforest. It grips the bark with sharp claws and long toes as it flushes out beetles, moths and flies from crevices, then darts out after them. The tail is fanned in flight. Sedentary. Breeds October–January. Nests between two vines 1–10 m high. Endemic. **Female**: *black breaks the white collar on sides of neck.* **Immatures**: *grey-brown mantle and breastband.* **Voice**: *like Frilled Monarch.*

Male

36 BROAD-BILLED FLYCATCHER

Myiagra ruficollis 16 cm

In mangroves, monsoon forest or scrub, it perches for long periods on a branch, characteristically quivering its white-shafted tail. It will fly out to take insects on the wing, or from leaves. Out of breeding season, it is solitary. Sedentary. Breeds October–February. Nests in a mangrove tree close to water, or in a paperbark up to 7 m high. Also in New Guinea and islands to the north. **Recognition**: *can be confused with the female Leaden and Satin Flycatchers.* **Sexes**: *similar but female paler and duller.* **Immatures**: *like females.* **Voice**: *loud buzzing, croaking, whistling.*

25

26

27

28

29

30

31

32

33

34

35

36

10

37, 40 SHINING FLYCATCHER

Myiagra alecto 19 cm

The male is the only all black flycatcher. Its iridescence, and the glowing rufous and white of the female, make this a most handsome species. Active body movements such as quivering of the tail, crest-raising, and darting sallies to seize prey, are characteristic. Its habitat is mangrove and other dense vegetation associated with water. Here it takes insects from foliage, but more often it creeps along the mud in search of ants and tiny shellfish. Probably sedentary. Breeds August–March. Also in New Guinea. Shining Flycatchers nest low or to 3 m high on a branch or fork over water. The nest is parasitised by cuckoos. **Immatures**: *like adult females.* **Voice**: *extended whistling 'cheee'; melodious trilling by both sexes.*

Male above, female below

38, 41 SATIN FLYCATCHER

Myiagra cyanoleuca 17 cm

Glossy plumage separates the Satin from the Leaden Flycatcher. Another distinguishing feature where their ranges overlap is that the Satin Flycatcher prefers the treetops of wetter forests and mountain ranges. It is constantly active and vocal, and darts out to take insects, including grasshoppers and wasps on the wing. Breeds September–February. There is migratory movement from south to north after breeding, including to New Guinea and adjacent islands. Nests are placed openly on a tree branch, and perhaps because of this are much favoured by Bronze and Pallid Cuckoos. Nests may be at heights of 3–25 m. **Immatures**: *like adult females, but are difficult to separate from immatures of Leaden Flycatcher.* **Voice**: *deep 'chee-chee', grating note, nasal buzzing.*

Male above, female below

39, 42 LEADEN FLYCATCHER

Myiagra rubecula 16 cm

Sometimes called Frogbird from its croaking call. In dry scrub and woodland in the south, mangroves and wooded vegetation near water in the north, it is robustly active. It darts out to take its prey of a wide selection of insects including bees, termites and cicadas; on the perch it is equally busy, with rapid quivering of the tail, raising of the neck feathers and vivacious calling. Rare in Tasmania and the southern part of its range, southern birds migrate northwards in winter. Both sexes have a small crest which is constantly being raised during their daily activities and both sexes call from the nest. Breeds August–February. The nest is placed on a tree fork 3–25 m high. Also in New Guinea. **Immatures**: *like adult females.* **Voice**: *clear high whistled 'too-whee' or 'too-whit'; nasal buzzing.*

Male above, female below

43 RESTLESS FLYCATCHER

Myiagra inquieta 21 cm

A notable feature is the peculiar grating call from which derives its common name of Scissors Grinder. It is often confused with the Willie Wagtail which it resembles, but can be distinguished by its white throat and the absence of a white eyebrow. Frequenting scrub or woodland in the search for insects, it often hovers close to the ground like the Jacky Winter, with crest raised. Nomadic or seasonal movements. Breeds July–March. Nests near water 1–22 m high. Also in New Guinea. **Sexes**: *similar.* **Immatures**: *brown above.* **Voice**: *beautiful penetrating musical 'tu-whee tu-whee'; grinding notes.*

44 WILLIE WAGTAIL

Rhipidura leucophrys 21 cm

Willie Wagtail is one of Australia's best known and best loved birds. It is familiar and tame in parks and gardens, endearingly switching its tail from side to side while running about snapping up insects, and expanding its white eyebrows conspicuously when agitated. It is strongly territorial and bravely drives off much larger birds. Some nomadic or seasonal movement. Breeds June–February. Fibrous nest on horizontal branch. Also in New Guinea and other south-west Pacific Islands. **Sexes**: *alike.* **Immatures**: *buff feather tips.* **Voice**: *'sweet-pretty-creature'; matchbox rattling alarm call, often at night.*

45 RED-WHISKERED BULBUL

Pycnonotus jocosus 23 cm

Introductions of this jaunty bird to Sydney a century or more ago became established slowly, but it has now spread into country areas. It has also been sighted occasionally in Melbourne. At one time dealers sold it as a 'Persian Nightingale'. It lives in thick bush, parkland and gardens, taking insects and feasting on fruit. It calls from high vantage points. Breeds August–March. Nests in vegetation 2–3m from ground. **Sexes**: *similar.* **Immatures**: *duller.* **Voice**: *tuneful animated chattering; call of 'kink-a-jou' and other notes.*

46 GREY FANTAIL

Rhipidura fuliginosa 16 cm

This popular little bird, known as the Cranky Fan and a number of other local names, is inquisitive and unafraid. It is briskly active, and darts after flying insects in erratic flight. It is widespread in mangrove, rainforest, woodland, scrub and garden and is found at all levels of tree cover. It is nomadic or migratory from north to south and from high country in the south-east into the drier inland. Breeds August–January. Nests on slender fork, low to 6 m high. Also in New Guinea and South Pacific islands. **Sexes**: *alike.* **Immatures**: *duller.* **Voice**: *sweet musical trilling and 'cranky' chattering.*

47 NORTHERN FANTAIL

Rhipidura rufiventris 18 cm

Similar to the Grey Fantail, but with a grey-brown bib and less white in the plumage, the Northern Fantail is more subdued in its behaviour, indulging less in tail-fanning and restless flitting from perch to perch. Generally solitary, it is sometimes found in small companies of mixed birds feeding at the edges of rainforest, or in mangrove or open forest. Breeds August–January. Nests 2–20 m high. Also in islands to the north. **Sexes**: *alike.* **Immatures**: *duller.* **Voice**: *musical silvery-toned 'just-keep-away-you'; notes of 'chunk-chunk' or 'chip-chip', continued after dark.*

48 RUFOUS FANTAIL

Rhipidura rufifrons 16 cm

Taking its name from its lovely fanned flame-like tail, this delightfully tame little bird darts in quick jerky runs or flights in pursuit of insects. It lives in the lower levels of the dense foliage of rainforest gullies, and migrates northwards beyond the shores of the continent in winter; some straying into drier country or cultivated areas. Breeds October–February. Nests 1–10 m high. Also in New Guinea and islands of the south-west Pacific. **Sexes**: *alike.* **Immatures**: *like adults.* **Voice**: *high-pitched tinny but sweet song; call of 'chip'.*

37

38

39

40

41

42

43

44

45

46

47

48

10

49 CRESTED SHRIKE-TIT

Falcunculus frontatus 16 cm

Race: *whitei* (Northern form)

This uncommon, even rare bird, is found only in a restricted area of the Northern Territory and there have been few sightings. Graeme Chapman's photograph opposite is therefore of special interest. Compared with its nearest relative, the Eastern Shrike-tit, it is smaller, its upperparts are yellower and its black throat and foreneck have some white barrings. Endemic. **Voice**: *like nominate race.*

50, 51 CRESTED BELLBIRD

Oreoica gutturalis 23 cm

The Crested Bellbird lives mostly in lightly wooded country of Australia's arid areas. It is seen singly, in pairs, sometimes in loose flocks, and spends much of the time foraging on or near the ground – hopping rapidly as it goes. Its food consists of a variety of seeds, insects including beetles, and especially hairy caterpillars. It has the strange habit of maintaining a larder of these in a live but incapacitated state around the rim of its large nest. As many as 12 caterpillars have been counted, with their hairs often shed. The male chooses a bare vantage point to deliver his song, with crest raised – all the while turning his head from side to side to give a ventriloquial effect. Breeds August–January or in response to rainfall. Nests 1–3 m high. Endemic. **Immatures**: *like adult females.* **Voice**: *the Aboriginal name for the bird Pan-pan-panellella reflects its chief song of far carrying melodious notes; also distinctive liquid 'plonks'.*

Male left, female right

52, 53 CRESTED SHRIKE-TIT

Falcunculus frontatus 19 cm

Nominate race (Eastern form)

Strikingly handsome plumage, crested head, robust body and strong bill are features common to all three races of this species. The bill is used to open seed pods or to prise bark off trees to get at larvae. The sound of chunks of bark being ripped away can often be distinctly heard. Solitary or in family parties, it keeps mostly to the highest branches. Sedentary or nomadic. Breeds August–January. Nests in tree fork up to 30 m high. Endemic. **Immatures**: *browner wings.* **Voice**: *several low chattering phrases resembling 'knock-at-the-door'; plaintive piping.*

Male above, female below

56 CRESTED SHRIKE-TIT

Falcunculus frontatus 17 cm

Race: *leucogaster* (Western form)

The extreme south-western corner of the continent is the home of this form of the Crested Shrike-tit. A white belly separates it from the northern and eastern forms. Like them, this bird is often located by its calls and by the noise of shredding bark when it is searching for insects and spiders. Endemic. **Male**: *similar, except for black throat and more prominent crest.* **Immatures**: *like adult females.* **Voice**: *like nominate race.*

Female.

54, 57 GOLDEN WHISTLER

Pachycephala pectoralis 18 cm

This boldly marked and beautiful whistler is widespread in the Australian Faunal Region (New Zealand excepted), with a range extending to Indonesia, Malaysia and the Philippines. Some 65 races are recognised. Singly or in pairs, sometimes in small parties, it frequents the dense vegetation of brush, scrub or eucalypt forest, and moves purposely through foliage searching for insects and berries. It may move to open forest to winter. Both sexes are particularly fine and strong singers, especially during the breeding season. Immature birds may wander considerably. Sharp noises like thunderclaps evoke a quick vocal response. Breeds September–January. Nests in bush or tree to 6 m high. **Recognition**: *because the male Golden Whistler goes through various plumage changes before he finally and permanently retains his golden colouring, and may often breed in brown plumage, it is not always easy to distinguish the sexes.* **Immatures**: *like adult females with rufous margin on wing feathers.* **Voice**: *lovely clear whistles with a whipcrack ending: 'chee-chee-chee-chee-tu-whit', and variations.*

Male above, female below

55, 58 MANGROVE GOLDEN WHISTLER

Pachycephala melanura 17 cm

Very similar in appearance to its close relative the Golden Whistler, except for its black tail, longer bill and slightly smaller size. There are two races: *melanura* (Black-tailed Whistler) occurring along the north-west coast of WA to the Kimberley region; and *robusta* (Robust Whistler) occurring along the coasts of the Northern Territory and Queensland. Confined to mangroves and marginal vegetation, it feeds on grubs, ants and beetles, sometimes feeding on the ground. Other races live in New Guinea. **Sexes** (race *melanura*): *male – similar to robusta; female – brighter yellow below.* **Immatures**: *like adult females with rufous edge to wing feathers.* **Voice**: *similar to Golden Whistler.*

Race *robusta*
Male above,
immature female below

49

50

51

52

53

54

55

56

57

58

59, 62 RUFOUS WHISTLER

Pachycephala rufiventris 17 cm

Widely distributed throughout almost the whole of the continent except in heavy forest, the Rufous Whistler's outstandingly beautiful song is amongst the most joyously exuberant of all Australian birds. It will burst into song in response to a clap of thunder, hence its popular name of Thunderbird. It feeds arboreally mainly on insects, occasionally berries. in the southwest it is locally nomadic, and in the south it is migratory, arriving in August and departing around April. Breeds September–February. The female Rufous Whistler is as accomplished a vocalist as the male. Seen in pairs or small migrating groups. Nests low to 10 m high. Also in New Guinea. **Immatures**: *like adult females but with rufous wing edges.* **Voice**: *repeated ringing 'ee-chong' with whipcrack ending; rippling musical 'pee-pee-pee', or 'joey-joey-joey'.*

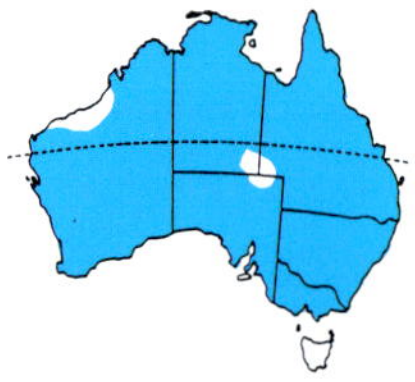

Male above, female below

60, 63 WHITE-BREASTED WHISTLER

Pachycephala lanioides 21 cm

Pairs of this striking whistler are to be seen searching for small crustaceans on the muddy floors of northern mangroves, or foraging among foliage for insects. It is an inquisitive and approachable bird, readily responding to imitations of its lively call. Sedentary in well-defined territories. Breeds March–November. Nests in dense mangrove fork 2–5 m above the mud. Endemic. **Recognition**: *where their habitats overlap, the female White-breasted Whistler may be differentiated from the female Rufous Whistler by its larger size and bigger bill, and less rufous coloration below.* **Immatures**: *male-brown washed olive, white throat; female-like adult female.* **Voice**: *rich and musical penetrating whistles 'per-weet', scolding alarm note.*

Male above, female below

61, 64 GILBERT'S WHISTLER

Pachycephala inornata 20 cm

A rather sad story attaches to the naming of this plainly-marked bird of southern Australia. In 1840, John Gould described a specimen collected in South Australia and named it *Pachycephala inornata* (meaning unadorned). In 1845, he also gave species rank to another specimen of the darker and more richly coloured Western Australian population, collected in the York district near Perth by John Gilbert, and named it *Pachycephala gilberti* as a tribute to his young and gifted assistant. Unfortunately, the two populations were later deemed to constitute a single species and the name *gilberti* lapsed under the international rule of priority – leaving Gilbert's name unrepresented among scientific bird names. (Also see article 'The Paradise Parrot and John Gilbert', page 70). The species lives in mallee and open forest. Usually in pairs, it quietly forages on or near the ground for insects, berries and seeds. Locally nomadic. Breeds September– December. Nests near the ground in a shrub or low tree. Endemic. **Recognition**: *black in front of eye; female greyish like small Grey Shrike- thrush.* **Immatures**: *brown eye instead of red.* **Voice**: *variety of rich notes, some ventriloquial; whistling, short 'chop-chop'.*

Male above, female below

65, 66 OLIVE WHISTLER

Pachycephala olivacea 22 cm

This species has Tasmanian and mainland races. In the north, it is uncommon in its haunts of beech forest at the tops of the highest mountains, where its call is wistful and elusive. In the south, it is more abundant in mountain and coastal thickets and in gardens. Here its lovely bursts of song are ringing, melodious and ventriloquial. Its calls are remarkably varied and it is probably the best songster of all the whistlers. It lives on insects and berries. Locally nomadic. Breeds September–January. Nests low. Endemic. **Immatures**: *like adult female with reddish wing tips.* **Voice**: *song of 'cheer-weet' or 'I'll-whit-choo' (south), peee-pooo (north).*

Male left, female right

67 RED-LORED WHISTLER

Pachycephala rufogularis 21 cm

This shy and solitary bird shares a restricted range with that of Gilbert's Whistler, with which it is sometimes confused. It frequents mainly mallee scrubs with dense cover and spends much time on the ground feeding unobtrusively on insects, larvae, seeds and berries. It is very vulnerable to clearance of its habitat. Sedentary. **Recognition**: *cinnamon in front of eye, face and throat, and buff-bellied.* Breeds September–December. Nests very low. Endemic. **Male**: *brighter.* **Immatures**: *eye brown.* **Voice**: *rich clear whistle followed by reedy sibilant notes 'see-ee-saw'.*

Female

68, 69 GREY WHISTLER

Pachycephala simplex 15 cm

This graceful bird is common in the monsoon coastal rainforests and mangroves of the Northern Territory and Cape York Peninsula. It is sedentary and lives mostly in the treetops, seldom descending to lower levels. It feeds quietly on a variety of insects. Usually solitary, but Cape York birds sometimes forage in flocks with other birds. There are several races in New Guinea; two in Australia.

Race *simplex* (Brown Whistler)
Recognition: *pale grey-brown above and dull white below.* Breeds all months and nests sometimes near the ground in a vine tangle, or high in the tree canopy. **Sexes**: *alike.* **Immatures**: *rufous brown.* **Voice**: *persistent melodious and varied song.*

Race *peninsulae* (Cape York form)
Recognition: *tinged greenish-olive above; pale lemon yellow below.* Breeds September–February. Nests low in vines or high in trees. **Sexes**: *alike.* **Immatures**: *rufous brown.* **Voice**: *persistent melodious and varied song.*

Race *simplex* left, race *peninsulae* right

70 LITTLE (Rufous) SHRIKE-THRUSH

Colluricincla megarhyncha 19 cm

(A) Race *rufigaster* (illustrated). (B) Race *parvula* (pale rufous below, whitish throat, black bill). A bird of rainforest, mangroves, gullies, parks and gardens. It forages at all levels of foliage, but mainly on the ground in rainforest, for insects and their larvae. It may appear in small feeding parties or in large mixed flocks. Sedentary. Breeds September–February. Nests 3–5 m high. Also in New Guinea. **Sexes**: *alike.* **Immatures**: *wing coverts edged brown.* **Voice**: *rich melodious 'to-whee-wot-wot'.* **Race parvula**: *'eeee, butch-butch-butcher'.*

59

60

61

62

63

64

65

66

67

68

69

70

10

71, 72 GREY SHRIKE-THRUSH

Colluricincla harmonica 23 cm

The rich melodious voice of this familiar and graceful bird puts it high on the list of the world's best songbirds. Its pure notes are far-carrying, and are often borrowed by lyrebirds and other mimics. The species also occurs in New Guinea, and there are several races – three in Australia. These are: *harmonica* (Grey Thrush) of the east and south-east; *brunnea* (Brown Thrush) of the far north; and *rufiventris* (Western Thrush) of the centre and south-west. Plumages vary progressively (clinally) from north to south and from east to west, without sharp divisions between the races. The natural habitat of the species is forest and open woodlands. In the southeast, it has adapted well to living alongside humans, and it is often seen and heard in parks and gardens. In the south-west however, it has yielded ground to encroaching cultivation. It is carnivorous, and eats a wide range of insects, lizards, small mammals, nestlings and sometimes carrion. Sedentary. Breeds July–February. Nests in a wide variety of sites including buildings. **Female**: *grey lores.* **Immatures**: *bright rufous eyebrow and wing coverts rufous.* **Voice**: *a glorious repertoire of melodious notes.*

***rufiventris* (male) left, *harmonica* (male) right**

73 SANDSTONE SHRIKE-THRUSH

Colluricincla woodwardi 26 cm

Named for its habitat in the sandstone gorges and cliffs of northern Australia, this is a terrestrial species, hopping with great agility over rocky outcrops, cliffs and caves in search of insects, especially grasshoppers and spiders. It often sings from a lofty crag, the song ringing and resounding through the natural rock amphitheatre. It is not rare, but rarely observed closely. Sedentary. Breeds November–January. Makes a nest of spinifex roots in rocky fissure. Endemic. **Sexes**: *alike.* **Immatures**: *darker wings and duller breast.* **Voice**: *rich muscial songs, clear whistling.*

74 BOWER'S SHRIKE-THRUSH

Colluricincla boweri 21 cm

Bower's Shrike-thrush is restricted to the highlands (above 400 m) and frequents dense rainforest. Sometimes visits nearby gardens. Perching quietly and inconspicuously, it watches for prey before making a sally, taking insects and their larvae in foliage or on the ground. Sedentary. Breeds October–January. Nests in tree or vine tangle 1–8 m high. Endemic. **Recognition**: *compared with the Little Shrike-thrush, it is a little larger, has a grey back and head, and is more streaked on the throat and breast.* **Sexes**: *alike.* **Immatures**: *browner and duller.* **Voice**: *varied repertoire of musical calls, silent in winter.*

75 GROUND (Scaly) THRUSH [White's Thrush]

Zoothera dauma 29 cm

The only natural representative in Australia of the widespread 'true' thrush family whose huge range extends even to Siberia. There are two Australian races: *lunata*, found over most of the range, including two isolated populations on the Atherton Tablelands and in SA; and *heinei* living in north NSW along the Great Dividing Range where the birds are rustier on the back and smaller. The species frequents cool, damp forest with dense ground litter and eats worms, insects and molluscs. It hops along, pausing to watch and listen for prey, and flies off swiftly to cover if alarmed. It wanders widely in the non-breeding season, and may be seen in loose flocks of 20 or more. Males sing particularly at dawn and dusk to proclaim territory in the breeding season. Beeds July–December. Nests in tree stump or fork to 15 m high. **Sexes**: *alike.* **Immatures**: *like adults.* **Voice**: *flute-like song.*

76 SONG THRUSH

Turdus philomelos 23 cm

This, the Throstle of poets, was introduced from Britain in 1863 to Melbourne where it has become fairly common in parks and gardens. Introductions elsewhere however all failed and the bird is not seen beyond Melbourne. A ground feeder, it pulls worms from lawns and is fond of snails using a rock as an 'anvil' to smash their shells. Breeds September–January and nests low in dense vegetation. **Sexes**: *alike.* **Immatures**: *like adults.* **Voice**: *a glorious song of repeated phrases which it pours out for hours on end, especially at dawn and dusk. In the words of Robert Browning:*

'. . . he sings each song twice over,
lest you should think he never could recapture
the first fine careless rapture!'

77, 78 BLACKBIRD

Turdus merula 25 cm

This all-black member of the thrush family, with a bright yellow bill, is one of the most abundant birds in Europe. It was introduced progressively into Australia, including releases by the Zoological Acclimatization Society of Victoria in Melbourne in 1862 and the Hobart Zoo subsequently in Tasmania. On the mainland, it has spread slowly northwards, and is still doing so, but the birds remain more numerous in the south. In Tasmania, it is now abundant. Introductions into New Zealand have adapted exceptionally well and the bird has even colonised the off-shore islands. Remarkably, many of these have become part albino, and are illustrated in the field guides as such. As in Europe, the Black-bird in Australia has adapted well to living in suburban areas and is a familiar sight in gardens, parks, orchards, as well as in natural habitats. It spends much time on the ground and is partial to lawns and short grass where it can probe for worms. In typical thrush fashion, it patters the ground with its feet (perhaps to emulate the sound of falling raindrops) then stops with head turned on one side to 'listen' for a worm. Insects, berries and fruit also form a part of its diet. Breeds September–January and nests in low bushes. **Immatures**: *more rufous brow than adult females.* **Voice**: *a true prima donna among bird songsters. From a high vantage point, it pours out its melodies in long recitals, especially early and late in the day, and its song carries over long distances. When disturbed, it rushes away protesting loudly with a torrent of scolding notes.*

Male left, female right

71

72

73

74

75

76

77

78

11

Wrens

Opposite: Splendid Blue Wren

He who shall hurt the little wren
Shall never be beloved by men
William Blake 1757-1827

Most of the species in this Section are linked by the name 'wren', and carry long cocked tails. They include the Fairy-wrens, Emu-wrens and Grasswrens of the family Maluridae; and the Fernwren, Scrubwrens, Heathwrens and Fieldwrens which form part of the warbler family Acanthizidae.

Historically, the name 'wren' belongs to a small brown bird with a stubby cocked-up tail widely distributed from Europe through Asia to Japan, affectionately known as the Jenny Wren in Britain and deep in legend and folklore. It must have surprised those who followed Colombus across the Atlantic during the following centuries to discover that this famous bird had a progenitor living in the New World, now familiar as the Winter Wren, and that it was there a member of a large family of sixty wrens (family Troglodytidae).

Australia's pert small birds of captivating ways with long tails cocked over their backs inevitably became known as wrens, and the name has even extended to a few species with wren-like characteristics whose tails are never cocked — such as the Scrubwrens. All evolved separately, however, and are not related to their namesakes elsewhere.

FAIRY-WRENS, EMU-WRENS, GRASSWRENS Family Maluridae

There are twenty-six species of this family, all living within the Australian Region. Twenty of these occur in Australia as endemic species, and divide into three generic groups.

Fairy-wrens, genus *Malurus* The male fairy-wrens are among the most exquisitely plumaged of the world's small birds, and their luminous patterning of blues, reds, black and white sets them apart in spectacular contrast to the almost uniformly brown females and immatures. The beauty of the males is reflected in some of the scientific names given to the species by John Gould and other ornithologists who first described them — such as *splendens* (splendid), *elegans* (elegant), *callainus* (coloured like turquoise), *amabilis* (lovely), *pulcherrimus* (very pretty) and *dulcis* (sweet). There are nine species comprising fourteen distinctive forms, each identified by well established names. They can conveniently be considered in four groups based on the characteristics of the male in breeding plumage:

— 'Blue' group, consisting of the Superb Blue, Splendid Blue, Turquoise and Black-backed Wrens;
— 'Chestnut-shouldered' group, consisting of the Variegated, Purple-Backed, Lavender-flanked, Lovely, Blue-breasted and Red-winged Wrens, in which the females can also be distinguished by their blue-grey tails;
— 'Bicoloured' group, consisting of the White-winged (the Black-and-white and Blue-and-white Wrens) and Red-backed Wrens;
— The quite distinctive Purple-crowned Wren.

The fairy-wrens are highly sociable and spend most of their lives in closely knit groups in or within the range of dense cover. They forage together for their main food of insects, roost tightly huddled together and frequently preen each other. They are strongly territorial, but may extend their foraging area in winter, sometimes combining two or more families in enlarged nomadic groups. One of the most intriguing ornithological experiences in Australia in the winter months is to encounter a family of fairy-wrens, tails cocked, busily flitting on the move in which a single blue male leads a retinue of perhaps a dozen, or even more, brown birds. As these can easily be mistaken for females, it used to be the popular, but erroneous, belief that the males ran harems, and this led to their being referred to as 'mormon wrens'. It is now known, thanks to several years of close observation by Ian Rowley and others on the Superb Blue Wren, that the brown birds comprise a majority of immatures from the nestings of the preceding breeding season in association with surviving males in eclipse winter plumage from progeny of previous years. The new immature males can be picked out by their black bills and light underparts, and the older males in eclipse plumage additionally by their blue tails. Shortly after nesting, the dominant female can often be identified by her tatty and frayed tail. As spring approaches, she drives her daughters away to find mates of their own, but one or two young males may stay. The female nests several times during a season, and she is able to do this by the relief afforded from parental duties in feeding and rearing the young by members of the family who assume the role of 'helpers'. These may be the older males who stayed or birds from earlier broods who, on reaching maturity, assist with the siblings of later broods. It may take four years or more before a male ceases to moult into eclipse winter plumage and wear full adult garb throughout the year. Breeding fairy-wrens when threatened are adept at feigning injury and leading intruders away from the nest by putting on the so-called 'rodent run'. Other fairy-wrens are believed to share the behavioural characteristics of the Superb Blue Wren. Fairy-wrens can adapt to all types of habitat, from the understorey of rainforests to the spinifex-clad rocky harsh interior, and one or more members of the family can be seen almost anywhere in Australia. 1-23

Emu-wrens, genus *Stipiturus* The diminutive emu-wrens have exceptionally long tails — about twice their body length — which have only six feathers, each of an open delicately filamented texture. These characteristics set the emu-wrens apart from all other Australian birds, and so few tail feathers are also rare among birds elsewhere. The structure of the feathers themselves is also unusual, since they lack the tiny interlocking hooks (barbicels) that keep a feather flat in most birds. This loose feather structure is shared with the Emu which — though a giant in comparison — explains the emu-wren's odd name. The weight of the Rufous Emu-wren, at about five grammes, is less than any other Australian bird. In nesting habits, diet and call-notes, emu-wrens closely resemble the fairy-wrens. They are usually timid and seldom emerge from the dense cover in which they live. When pressed, they fly to a vantage point to fix their bearings before suddenly disappearing again to creep mouse-like through the tangled undergrowth. They are poor fliers and, with tails streaming behind them, look like huge insects. There are two widely separated species, each adapted to a different type of habitat. Both are endemic. 24-26

11

Grasswrens, genus *Amytornis* The country in which Australia's eight species of grasswrens live is mostly so inhospitable and remote from human habitation that five of them were only discovered this century; the Black Grasswren in 1901 (unrecorded again until 1968), the White-throated in 1905 (unrecorded again until 1973), the Dusky and Carpentarian in 1914 and the Grey in 1921 — but only collected and described in 1967. The Striated Grasswren was discovered in 1883, and the Eyrean in 1874 but was 'lost' until rediscovered in 1972. The problems of photographing these elusive birds have also posed a special challenge, and there is a story of patience and enterprise to be told behind most of the photographs of this group. For instance, the first photographs of the adult Eyrean Grasswren, which were taken by the authors as members of a nine-person expedition to the Simpson Desert in 1977, involved four four-wheel-drive vehicles and a spotter plane. The palm must, however, go to the intrepid Bert and Babs Wells who were dropped by helicopter in the heart of the Kimberley region in Western Australia in 1974 to become the first to photograph the Black Grasswren, and who were lucky to escape with their lives after being trapped by a bushfire which swept the area. 27-34

OTHER WREN-LIKE BIRDS Family Acanthizidae (part)

Thirteen wren-like birds of the warbler family Acanthizidae are included here.

Scrubwrens, genus *Sericornis* There are six species of scrubwren in Australia, including the somewhat similar Scrubtit. All are endemic, except the Little Scrubwren of the far north of Cape York Peninsula which occurs also in New Guinea, where is is widespread. Scrubwrens mostly frequent thick undergrowth in moist places and forage mainly on the ground or low in the vegetation. Several have white eyebrows and pale eyes. The Scrubtit lives only in Tasmania and a race of the White-browed Scrubwren also occurs there. On the mainland, the latter species ranges right around the eastern, southern and western coastal belts from Cairns to Carnarvon, WA. There is considerable geographic variation, particularly in the breast plumage. The tails are not cocked. 36-38, 40-41, 44, 47

Fernwren, genus *Crateroscelis* In some ways, the endemic Fernwren resembles the scrubwrens, but the differences between them are sufficiently distinct to entitle the Fernwren to a genus of its own. It is confined to the rainforests of the Atherton area of northern Queensland, and is one of Australia's least-known birds. 35

Heathwrens and Redthroat, genus *Sericornis* The nesting habits and general behaviour of the two heathwrens (formerly genus *Hylacola*) and the Redthroat (formerly genus *Pyrrholaemus*) are similar. They are among the most accomplished songsters of Australia's birds and are excellent mimics. A distinction is that the heathwrens carry their tails cocked and the Redthroat does not. All are endemic. 39, 45-46

Fieldwrens, genus *Sericornis* The two endemic fieldwrens (formerly genus *Calamanthus*) are small, shy, ground-frequenting birds dispersed over the southern part of the continent and in Tasmania. Both species often carry their tails cocked, and are robust songsters. They are also notable for their many geographical colour variations — ranging from greenish-grey above and yellow-white below in south-eastern Australia and Tasmania, to the smaller western form, which is greyer above and whiter below. No less than fifteen races and twelve colloquial names have been listed at various times for them — a record for any Australian bird. 42, 43

Opposite: Eyrean Grasswren

PHOTOGRAPHERS: H. & J. Beste 6, 13, 16, 34, 35; G. Chapman 10, 12, 14, 17, 22, 23, 27, 29, 30, 33, 42; A.T. Foster 7; T. & P. Gardner 11, 38; P.N. Green 2; W.J. Labbett 26, 45; T.G. Lowe 20; I. Morris 28; F. & B.J. Park 25; A.Y. Pepper p. 105; J. Purnell 8, 9, 47; L. Robinson 32; A.H. Tolhurst 4; T.A. Waite 3, 43; J.D. Waterhouse 37, 40; G. Weber 1; C. Webster 15, 18; B.A. & A.G. Wells 5, 19, 24, 36, 39; Authors p. 107; 21, 31, 41, 44, 46

11

1, 2 SUPERB BLUE WREN

Malurus cyaneus

(Mainland form) 13 cm

Popularly known simply as the Blue Wren, this often tame and endearing small bird lives in low dense bushy vegetation and often frequents parks and gardens. It may be seen in family parties of perhaps a dozen birds in winter — an interesting characteristic which is further described in the Introduction. It is strongly territorial and, when vocal warnings of the male to competing males do not suffice, it may engage in aerial battles. It will also attack its image in windows. It often bathes in a shallow pool, or skims over deeper water, and delights in showering in the rain or under a garden sprinkler. Normally sedentary, but some individuals become nomadic in winter. Breeds July-March and may raise several broods. Usually nests low in a bush or tangled thicket. Endemic. **Immatures and young males:** (see Introduction). **Voice:** *light trilling like the tinkling of a small alarm clock.*

Male left, female right

3 SUPERB BLUE WREN

Malurus cyaneus (Tasmanian form) 14 cm

The Tasmanian bird is larger, has a longer tail, and is a deeper blue; its bib is also blue instead of black. It was first described by William Anderson, Captain Cook's friend and master surgeon on his third and tragic voyage on the *Resolution* (1776-1780). (Also see p. 124) In his journal, he noted it as a 'small bird with a pretty long tail has part of the head and neck of a beautiful azure colour'. The surgeon's mate, William W. Ellis, painted it from life, dating the work 1777. Its habits resemble those of its counterpart on the mainland.

4, 5 SPLENDID BLUE WREN

Malurus splendens

(Nominate race) 13 cm

This brilliant blue wren, the south-western counterpart of the Superb Blue Wren, was thought by John Gould to be more gorgeous than any other of its race. The male, and that of its close relative the Black-backed Wren (6), are distinguished by their bright violet-blue underparts crossed with a black band on the upper breast. The species was first collected in 1826 at Albany in WA by the French naturalists Quoy and Gaimard on the voyage of the *Astrolabe.* It inhabits dense bushy woodland. In most respects its behaviour is similar to that of the Superb Blue Wren. It is however markedly shyer, especially the male. Maturing young stay with the family for one or more seasons, and territories are jealously guarded. Sedentary. Breeds September-January. Nests within 1m of the ground. Endemic. **Female:** *note bluish tail.* **Immatures:** *like females, but tail brown.* **Voice:** *trilling, and a single sharp note.*

Male left, female right

6 SPLENDID BLUE WREN

Malurus splendens 13 cm

Race: *melanotus* (Black-backed Wren)

The Black-backed Wren resembles its close relative the Splendid Blue Wren in almost all respects except for a black lower back and rump. It lives as a widely separate population in mallee, woodland and scrub. It is also shy and relatively rarely-seen. Sedentary. Breeds September-January. Nests low in a bush. **Female**: *brown, tail dull blue, bill light brown.* **Immatures**: *like adult female.* **Voice**: *high-pitched trilling.*

Male

7 SPLENDID BLUE WREN

Malurus splendens 14 cm

Race: *callainus* (Turquoise Wren)

The dun colours of mulga scrub, spinifex and saltbush of the arid centre habitat of the Turquoise Wren, are the antithesis of the luminous blue-green colouring of this lovely wren. Within the context of differing environments, its habits are similar to the others of the species, but it is somewhat more nomadic. Breeds variably, mainly November-April. Nests on or close to the ground, well-hidden. Endemic. **Female**: *like non-breeding male, bill light brown, tail dull blue.* **Immatures**: *like females.* **Voice**: *trilling.*

Male

8, 9 VARIEGATED WREN

Malurus lamberti

(Nominate race) 14 cm

Living to the east of the Great Dividing Range, the Variegated Wren is the best known of the chestnut-shouldered wrens. It is shyer than the Superb Blue Wren, with which it shares territory, and so is less frequently seen. Families up to 10 frequent heathland and open forested country, occasionally parks and gardens. Sedentary. Breeds August-January. Nests on or near ground in dense cover. **Female:** *distinguished by a chestnut bill, and a chestnut patch ringing the eye. The tail is dull grey-blue and longer than that of the female Superb Blue Wren.* **Immatures:** *like adult female.* **Voice:** *brief high-pitched trills, softer than the Superb Blue Wren, but more metallic.*

Male left, female right

10, 11 VARIEGATED WREN

Malurus lamberti

Race: *assimilis* (Purple-backed Wren) 14 cm

The most widely-distributed of the chestnut-shouldered wrens and common in the drier inland areas of all States. Its habitat includes mallee, lightly timbered grassland and a wide variety of low-growing vegetation. Usually in small family parties. Sedentary. Breeds variably, mainly September-January. Nests on or close to the ground. Endemic. **Recognition:** *the male has deeper violet-blue tonings than the males of the nominate race of the Variegated Wrens, but the differences are slight and difficult to detect in the field.* **Female:** *like males in eclipse plumage, but bill chestnut.* Immatures: *like adult females.* **Voice:** *like nominate race.*

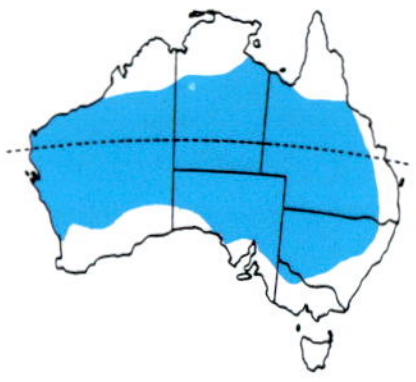

Male left, female right

12 VARIEGATED WREN

Malurus lamberti 14 cm

Race: *dulcis* (Lavender-Flanked Wren)

This race of the Variegated Wren is distinguished by the lavender flanks of the male (although not always obvious) and the blue-grey upperparts of the female. It lives among the rugged sandstone ridges of the Kimberley and Arnhem Land, in low vegetation. It is continually on the move usually in small parties. Breeds October-December. Nests low. **Female**: *blue-grey upper parts and tail.* **Immatures**: *like females.* **Voice**: *short sharp notes and trills.*

Male

1

2

3

4

5

6

7

8

9

10

11

12

11

13, 16 LOVELY WREN

Malurus amabilis 14 cm

Restricted to the humid coastal parts of Cape York Peninsula south almost to Townsville, the Lovely Wren frequents the edges of mangroves, rainforest and streams. In pairs and family parties, it forages on or close to the ground, usually in dense cover. Sedentary. Breeds September-December. Nests low or on the ground in dense cover. Endemic. **Recognition**: *the female, unlike the females of the other blue wrens, is almost as brightly coloured as the male.* **Immatures**: *like adult females; young males have black bills.* **Voice**: *faint trilling, more feeble than the others of the species.*

Male above, female below

14, 17 BLUE BREASTED WREN

Malurus pulcherrimus 14 cm

The indigo-blue breast of the male, when caught in strong light, beautifully sets off the chestnut shoulders, and distinguishes him apart from the black-breasted male of the very similar Purple-backed Wren whose range partially overlaps. It lives in scrubby mallee, heathland or swampy areas of southern Western Australia and in South Australia. Its behaviour is similar to but quieter than the other fairy wrens. Typically the male is shyer and keeps in the background. Gregarious, it forages in small parties on or near the ground. Sedentary. Breeds August-September. Nests well hidden low in a small bush. **Immatures**: *like female, but bill brown.* **Voice**: *thin trilling.*

Male above, female below

15, 18 RED-BACKED WREN

Malurus melanocephalus 13 cm

The Red-backed Wren is one of the smallest of the fairy wrens, and the name Elfin Wren is among its alternative names. It is also unique in the complete absence of blue in its plumage, but its flashing scarlet 'saddle' nevertheless makes it a spectacular little bird. The toning of this red is however variable and may fade at different seasons. Sometimes too, the male breeds in brownish plumage. It is seen in pairs or family parties and favours thick grassland associated with a wide variety of other vegetation. It is common within its range. Sedentary. Breeds August-January. Nests low. **Female**: *pale-brown above, with no chestnut around eye.* **Immatures**: *like adult females.* **Voice**: *trilling.*

Male above, female below

19 WHITE-WINGED WREN

Malurus leucopterus 13 cm

Nominate Race: (Black-and-white Wren)

The black-and-white form of the species was the first to be discovered — by the French naturalists Quoy and Gaimard in 1818 on Dirk Hartog Island off the west coast of Australia. It is also to be found on Barrow Island some 500 km to the north. It moves in pairs and family groups through the vegetation of the islands. Sedentary. Breeds July-September. Nests in dense bush close to the ground. Endemic. **Female**: *light brown above, paler below.* **Immatures**: *like females.* **Voice**: *weak trilling.*

Male

20, 21 WHITE-WINGED WREN

Malurus leucopterus

Race: *leuconotus* (Blue-and-white Wren) 13 cm

On the mainland, the plumage of the species is always blue-and-white. It has a very wide range over the drier parts of the continent, and frequents treeless plains with shrubs, spinifex and grass thickets and is often the dominant bird. In the words of Arnold McGill in his *Australian Warblers*, 'the males in full plumage look like large blue-and-white butterflies as they fly swiftly from one patch of herbage to another, or perch for a brief time on top of a clump of spinifex or salt-bush'. Habits are similar to those of other fairy wrens. The male is especially wary in the family party, which may number 20 birds, and many observers have commented on the ability of a whole family group to melt into the vegetation with astonishing speed when alarmed. Sedentary or locally nomadic. Breeds mainly July-September. Nests low. Endemic. **Immatures:** *like adult females, tail brown.* **Voice:** *like the sound of a fishing reel being wound.*

Male left, female right

22 RED-WINGED WREN

Malurus elegans 14 cm

Once known in Perth, the range of this man-shy wren has retreated to the south-west ern corner of the continent. Also known as the Elegant Wren. It lives in dense wet thickets fringing freshwater swamps and creeks and the undergrowth of jarrah and karri forests and is difficult to observe. Its name describes the rufous wing area common to all the chestnut-shouldered group. Slightly larger than its relatives. Gregarious and sedentary. Breeds August-January. Nests close to ground near water. **Female**: *grey-brown above, pale below; tail dull blue.* **Immatures**: *like females, tail brown.* **Voice**: *trilling.*

Male

23 PURPLE-CROWNED WREN

Malurus coronatus to 15 cm

Found only in two isolated areas of the far north of Australia, it is the least known and most unusual of the fairy wrens, being larger than the others. Its lovely lilac coloration, so distinctive, is not only unique among the wrens, but very rare in all Australian bird species. It was named by Gould from specimens collected on A.C. Gregory's expedition across north Australia in the 1850s. Sedentary. Family parties search for insects and their larvae, moving swiftly through dense riverside vegetation, keeping mainly in the shadows, but sometimes catching insects on the wing. Females and immatures are inquisitive; as usual, the male is more retiring. Breeds most of the year. Nests on or near the ground near water, bulkier than other wren nests. **Immatures:** *similar to adult females, but crown brown.* **Voice:** *loud sustained 'chicka- chicka- chickaa-chicka' often as a duet.*

Male left, female right

13

14

15

16

17

18

19

20

21

22

23

11

24, 25 SOUTHERN EMU-WREN

Stipiturus malachurus 20 cm

The eastern population of the Southern Emu-wren inhabits dense undergrowth in swampy areas, coastal dunes and heath, while that of the south-west frequents low arid scrub. Like the fairy-wrens, the species lives in pairs and family groups. It is unusually timid and, except when flushed, seldom exposes itself. It will then quickly drop to cover and scurry away mouse-like to disappear in the undergrowth. At the time of settlement, the Southern Emu-wren was common around Sydney, but it is now much rarer, and always difficult to spot. Sedentary. Breeds July-January. Nests near ground. Endemic. **Immatures:** *duller.* **Voice:** *thin very high-pitched trilling.*

Female left, male right

26 RUFOUS-CROWNED EMU-WREN

(A) *Stipiturus ruficeps ruficeps* 14 cm
(B) *S.r. mallee* (Mallee Emu-wren)

The separated populations of this emu-wren are here treated as northern and southern races of a single species. The former lives in spinifex and other low vegetation; the latter in understorey of mallee eucalypt country. **Recognition**: *both races smaller with shorter less filamentous tails than Southern Emu-wren. (A) — bright-rufous crown without streaks. (B) — dull-rufous crown slightly streaked.* **Females**: *blue bib absent.* **Immatures**: *like females.* **Voice**: *like Southern Emu-wren.*

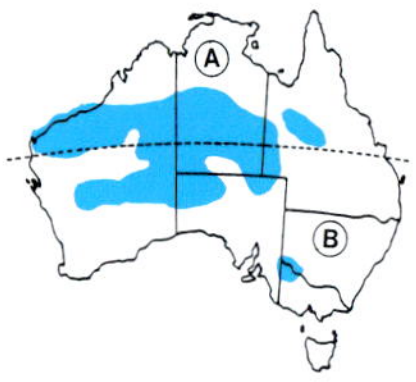

Male (B)

27 THICK-BILLED GRASSWREN

Amytornis textilis 18 cm

There are three forms: *(A) race textilis*, (B) race *modestus* and (C) race *myall*. The species inhabits thick saltbush on inland plains and around salt lakes. Populations have declined alarmingly. Timid and elusive, it hides within base of bush during heat of day and is difficult to flush. Bounces hastily from bush to bush. In pairs or small groups it feeds on ants, beetles, or berries. Sedentary or nomadic. Breeding related to rainfall. Nests low. **Female**: *rufous patch on flanks.* **Immatures**: *duller, less distinct markings.* **Voice**: *silvery melodious song; reeling or buzzing call, squeaky alarm.*

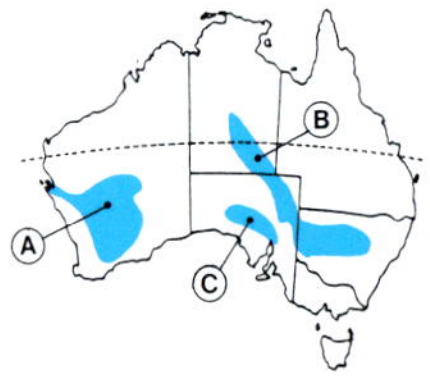

Male (A)

28 WHITE-THROATED GRASSWREN

Amytornis woodwardi 21.5 cm

Inhabiting the rugged sandstone escarpments of west Arnhem Land, it scurries secretively over rocky outcrops with head and tail lowered. Sometimes it watches intruders curiously from cover in rock crevices, or sits on an open perch to sing with the head thrown back. In pairs or parties it feeds in damp gullies on insects and seeds. Sedentary. Breeds December-March. Nests in porcupine grass tussock. Endemic. **Female**: *more rufous underneath.* **Immatures**: *duller, with buff abdomen.* **Voice**: *rich far-carrying trills; 'churring'.*

Male

29 CARPENTARIAN GRASSWREN

Amytornis dorotheae 17 cm

Discovered only in 1913, this grasswren was named *dorotheae* after the daughter of the distinguished ornithologist H.L. White. Its habit of keeping low in the spinifex of the remote sandstone ranges of the Gulf country and sheltering among the rocks makes it one of Australia's most elusive and least known birds. Sedentary. Breeds October-March. Nests low in grass or bush. Endemic. **Female**: *chestnut underparts.* **Immatures**: *duller.* **Voice**: *musical song; 'ticking'.*

Male

30 STRIATED GRASSWREN

Amytornis striatus 17.5 cm

The Striated Grasswren has the widest range of the grasswrens but its distribution is patchy. It frequents dense spinifex of inland sandy plains and mallee and exhibits extraordinary agility in threading its way through prickly porcupine grass. Singly or in pairs or small groups it bounces along rapidly near cover when feeding on seeds, ants and beetles. Shy but inquisitive during breeding. Sedentary or nomadic. Breeds variably according to rainfall. Nests in grass tussock. Endemic. **Female**: *rufous flanks.* **Immatures**: *duller.* **Voice**: *silvery trilling, cheeping.*

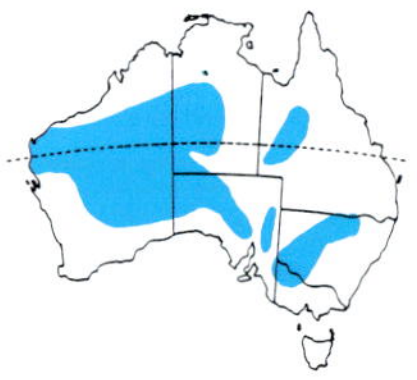

Male

31 EYREAN GRASSWREN

Amytornis goyderi 14 cm

The Eyrean Grasswren mainly lives and nests in the sandhill canegrass tussocks which grow along the tops of the long ridging dunes of the Simpson Desert, from which it is hard to flush. It eats insects and canegrass seed, hulling it with its exceptionally strong finch-like bill. Probably highly nomadic, seeking areas where canegrass is in seed following a chance fall of rain. Rainfall may also trigger its breeding. It gets moisture from insects and seems to show no interest in water.
It characteristically hops with one foot in front of the other. **Sexes**: *similar.* **Immatures**: *duller.* **Voice**: *high-pitched notes.*

32 GREY GRASSWREN

Amytornis barbatus 20 cm

Confined to a few relatively small remote areas on the Bulloo, Diamantina and Cooper River systems, it lives in tall dense clumps of lignum and swamp canegrass, and eats seeds, insects and some water snails. It hops briskly, tail cocked, when feeding, chirping constantly. When calling, it often perches prominently. Dashes to cover when alarmed, tail trailing. Often wanders in groups of up to 15 out of breeding season. Flies low and swiftly. Breeds July-August. Nests low. Endemic. **Female**: *less distinct breast markings.* **Voice**: *prolonged cricket-like twittering from thicket, especially when disturbed.*

Male

33 BLACK GRASSWREN

Amytornis housei 23 cm

Confined to the Kimberley region of rocky outcrops, gullies and tumbled rocks interspersed with dense spinifex. It forages in small parties for insects and seeds, and darts over the ground with head and tail lowered. It makes short low flights to a rock perch, flicking its tail on alighting. Climbs up and down rocks instead of flying, keeping mostly to the shadows. Sedentary, but small groups wander when not breeding. Breeds probably in the wet season, December-April. Nests in porcupine grass tussock. Endemic. **Female**: *chestnut below.* **Immatures**: *duller.* **Voice**: *trilling, with bell-like notes.*

Male

34 DUSKY GRASSWREN

Amytornis purnelli 17 cm

Very like the Thick-billed Grasswren except for its thinner bill; it differs chiefly in its preferred habitat of spinifex on rocky hillsides. While grasswren scampering is often described as mouselike, it is most unrodent-like when it perches, tail cocked, on a rock or bush to sing musically, sometimes in duet. Often appearing in parties, it is shy but inquisitive. Distraction displays draw attention away from the nest. Sedentary or locally nomadic. Breeding related to rainfall. Nests low. Endemic. **Female**: *chestnut flanks.* **Immatures**: *duller streaking.* **Voice**: *silvery song ending with a quick jangle of notes.*

Male

35 FERNWREN

Crateroscelis gutteralis 14 cm

Dusky brown-green plumage helps the Fernwren blend perfectly with its dense shady environment in remote highland rainforest. It dwells on the forest floor, searching amongst leaf litter for insects, spiders and snails, constantly on the move, hopping, bowing the head, and flicking the tail. It is calmly tolerant of observation. Usually alone or in pairs. Sedentary. Breeds August-February. Nests low. Endemic. **Recognition**: *its white collar is distinctive.* **Female**: *duller.* **Immatures**: *brown face and throat.* **Voice**: *single high repeated whistle, cheeping and chattering.*

Male

24

25

26

27

28

29

30

31

32

33

34

35

11

36, 37 WHITE-BROWED SCRUBWREN

Sericornis frontalis 12 cm

The bold and friendly White-browed Scrubwren is abundant throughout its range and is one of the most popular of the small birds. It lives in undergrowth in a wide variety of low vegetation and is constantly on the move in pairs and parties briskly foraging for insects such as beetles and moths — and some seed in Tasmania. It advertises its presence with incessant 'zitting'. There are many geographical plumage variations, with four main forms: (A) south-eastern race *frontalis* (illustrated at right); (B) Queensland race *laevigaster* (with buff underparts and plain throat); (C) Tasmanian race *humilis* (which is larger, plainer and darker): and (D) Western race *maculatis* (illustrated at left) — previously known as the spotted Scrubwren. Sedentary. Breeds July-January. Nests low. Endemic. **Female***: duller.* **Immatures***: duller and browner.* **Voice***: an incessant 'zit-zit' and some oft-repeated musical notes.*

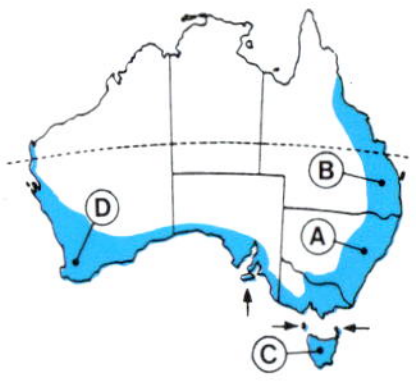

38 ATHERTON SCRUBWREN

Sericornis keri 13 cm

Only in recent years has this resident of a very restricted range on the Atherton Tablelands been declared a separate species from the Large-billed Scrubwren, which it closely resembles, and the minimal differences make them different to identify in the field. Usually solitary, rarely in pairs. Sedentary. Breeds possibly July-February. The only nest so far reported was close to the ground. Endemic. **Recognition***: (cf the Large-billed Scrubwren): slightly larger, longer legs and bill, forehead and face darker, feeds on the ground.* **Sexes***: alike.* **Immatures***: bright yellowish underparts.* **Voice***: chatter.*

39 REDTHROAT

Sericornis brunneus 12 cm

A small chestnut throat patch distinguishes this otherwise drab bird. It is widespread on inland plains and semi-arid areas, and is common in WA. It frequents thick scrub and saltbush, also mulga and mallee. Often calls from the top of a bush. In pairs and family groups it searches low vegetation for insects. Sometimes in flocks of up to 30. Sedentary. Breeds August-December. Nests in a bush clump close to the ground, sometimes used by cuckoos. Endemic. **Female***: lacks throat patch.* **Immatures***: paler than adult female.* **Voice***: twittering and chattering, a rambling melodious song, and a splendid mimic.*

Male

40 YELLOW-THROATED SCRUBWREN

Sericornis citreogularis 14 cm

An outstanding songster and an accomplished mimic, incorporating into its song imitations of dozens of other birds. It sings especially at daybreak, generally while moving over the forest floor. It lives deep in rainforest in dense moist gulllies. Its long pendulous nest is nearly always slung over water in a dark, sunless spot, a habit which accounts for its being known as 'devil-bird' in Queensland. Sedentary. Breeds August-March. Endemic. **Female***: brown face patch.* **Immatures***: like females.* **Voice***: range of beautiful melodious notes including mimicry; also a sharp 'tchit'.*

Male

41 LARGE-BILLED SCRUBWREN

Sericornis magnirostris 12-13 cm

Soberly-plumaged, it is a resident of the same sort of damp forest as the Yellow-throated Scrubwren. However, it feeds in middle or upper foliage levels, not on the ground. It is active and agile, flushing out prey by much fluttering, twittering and chattering. It is seen in groups of up to ten but in the more northerly part of its range it may join mixed flocks of similarly-occupied birds. Sedentary. Breeds July-January. Nests to 10 m high, or takes over nest of Yellow-throated Scrubwren, often in turn parasitised by cuckoos. Endemic. **Sexes***: alike.* **Immatures***: paler.* **Voice***: harsh 'chew', twittering; minor mimic.*

42 RUFOUS FIELDWREN [Calamanthus]

Sericornus campestris 13 cm

This shy little bird is patchily spread through arid and semi-arid country and lives in a variety of low vegetation. It hops on the ground, tail erect, when searching for insects and seeds, and scurries mouse-like to cover when alarmed. Usually in pairs and small parties. The male may sing from the top of a shrub, particularly at dawn and dusk. There are many geographical colour variations, but birds in the south-west tend to be greyer. Sedentary. Breeds July-December. Nests on the ground. Endemic. **Sexes***: alike.* **Immatures***: duller and more finely streaked.* **Voice***: 'whirr-whirr-chick-chick-whirr-ree-ree'.*

43 STRIATED FIELDWREN [Calamanthus]

Sericornus fuliginosus 13 cm

Confined to the south-eastern corner and Tasmania, lives in moist coastal swamp and heath. It is seen singly, in pairs and small parties and creeps mouse-like through the undergrowth. It feeds on insects and seeds — hopping on the ground with tail cocked. Males sing from a conspicuous perch sometimes wagging their cocked tails from side to side. Sedentary and strongly territorial. Breeds July-December. Nests on or near the ground in dense cover. Endemic. **Female***: buff-yellow throat tinge.* **Immatures***: like adult females.* **Voice***: similar to 'Rufous Fieldwren.*

Male

44 LITTLE [Tropical] SCRUBWREN

Sericornis beccarii 11 cm

This small scrubwren of New Guinea and adjacent islands, occurs in Australia only in the northern rainforests of Cape York Peninsula. In pairs and parties, the species actively forages in middle levels of foliage for insects. Probably sedentary. Breeds October-December. Nests close to ground in dense cover. **Sexes***: similar.* **Immatures***: rufous wash on flanks.* **Voice***: musical warblings, grating calls.*

45 SHY HEATHWREN [Hylacola]

Sericornus cautus 14 cm

The Shy Heathwren in the east lives in mallee, a habitat which separates it from the Chestnut-rumped Heathwren. It is in serious danger from mallee clearance. In the west, it lives in dry heathland. Despite its name, it does not appear to be any shyer than its relative — both are timid and elusive. Sedentary. Breeds August-November. Nests near the ground. Endemic. **Female***: duller.* **Immatures***: fawn wash on throat and breast.* **Voice***: 'chee-chee-chick-a-dee', some mimicry.*

Male

46 CHESTNUT-RUMPED HEATHWREN [Hylacola]

Sericornus pyrrhopygius 14 cm

This small bird with a cocked tail, lives in small parties in thick undergrowth of heath, woodland and forest. It is not only an accomplished songster in its own right but has exceptional skill in mimicking the calls of many other birds: one observer identified in a continuous song the voices of fifteen species, ranging from the small Yellow Robin to the large Yellow-tailed Black Cockatoo. Sedentary. Breeds June-November. Nests low. Endemic. **Female***: duller.* **Immatures***: no streaking on underparts.* **Voice***: long variable melodic song with mimicry of other birds.*

Male

47 SCRUBTIT

Sericornis magnus 12 cm

Confined to Tasmania, the Scrubtit is found mostly in the quiet ferny gullies, forests and thickets of the west and south-west of the State. Retiring, but it can be 'chirped-up'. It is a most active little bird, scrutinising leaf foliage for insects, and hopping up tree trunks in the manner of treecreepers. Sedentary. Breeds September-January. Nests low, often among ferns. Endemic. **Recognition** *(cf the White-browed Scrubwren): longer curved bill, underparts pale lemon-yellow.* **Sexes and immatures***: similar.* **Voice***: whistled 'to-wee-to'.*

36

37

38

39

40

41

42

43

44

45

46

47

12 Warblers

Included in the Section are groups of small rather plain songbirds belonging to the families Acanthizidae and Sylviidae most of which are commonly known as warblers.

THORNBILLS AND ALLIES Family Acanthizidae (Part)

Speckled Warbler, genus *Sericornis* (formerly *Chthonicola*) A distinctly marked ground-frequenting warbler, not unlike a heathwren in appearance, but smaller. 1

Rock Warbler, genus *Origma* The sole member of the genus and unique among warblers from its habit of always nesting in caves or other dark places in rock formations. 2

Fairy Warblers, genus *Gerygone* The generic name for this group, meaning 'born of sound' in Greek, is well chosen for, as Arnold McGill writes in his *Australian Warblers* (1970) 'Most of the nine Australian species pour forth, in spring and summer, fairy-like melodies from the treetops, and because these small birds are so active in movements, the pleasing call-notes are mostly the only indication of their presence'. The White-throated Warbler (or Bush Canary) typifies this engaging group of small birds. Odd man out is the Brown Warbler who only warbles in the north of its range. In the south, it advertises its presence with busy repetitions of '*which-is-it*' from which it is easily identified. There are twenty species of fairy-warblers ranging from South-east Asia to New Zealand. Nine of these occur in Australia of which five extend into New Guinea. Fairy-warblers are essentially arboreal in habits and live among the foliage usually singly or in pairs. Their nests are elaborate suspended structures with side entrances under a substantial hood. Some of the species build their nests close to those of wasps and hornets — for reasons that can only be guessed at. 3-12

Weebill, genus *Smicrornis* The sole member of the genus and Australia's smallest bird. It is arboreal in habits and often associates with thornbills in mixed feeding flocks. 13-14

Thornbills, genus *Acanthiza* Of the thirteen species of thornbills, twelve are endemic and one occurs in New Guinea. From the rainforests along the eastern sector to the spinifex-studded dry interior, thornbills in one form or other have learned to adapt. The generic name *Acanthiza*, latinized from the Greek, could be interpreted as 'I live in thorny places', and might apply to those thornbills that move so adroitly through tangled undergrowth, such as the Brown Thornbill. While similar in size to the fairy-warblers, they are more distinctively marked. They do not however compete as songsters. Unlike the fairy-warblers, they are prone to live in closely knit flocks, and some are ground feeders; these include the Buff-rumped, Yellow-rumped, Chestnut-rumped and Samphire Thornbills. 15-26

Whitefaces, genus *Aphelocephala* The size, feeding habits and general behaviour of the three whitefaces closely resemble those of the terrestrial thornbills. Although their ranges converge in northern South Australia, they do not appear to interbreed. 28-30

OLD WORLD WARBLERS Family Sylviidae (Part)

This large family of some 350 species, ranges widely mostly over Old World countries from Europe to Asia. They are mostly small birds with slender bills and weak flight. Their plumage is usually plain brown or green and they are insect eaters. Many are excellent singers. Collectively, they are sometimes referred to as grass warblers. Nine species occur in Australia — seven of which are included in this Section.

Spinifexbird, genus *Eremiornis* A lone genus and species. As the name implies, the bird is especially at home in a spinifex environment which can give it protection. 27

Grassbirds, genus *Megalurus* Two of the five species of grassbirds live in Australia; one is endemic. 31-32

Reed Warblers, genus *Acrocephalus* There are about a dozen species of reed warblers in Europe and Asia. Many are vociferous vocalists and, once heard, the call is unforgettable. Australia's single representative has a glorious voice. 33

Fantail Warblers, genus *Cisticola* There are some seventy-five species of the fantail, or grass warblers, of this genus. They range widely from Africa, which is their stronghold, eastward through southern Asia as far as Japan and Australia. Here two species occur: *Cisticola juncidis*, the Common Fantail Warbler, and *Cisticola exilis*, the Golden-headed Fantail Warbler, traditionally known in Australia as the Tailorbird. (There are also nine species of tailorbirds in Asia, but these belong to the genus *Orthotomus*, and are not immediately related.) The first named, known as the Zitting Cisticola in Africa, occurs only sparsely in Australia. The latter however is relatively common and, happily, can still be seen within an afternoon's jaunt from Sydney in the reed beds of Dee Why Lagoon and on Kurnell Peninsula. Its pretty and distinctive call discloses its whereabouts unerringly and there is a bonus if its marvellous nest is seen in which living growing leaves are skilfully stitched into its construction. On the subject of bird's nests, Alec Chisholm, in his book *Bird Wonders of Australia* (1934), quotes from an essay written by a small Queensland boy: " 'Who taught the birds to build their nests?' he wrote 'You would think that they would have to be taught how to make them. But, no. It is Nature, or we think it is. Without they are taught in a tongue of which we do not know, or God whispers it to them at night'." 34-36

Opposite: White-throated Warbler

PHOTOGRAPHERS: H. & J. Beste 15; G. Chapman 5, 23, 27, 30, 36; J.E. Dart 20; J.A. Estbergs 14; A.T. Foster 32; T. & P. Gardner 22, 34; R.A. Graves 26; R.I. Green 21; K. Ireland 9; R. McLennan 33; I.L. Morgan 19; I. Morris 8; J. Purnell 17, 18, 31; K. & B. Richards 13, 25; G. Rogerson 7; E.C.J. Smith p. 117; J.D. Waterhouse 3; G. Weber 1, 2, 16; C. Webster 35; E.E. Zillman 4, 6, 10, 24; Authors 11, 12, 28, 29.

12

1 SPECKLED WARBLER
Sericornis saggitatus 12 cm

This small ground-frequenting warbler lives mostly in open woodland. Usually in family parties, often in company with thornbills, it hops briskly over the ground foraging for insects and seeds among grass and leaf litter. It is often parasitised by the Black-eared Cuckoo with matching chocolate-red eggs coloured with a rub-off pigment. Breeds September-January. Domed nest on ground near cover. Endemic. **Male**: *eyebrow black instead of red-brown.* **Immatures**: *less streaking on underparts.* **Voice**: *subdued warbling song, also short 'zzt' contact call; accomplished mimic.*

Female

2 ROCK WARBLER
Origma solitaria 14 cm

The Rock Warbler is the only Australian bird restricted to New South Wales, where it lives in coastal sandstone between the Hunter River and the Clyde River. It is unique among Australian warblers in building its nest in a cave or deep rock crevice, sometimes even in the dark places of man-made structures. It feeds on insects and seeds. Breeds August-December. The spherical nest with side entrance is suspended from the rock undersurfaces with spider's web. Endemic. **Sexes**: *similar.* **Immatures**: *duller.* **Voice**: *sad shrill 'goodbye' also a chattering note; minor mimic. Sings even when being handled by bird banders.*

3 BROWN WARBLER
Gerygone mouki 10 cm

Common in pairs and small parties usually feeding in middle storey of forest. The birds north of Mackay (race *mouki*) are greyer above and buffer below than those south of Queensland (race *richmondi*). An intermediate race *amalia* is named for Frau Amalie Dietrich, the German naturalist who spent the years 1864-72 in the Queensland bush collecting for a Hamburg museum. Breeds September-February. Intricate, beautiful hanging nest near water. Endemic. **Sexes**: *alike.* **Immatures**: *like adults.* **Voice**: *(south) busy repetitive 'what-is-it', 'which-is-it'; (north) musical 'having-a-good-time'; minor mimic.*

4 LARGE-BILLED WARBLER [Gerygone]
Gerygone magnirostris 12 cm

In Australia, the Large-billed Warbler, which also occurs in New Guinea and adjacent islands, is often called the Floodbird because of its habit of building a nest hanging over water skilfully camouflaged as flood debris. It frequents coastal moist vegetation and forages on middle and lower levels of foliage for insects. Usually seen singly or in pairs, sedentary. Breeds September-April. Nest is a loose slender dome with ragged tail 50 cm long, and often near a wasp nest. **Sexes**: *similar.* **Immatures**: *similar to adults, brown eye.* **Voice**: *long cadence of repeated notes gradually slowing and descending like a spring unwinding.*

5 DUSKY WARBLER [Gerygone]
Gerygone tenebrosa 12 cm

This, the largest of the warblers, frequents the mangroves, swamps and scrubby creeks of the north-western coasts and is one of the least known. Its habits are apparently similar to those of other warblers, including working over foliage to snap up insects, sometimes in small loose feeding groups. Breeds October-March. Domed suspended short-tailed nest. Endemic. **Sexes**: *alike.* **Immatures**: *underparts washed pale yellow.* **Voice**: *a hesitant four-note warble.*

6 MANGROVE WARBLER [Gerygone]
Gerygone laevigaster 11 cm

The Mangrove Warbler, also in New Guinea, occurs in Australia as two races: *race laevigaster* along the north coast, smaller and whiter below, and race *cantator* on the east coast, rustier above. The latter has slowly extended to south of Sydney in the past twenty years. Usually in pairs mostly in coastal mangroves and margins of rivers and creeks and forages for insects. Friendly, tame and extremely active. Breeds September-April, longer in the north. Domed suspended nest 1-8 m above ground. **Sexes**: *alike.* **Immatures**: *underparts pale yellow wash.* **Voice**: *sweet jaunty warble descending the scale.*

7 WHITE-TAILED (Western) WARBLER [Gerygone]
Gerygone fusca 10 cm

Traditionally called the Western Warbler, this species occurs in several isolated populations across the continent. It differs from all the other warblers of the genus in its adoption of the drier inland areas where it lives in open forest and mallee scrub. Singly or in pairs, it forages mostly in the canopy for insects. Breeds September-January. Nest slender-tailed suspended from twig. Endemic. Recognition: *broad band of white to base of tail.* **Sexes**: *alike.* **Immatures**: *underparts pale yellow wash.* **Voice**: *similar but slower than White-throated Warbler.*

8 GREEN-BACKED WARBLER [Gerygone]
Gerygone chloronota 10 cm

Widespread in New Guinea, this species occurs in Australia only in the northern monsoon rainforests and mangroves from the Kimberley region to Arnhem Land, where it is abundant. It is a common garden bird in Darwin and is fearless. In pairs and loose flocks it forages for insects in the outer foliage. Breeds mainly October-April. Nests in small trees up to 10 m. **Sexes**: *similar.* **Immatures**: *duller.* **Voice**: *delicate warbling song similar to the White-tailed Warbler.*

9 WHITE-THROATED WARBLER [Gerygone]
Gerygone olivacea 11 cm

This warbler differs from all others because of the migratory movements of the southern birds and, particularly, for its distinctive song from which it is popularly known as the Bush Canary. It is one of the first to be heard in spring. It is quiet and unobtrusive during the non-breeding period. Habits are similar to other warblers. Also occurs sparsely in New Guinea. Breeds sporadically in the north and often nests in ant-infested trees up to 12 m. **Sexes**: *similar.* **Immatures**: *yellow on throat.* **Voice**: *repeated cascades of falling silvery notes.*

10-12 FAIRY WARBLER [Gerygone]
Gerygone palpebrosa 11 cm

The range of the Fairy Warbler spreads from New Guinea into eastern Australia, where two races are recognised — (A) *personata* (Black-throated Warbler) in Cape York to about Cooktown and (B) *flavida* to the south. The former is characterised by the black/brown throat of the male which progressively disappears where the two populations intergrade in the region of Cooktown. The females of the two races are indistinguishable and the markedly different male plumage of race *personata* is the only instance in Australia where a warbler is sexually dimorphic. The species frequents coastal rainforests and mangrove and forages in the outer foliage for insects. It is seen in pairs during the breeding season and is strongly territorial. It gathers in family parties in the non-breeding period sometimes with other species. Breeds sporadically throughout the year, especially October-March, and has the remarkable habit of siting its nest close to a wasp nest. It plays unwilling host to egg-laying by bronze-cuckoos. **Sexes**: *race flavida similar.* **Immatures**: *yellow wash on throat* **Voice**: *twitters constantly, tinkling undulating warble.*

Race *flavida* left; race *personata* male centre; race *personata* female right

1

2

3

4

5

6

7

8

9

10

11

12

12

13, 14 WEEBILL

Smicrornis brevirostris 8 cm

The smallest bird in Australia, with an unusually short bill — hence Weebill and *brevirostris* which means 'short bill'. Widespread almost throughout the continent. Plumage grades regionally from brown in the southern populations to yellow in the interior, the north and the west of the continent — with many intermediate variations. The brown and yellow birds were formerly recognised as two species S. *flavescens* (Yellow Weebill) and S. *brevirostris* (Brown Weebill). In pairs or in loose parties, it frequents a wide range of habitats, with a preference for the drier eucalyptus forests and woodlands. It behaves much like the thornbills in flitting and hovering in and over foliage in search of insects. Breeds August-February, south; October-May, north. Neat domed nest with narrow spout-like entrance suspended in tree 1-10 m or more from ground. Endemic. **Sexes**: *alike.* **Immatures**: *duller.* **Voice**: *short clear whistling call 'wee-winnie-wield', sometimes sustained song.*

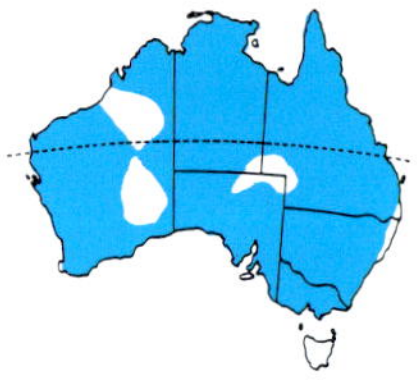

Brown form left, yellow form right

15 MOUNTAIN THORNBILL

Acanthiza katherina 10 cm

This species is confined to rainforest and nearby stream vegetation above 400 m between Townsville and Cooktown. It feeds actively on insects in high foliage in pairs and small parties. Although its plumage is normally yellow-green, this varies according to locality, becoming less yellow at increasing altitudes. Breeds September-January. Domed nest with slightly-hooded entrance, of grass and stalks faced with moss. Endemic. **Recognition**: *although closely resembling the Brown Thornbill, the most distinguishing characteristic is a creamy-white eye.* **Sexes**: *similar.* **Immatures**: *duller.* **Voice**: *sweet warbling similar to Brown Thornbill.*

16 BROWN THORNBILL

Acanthiza pusilla 10 cm

Singly and in pairs, it frequents wet areas with dense undergrowth and sometimes seen in gardens. An acrobatic feeder in the foliage while looking for insects and spiders. Five races are recognised, the northern birds being brighter, the northernmost race *mcgilli* (from Eungella) is very yellow. In western Victoria it intergrades with the Inland Thornbill. Breeds August-December. Rather an untidy nest of fibre and moss close to the ground. Endemic. **Recognition:** *large dark-red eye and streaked breast.* **Sexes**: *alike.* **Immatures**: *duller.* **Voice**: *varied musical notes and 'tzit-tzit' calls. Accomplished mimic and, like the Rock Warbler, is reported to sing even while being banded.*

17 INLAND THORNBILL

Acanthiza apicalis 11 cm

Also called the Red-rumped and Broad-tailed Thornbill, replacing the Brown Thornbill in the drier country over much of the continent. It also inhabits wet eucalyptus forest in the west. Singly or in pairs it forages for insects mainly in low shrub, but also in trees. There are variations in plumage and rump colour and three races are recognised. Breeds July-December. Nests in low branches of tree or shrub. Endemic. **Recognition**: *rump brighter red and forehead more scalloped than Brown Thornbill.* **Sexes**: *alike.* **Immatures**: *duller.* **Voice**: *strong song with mimicry.*

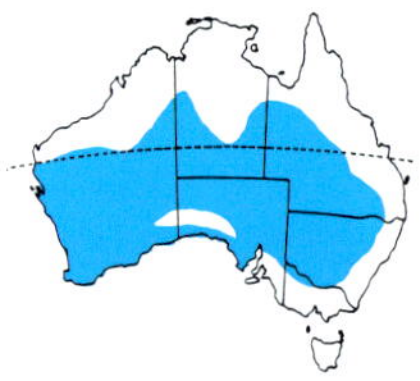

18 TASMANIAN THORNBILL

Acanthiza ewingi 10 cm

Confined to Tasmania and the Bass Strait Islands, it is difficult to distinguish from its close relative, the Brown Thornbill, which also occurs in Tasmania. It, however, prefers the wet forests while the latter keeps to dry forest. They co-exist where their ranges meet without inter-breeding. Breeds August-December. Nests in outer branches of small tree or shrub. Endemic. **Recognition**: *longer legs, smaller bill and paler underparts than the Brown Thornbill.* **Sexes**: *alike.* **Immatures**: *paler.* **Voice**: *musical warbling.*

19 CHESTNUT-RUMPED THORNBILL

Acanthiza uropygialis 10 cm

The Chestnut-rumped Thornbill, whose bright rufous rump is very apparent in flight, is seen in family parties and small nomadic flocks. Its range includes the drier inland areas. It frequents a variety of habitats including woodland, scrub and garden and is often seen near buildings. It feeds on the ground on insects and seeds. Breeds July-December. Nests in variety of places including hollow timber and buildings. Endemic. **Recognition**: *(cf Inland Thornbill) pale eyes.* **Sexes**: *alike.* **Immatures**: *duller.* **Voice**: *animated twitter and occasional mimic.*

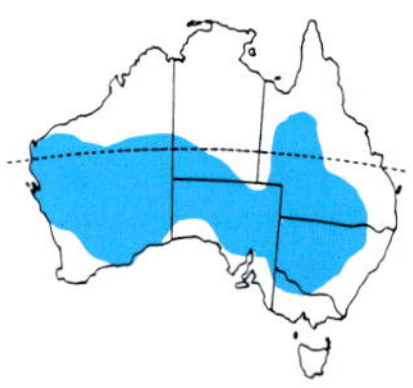

20 SLATE-BACKED THORNBILL

Acanthiza robustirostris 9 cm

Dense scrub and mulga in isolated arid inland territory is the home of this small thornbill. Probably seasonally nomadic. It usually forages in pairs arboreally for insects and caterpillars, also on the ground. Breeding is rainfall-related. Nests in slender branches, usually about 2 m high. Endemic. **Recognition**: *iris dark red-brown; forehead streaked black (cf Inland Thornbill) breast is not streaked.* **Sexes**: *alike.* **Immatures**: *like adults.* **Voice**: *loud and vigorous with call like 'wi-pu-chew'; also sibilant 'ssew' like Chestnut-rumped Thornbill.*

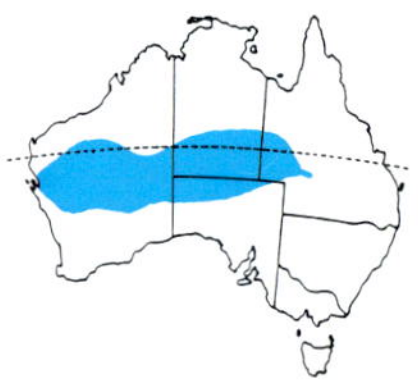

21 WESTERN THORNBILL

Acanthiza inornata 10 cm

This plain bird is common in the woodland areas of south-west Western Australia, including Kings Park in Perth. It forages in pairs or small parties of 5-15 birds for tiny insects in the foliage or on the ground. Breeds July-December. Nests close to the ground in low shrub. Endemic. **Recognition:** *plain plumage, short tail, pale eye.* **Sexes**: *alike.* **Immatures**: *paler.* **Voice**: *feeble tinkling twittering, with mimicry.*

22 BUFF-RUMPED THORNBILL

Acanthiza reguloides 11 cm

The Buff-rumped Thornbill varies in appearance and habit throughout its range. The most distinctive variant is the race *squamata,* ranging from the Atherton Tablelands south to Mackay, which has much brighter yellow underparts and rump. The northern birds, also known as the Varied Thornbill, frequent the tree tops as often as the ground, while the southern birds are mainly terrestrial. The species is fairly common and usually seen in small flocks of up to 15 birds. Breeds August-December. Variety of nest sites less than 2 m from ground. Endemic. **Sexes**: *alike.* **Immatures**: *duller.* **Voice**: *very similar to Yellow-rumped Thornbill, less melodious.*

23 SAMPHIRE (Slender-billed) THORNBILL

Acanthiza iredalei 9 cm

Sparsely distributed across its range, with some geographic plumage variations, it characteristically frequents samphire and saltbush on margins of inland salt lakes, saltbush or dense arid mallee heath and is usually seen in small parties searching for insects. It is shy and groups quickly scatter to cover. Breeds mainly July-November. Nests in low vegetation. Endemic. **Sexes**: *alike.* **Immatures**: *like adults.* **Voice**: *quiet twitter, warbles especially in flight.*

24 YELLOW-RUMPED THORNBILL

Acanthiza chrysorrhoa 10 cm

The best known and most widespread of the terrestrial thornbills. In pairs and small parties, often with other thornbills, it is seen in all foliaged areas including parks and gardens. It is relatively unafraid of humans. Breeds June-December, or after rain. Nests in drooping foliage 2—10 m high, sometimes in nests of other birds. For reasons unknown, it builds an open cup false nest over the domed structure in which the eggs are laid. Endemic. **Recognition:** *bright yellow rump and black spotted crown.* **Sexes:** *alike.* **Immatures:** *like adults.* **Voice:** *described as a showering silvery cadence of animated notes*

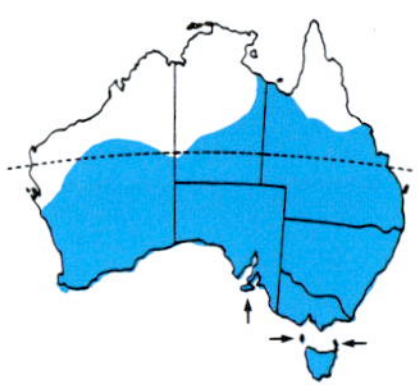

13

14

15

16

17

18

19

20

21

22

23

24

12

25 YELLOW (Little) THORNBILL

Acanthiza nana 9 cm

This, the yellowest of the thornbills, frequents a wide variety of habitats ranging from forest to parks and gardens where it readily accepts the proximity of people. It feeds on insects on outer foliage especially of acacia and casuarina and is rarely on the ground. Inland birds are paler. Breeds August-December. Nest suspended 3-12 m up from outer branches of trees. Endemic. **Recognition**: *(cf Weebill) bill longer, pale eyebrow absent.* **Sexes**: *alike.* **Immatures**: *like adults.* **Voice**: *constantly repeated sharp metallic 'ts-ts', 'ts-ts-ts' with a thin note through it.*

26 STRIATED THORNBILL

Acanthiza lineata 10 cm

Forages mainly in the canopy of eucalypts for leaf-dwelling insects. It is sociable and is at home in suburban gardens. There is some geographic variation and four races are recognised. Small flocks of 15-20 birds may gather in the non-breeding season often with other species. A banded bird, retrapped after an interval of 11 years suggests that the species has a long life span. Breeds July-December. Nest is suspended from a branch from 3-10 m up. Endemic. **Sexes**: *alike.* **Immatures**: *darker and duller.* **Voice**: *extremely high-pitched continuous chatter, 'zit-zit-zit'.*

27 SPINIFEXBIRD

Eremiornis carteri 15 cm

In reluctant feeble fluttering flight, the ground-frequenting Spinifexbird seems to be hindered by its heavy wedge-shaped tail (a diagnostic feature). It hops quickly across open ground with tail erect and moves through clumps of spinifex by clinging to the stems, with its tail used for balance. It feeds on insects and plant matter. Breeds August-April. Nests in a spinifex clump. Endemic. **Sexes**: *alike.* **Immatures**: *paler.* **Voice**: *a rapid chirrup, with repeated variations 'chidlee-dt', 'chpee-it'; contact call a repeated 'tk'.*

28 SOUTHERN WHITEFACE

Aphelocephala leucopsis 10 cm

This chubby bright-eyed bird is familiar still around country homesteads, as it was once in the outskirts of towns of the south before domestic and feral cats helped reduce its numbers. In family or mixed parties, it hops busily over the ground in search of grass seeds, often with thornbills. There is minor variation of flank colour progressively across Australia and three races are recognised. Breeds July-December. Nests in shrubs and low trees, and in any convenient repository around farm buildings. Endemic. **Sexes**: *alike.* **Immatures**: *like adults.* **Voice**: *twitters and sweet silvery song.*

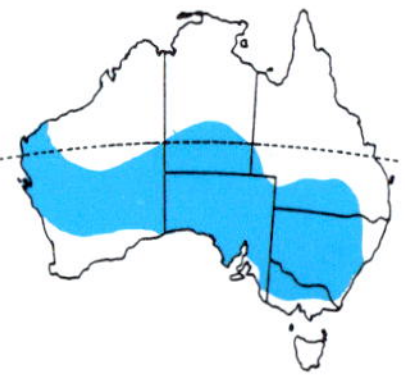

29 BANDED WHITEFACE

Aphelocephala nigricincta 11 cm

Widespread in the arid inland, it lives in spinifex or on stony flats with scattered shrubs, and behaves very much like other whitefaces. In pairs and small parties, it hops over open ground foraging for seeds and moist-bodied insects. It moves seasonally as conditons dictate. Breeds February-August. Bulky domes nest in prickly bush, lined with flowers and feathers, and with narrow entrance tunnel also lined. Endemic. **Sexes**: *alike.* **Immatures**: *browner and paler.* **Voice**: *twittering when feeding; musical trill sometimes in flight.*

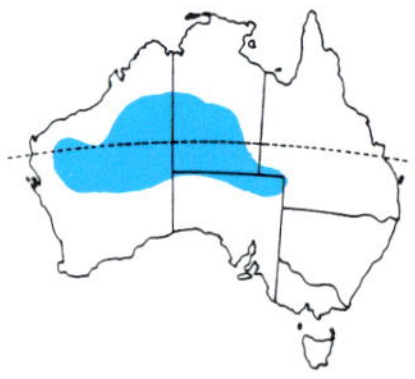

30 CHESTNUT-BREASTED WHITEFACE

Aphelocephala pectoralis 10 cm

In remote gibber plains and desert tablelands studded with low shrubs in South Australia, this bird lives out its life away from the eyes of all except those who set out to track it down. In pairs and small companies, often with other whitefaces, it hops over the ground in search of seeds and insects. It is nomadic, according to seasonal conditons, and is shy, flying to cover if disturbed. Breeds August-September, sometimes seasonally. Loose globular nest with side entrance. Endemic. **Sexes**: *similar.* **Immatures**: *paler chest band, with some dark spots.* **Voice**: *softer, more plaintive than Banded Whiteface.*

31 TAWNY GRASSBIRD

Megalurus timoriensis 17 cm

The Tawny Grassbird, whose range extends through New Guinea to the Philippines, is represented in Australia by the race *alisteri*. During the nesting season, the male performs an impressive display flight, soaring to hover in a crescendo of song to proclaim its territory. It frequents rank grassy vegetation or open forest with thick understorey. It is extremely secretive, and rarely shows itself except when climbing briefly to stem top to observe intruders. Sedentary or occasionally nomadic. Breeds August-April. Cup shaped nest in deep grass. **Sexes**: *alike.* **Immatures**: *darker brown.* **Voice**: *(in cover) persistent 'tsi-lik'.*

32 LITTLE GRASSBIRD

Megalurus gramineus 15 cm

Like the Tawny Grassbird, this smaller grassbird keeps close to the cover of moist reedbeds near the edge of water and is equally difficult to see, even when its song comes from a point close by. It can, however, be coaxed from cover by mimicry. Sometimes scurries over mud in search of insect prey. Calls from top of bush. Sedentary but may make long flights to temporary inland swamps. There are three races with minor plumage variations. Endemic. **Sexes**: *alike.* **Immatures**: *paler.* **Voice**: *plaintive, drawn-out three-note 'pip-pooo-pooo'.*

33 REED WARBLER

Acrocephalus stentoreus 17 cm

The Reed Warbler, notable for its glorious song, is found extensively through New Guinea to southern Asia and Africa. In Australia, it lives in moist wetlands where reeds grow, but is scarce in Tasmania. Southern populations in eastern Australia migrate northwards after breeding. Breeds September-February. Nests in skilfully-woven cups 'tied' to reed stems or willow branches over water. **Sexes**: *alike.* **Immatures**: *darker brown.* **Voice**: *its song is poured forth constantly in the breeding season, at night too, and is unsurpassed even by the European Nightingale.*

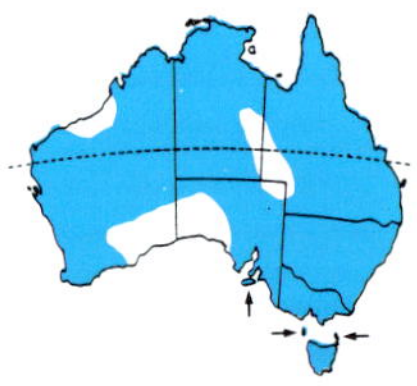

34, 35 TAILORBIRD OR GOLDEN-HEADED FANTAIL WARBLER [Golden-headed Cisticola]

Cisticola exilis 10 cm

It nests in vegetation where it can cobble living leaves into the structure. These are stitched together along their edges in combination with grasses using a thread of spider-web or fibre — and even cotton, rayon and wool where the birds live near inhabited areas. The female uses her beak as a needle and the male gathers the 'thread' for her. The number of leaves used varies with their size and counts have ranged from three to forty-five in a nest built in a verbena bush. Nests are entered from the side and are softly lined. The male sports a golden crown only during the breeding season. At other times, his head becomes streaked like that of the female and he grows a longer tail. The species frequents reed beds and swampy vegetation in open areas and is often abundant. It feeds on insects. Breeds September-March. Nests in thick vegetation close to the ground. Also widespread in India, China, south-east Asia and New Guinea. **Female**: *lacks golden crown of breeding male.* **Immatures**: *yellow underparts.* **Voice**: *incessant buzzing and pretty ventriloquial song of liquid notes usually delivered briefly on popping up to a reed top or higher vantage point.*

Male left, female right

36 COMMON FANTAIL WARBLER [Zitting Cisticola]

Cisticola juncidis 10 cm

Although widespread and common in southern Asia, Africa and southern Europe, it has only recently been reported in Australia from a few isolated coastal areas. Similar in habits to its Golden-headed relative and difficult to separate except when male head markings differ in the breeding season. Breeds December-April. Builds nest in grass clump near the ground with green leaves occasionally stitched in. The calls however differ. **Female**: *crown more streaked.* **Immatures**: *similar to female but duller.* **Voice**: *male — 'zit-zit' or 'lik-lik'; female calls infrequently.*

25

26

27

28

29

30

31

32

33

34

35

36

13 Honeyeaters

Opposite: Brush Wattlebird

Honeyeaters are as much a feature of Australia's natural environment as are the parrots, with one or more of the sixty-seven species occupying every niche of the country from the tree-lined streets of the cities to the rainforests and deserts. Fifty-four of them are endemic. Honeyeaters constitute the family Meliphagidae which is almost entirely confined to the Australian Region and comprises 170 species. They are notable for the great diversity of colour and patterns of plumage and the wide range of sizes — from the 11cm Scarlet Honeyeater of eastern Australia to the 48 cm Yellow Wattlebird of Tasmania. Their bills, which are mostly decurved, vary in length according to feeding habits, but all have brush tongues which enable them to sip nectar from blossom.

Nectar bearing flowering trees and plants grow in profusion across the continent and in great variety, from the tallest trees of the rainforests to the low vegetation of heath and desert. Plant species which produce large quantities of nectar include many of the eucalypts, grevilleas, melaleucas (paperbarks), banksias and eremophilas (particularly the red-flowering Emu Bush). Some drip nectar to the extent that an unwary camper under such a tree can wake up in a very sticky sleeping bag. Flowering is triggered by seasonal and climatic influences and tends to be sporadic and unpredictable. Because a number of honeyeaters tend to follow the blossom, they are often highly nomadic and sometimes congregate on trees in large numbers when flowering is abundant. Honeyeaters are, however, versatile feeders and are not solely dependant on nectar; some feed on nectar hardly at all. Fruit is also eaten and they are particularly fond of insects. These are gleaned from foliage, prised from bark and crevices, and taken from the air. A major diet for some species, such as the miners, is honeydew, manna and lerps. (Honeydew is the sugary secretion produced by many sap-sucking insects of which the Cicada is an example. Whitish, sugary deposits which gather on leaves when honeydew evaporates are generally termed manna; it is sometimes produced in such quantity that it falls from a tree like rain. Lerps are small protective sugary coverings built on leaves by small insects called *Psyllids* to protect their nymphs which grow and feed on the tissue below. They are easy prey and some birds appear to eat the lerps as well.)

The White-naped Honeyeaters, genus *Melithreptus* The most marked characteristic of these seven predominantly black-headed honeyeaters is the white nape band on all except Tasmania's Black-headed Honeyeater. Apart from the White-throated Honeyeater, whose range extends to New Guinea, the group is endemic. 1-7

The 'Myzomela' group Of about twenty-two species of the genus *Myzomela*, which extends to Fiji, Sulawesi and islands north of New Guinea, three are represented in Australia and are typified by the Scarlet Honeyeater. 8-10

Honeyeaters with well-marked breasts, genus *Ramsayornis* The breasts of these two honeyeaters are barred or scalloped. The Bar-breasted Honeyeater is endemic. 11-12

The Spinebills, genus *Acanthorhynchus* The long awl-shaped bills of these two endemic species, in comparison to their size, is greater in length than that of any other Australian passerine, and they are appropriately included among the honeyeaters. 13-14

The monotypic honeyeaters These eight 'orphaned' species of honeyeaters that differ from their relatives to the extent that each enjoys the status of a separate genus. 15-19, 58-60

The 'Conopophila' group A group of three small honeyeaters of the genus *Conopophila* — all endemic except for an extension of range of the Rufous-banded honeyeater into New Guinea and the Aru Islands. 20-22

The dimorphic honeyeaters, genera *Certhionyx* and *Cissomela* Three endemic species characterised by well-marked differences in plumage between the sexes of the Pied and Black Honeyeaters and between the adult bird and the immature in the Banded Honeyeater. 23-27

The 'Meliphaga' group, (including the genus *Xanthotis)* Twenty-five of the thirty-six species of these predominantly grey or green birds are represented in Australia, of which twenty are endemic. Affinities exist among the Lewin's and Lesser Lewin's and the Graceful Honeyeater (30-32) and between the five species with plumes (44-48). 28-52

Genus *Phylidonyris* A group of five endemic honeyeaters which includes the familiar trio of yellow-winged honeyeaters, the Crescent, White-cheeked and the New Holland, the first honeyeater to be discovered. 53-57

The Miners, genus *Manorina* The four miners, characterised by the Noisy Miner, are endemic, and are among the most successful and abundant of Australia's honeyeaters. The Indian Mynah is unrelated. 61-64

The Wattlebirds, genus *Anthochaera* The three species of this large honeyeater are endemic. They take their name from the conspicuous wattles of the Red Wattlebird and the large Yellow Wattlebird of Tasmania. 65-68

The Friarbirds, genus *Philemon* Of about seventeen species of friarbirds in the Australian Region, six occur in Australia. Some have largely unfeathered heads of blackish bare skin, and so these 'shaven headed' birds became known as 'friarbirds'. 69-72

WILLIAM ANDERSON

The New Holland Honeyeater was discovered by William Anderson, surgeon and naturalist on Cook's last two voyages, following a call at Adventure Bay in Van Diemen's Land from 26 to 30 January 1777. An entry in his journal of 29 January includes descriptions in Latin and English of three birds he named *Motacilla azurea, Motacilla canescens* and *Tanagra diemeniana* — known today respectively as the Superb Blue Wren (see historical reference also in Section 11), New Holland Honeyeater and Striated Pardalote. The gifted Anderson died at sea on 3 August 1778 off the north-west coast of America, and Cook named an island in the vicinity after him. His names for the three birds did not survive as the accepted scientific names, but his scholarly descriptions of them did, and this gives him a secure place in Australian ornithological history as being among the first to provide descriptions of this calibre of birds in Australia.

PHOTOGRAPHERS: G.D. Anderson 19; H. & J. Beste 49; G. Chapman 11, 21, 23, 25, 36, 42, 51, 56, 58, 60, 65, 67; J.A. Estbergs 20; J. Frazier 32, 62; T. & P. Gardner 9, 34; R. Garstone 24; G.J. Glass 5; P.N. Green 61; E. & D. Hosking 69; K. Ireland 8, 38; F.F. James 48; R. & D. Keller 1, 2; T.R. Lindsey 35; M. McNaughton 57; J. Purnell 39, 59, 63; K. & B. Richards 3, 4, 31, 44; L. Robinson 29, 30, 41; E.C. Slater 43; T.A. Waite 6, 7, 14, 33, 55, 68; G. Weber 16, 47; B.A. & A.G. Wells 12, 13, 22, 26, 27, 52; E.E. Zillmann 17, 46, 72; Authors p. 125; 10, 15, 18, 28, 37, 40, 45, 50, 53, 54, 64, 66, 70, 71

13

1 WHITE-NAPED HONEYEATER

Melithreptus lunatus 14 cm

The birds in the south-west are larger with longer bills and some have a greyish-white eye skin. In the east and south-east the eye skin is orange-red, and the white nape band fainter. In wet and dry forest it forages in pairs and parties for insects, keeping mainly to high foliage, and amidst blossoms for nectar and pollen. In the south-east it is migratory, moving northwards in autumn in noisy flocks, returning in spring. Breeds July-January. Nests 5-16 m from ground, sometimes host to cuckoos. Endemic. **Sexes**: *alike*. **Immatures**: *duller, nape crescent indistinct*. **Voice**: *subdued, leisurely piping call also a hissing note.*

2 WHITE-THROATED HONEYEATER

Melithreptus albogularis 14 cm

Replaces the White-naped Honeyeater in most of tropical Australia. Although the two species overlap from northern NSW to the Atherton Tablelands, the White-throated differs in having bluish-white skin above the eyes, and the white nape crescent reaches to the eyes. In the Kimberleys they are smaller and paler. Inhabits forest and woodland, parks and gardens. Breeds January-October. Nest 5-16 m from the ground. Also in New Guinea. **Sexes**: *alike* **Immatures**: *duller*. **Voice**: *rapid hissing notes; loud clear whistling.*

3 BROWN-HEADED HONEYEATER

Melithreptus brevirostris 13 cm

Living in mallee, open forest, arid scrub, parks and gardens, it overlaps much of the range of the White-naped Honeyeater, whose immatures it resembles. Populations westward have increasingly dark head plumage. Acrobatically active in foliage, it investigates bark crevices for insects; it also feeds on nectar. Occurs in small stable flocks when not nesting. Sedentary or locally nomadic. Breeds August-December. Nests 3-15 m from the ground. Endemic. **Sexes**: *alike*. **Immatures**: *eye patch pale blue, nape crescent duller* **Voice**: *loud and noticeable; calls of 'chick-chick'.*

4 BLACK-CHINNED HONEYEATER

Melithreptus gularis 16 cm

In the south-east this honeyeater has an olive-green back and bluish eye crescent. In savannah woodland and dry eucalptus forest it actively keeps to the upper foliage, feeding on insects and nectar. Like the White-eared Honeyeater it collects hair for its nest, apparently preferring blondes! It usually occurs in largish parties, sometimes in mixed feeding flocks. Sedentary or locally nomadic. Breeds July-December. Nests high in outer foliage. Endemic. **Sexes**: *alike*. **Immatures**: *head brown, bill and legs orange*. **Voice**: *various high-pitched calls including 'chirrup'; prolonged bursts of singing.*

5 GOLDEN-BACKED HONEYEATER

Melithreptus gularis laetior 16 cm

This is the northern race of the Black-chinned Honeyeater, and distinguished from it by a striking golden-yellow back and yellow-green eye crescent. It lives in a variety of habitat from open woodland to desert scrub. Often in flocks, it feeds on nectar and insects. Sedentary, nomadic in drier regions. Breeding probably related to rainfall. Very few nests have been found; one in a bauhinia tree was 3 m from ground. Endemic. **Sexes**: *similar*. **Immatures**: *duller, head brown*. **Voice**: *like Black-chinned Honeyeater; loud rollicking 'o-swallow-swallow' repeated.*

6 STRONG-BILLED HONEYEATER

Melithreptus validirostris 18 cm

Confined to Tasmania and the Bass Strait Islands, where it is common in forest and scrub and is usually seen in pairs and flocks. It feeds mainly on insects, their larvae and spiders. These are often prised from trunks of trees and flocks splitting and stripping bark in a patch of forest, calling as they go, set up quite a noise. Falling debris also marks their presence. It is seldom seen in blossom seeking nectar. Breeds July-December. Nests 5-25 m from the ground, possibly in loose colonies. Endemic. **Female**: *smaller*. **Immatures**: *bill, legs and eye patch orange*. **Voice**: *loud rattling noises; call of 'cheep'.*

7 BLACK-HEADED HONEYEATER

Melithreptus affinis 14 cm

Confined to Tasmania and the Bass Strait Islands where it is common. It is the only bird of the *Melithreptus* group to lack a whitish nape. It lives in forest, woodland and heath, and forages in noisy flocks in middle and outer foliage for a variety of insects. Occasionally takes nectar. Nomadic in winter. Breeds October-January. Nests to 20 m high in outer foliage; host to cuckoos. Endemic. **Female**: *smaller*. **Immatures**: *paler; throat whitish yellow*. **Voice**: *sharp whistling; soft 'chip'.*

8 SCARLET HONEYEATER

Myzomela sanguinolenta 11 cm

A brilliant bird, and smallest of the honeyeaters, lives in tall flowering trees and briskly works over the canopy among the blossoms in loose parties, often with other honeyeaters, probing for nectar and taking insects with darting and hovering movements — singing as it goes. It is nomadic following the blossoms; the southern birds migrate northwards to winter. Breeds July-January. Nests in foliage to 10 m high. Also on islands from Sulawesi to New Caledonia. **Female**: *upperparts olive-brown, underparts off-white*. **Immatures**: *like females*. **Voice**: *male's song is a clear and beautiful warble.*

Male

9 BROWN-BACKED HONEYEATER

Ramsayornis modestus 12 cm

The small Brown-backed Honeyeater lives in coastal paperbark swamp and mangrove, and feeds on nectar from flowering trees, sometimes in mixed flocks. Also takes insects on the wing. Southern birds migrate northwards to winter; some possibly to New Guinea, where there is also a resident population. They leave in April and return in August. Breeds August-March. Its nest is like the Bar-breasted Honeyeater's. **Sexes**: *alike*. **Immatures**: *breast more streaked*. **Voice**: *typical call of 'mick-mick'; chattering.*

10 DUSKY HONEYEATER

Myzomela obscura 14 cm

This plain long-billed honeyeater is common in humid tropical rainforest and coastal vegetation of the north-east. Singly and in pairs, it feeds high in the blossoms for nectar, often with other honeyeaters. It also takes flying insects on the wing. Locally nomadic. Breeds September-January. Nests in tree tops. Also in New Guinea and Indonesia. **Sexes**: *alike*. **Immatures**: *duller*. **Voice**: *squeaky trilling chatter 'chrip-chrip-chrip', and harsher notes.*

11 RED-HEADED HONEYEATER

Myzomela erythrocephala 13 cm

The Red-headed Honeyeater lives in mangroves and coastal eucalypts of the tropical north. It flits and darts tirelessly among the blossoms in pairs or small flocks feeding on nectar, sometimes in mixed flocks. It also takes insects. Breeds March-September. Nests low in mangrove fork. Also in New Guinea and Indonesia. **Female**: *browner, red wash over forehead, cheeks and chin; upperparts grey-brown*. **Immatures**: *like females*. **Voice**: *brisk hard 'chirrups'; jingling song.*

Male

12 BAR-BREASTED HONEYEATER

Ramsayornis fasciatus 15 cm

A honeyeater of tropical rainforest, woodland near water, and mangrove swamp. Usually in pairs, it works over flowering trees for nectar and insects. Large flocks sometimes gather when blossom is abundant. Breeds mainly October-March. Nest is usually built overhanging water, and its distinctive feature, shared alone with its close relative the Brown-backed Honeyeater, is a side entrance. **Female**: *fainter markings*. **Immatures**: *head brown not scalloped; bold brown streaking on breast*. **Voice**: *rapidly repeated shrill piping note.*

1

2

3

4

5

6

7

8

9

10

11

12

13 WESTERN SPINEBILL

Acanthorhynchus superciliosus 15 cm

Restricted to the woodlands, heath and gardens of south-west WA, and frequents Kings Park, Perth. Usually in pairs, it works over a range of flowering shrubs for nectar, such as dryandras, banksias, myrtles, and particularly kangaroo paws. Pollen rubbed off from its hanging flowers onto the spinebill's back is thus carried to pollinate other kangaroo paws. It also takes insects. Breeds September-January. Nests in a low bush or tree. Endemic. **Female**: *upperparts olive-grey, throat buff-brown, paler orange collar* **Immatures**: *like females* **Voice**: *staccato piping 'kleet-kleet'.*

Male

14 EASTERN SPINEBILL

Acanthorhynchus tenuirostris 16 cm

The awl-shaped bill typical of honeyeaters attains its greatest length in the spinebill, and enables it to reach into long tubular flowers. Like hummingbirds, it hovers while feeding, with audible beatings wings. Singly or in small groups, it frequents forest, woodland and suburban gardens wherever there are nectar-bearing blossoms. Also takes insects. Breeds October-January. Nests in bush or low tree. Endemic. **Female**: *duller, crown grey.* **Immatures**: *upperparts grey-olive, underparts fawn.* **Voice**: *rapid succession of high piping notes and a soft 'chee-chee'.*

Male

15 GREEN-BACKED HONEYEATER

Glycichaera fallax 12 cm

This predominantly New Guinea honeyeater is confined in Australia to a small area of rainforest at the top of Cape York Peninsula. Usually in pairs, it moves non-stop from the tree-tops down almost to the ground, actively foraging for insects in the foliage. In June 1981, adults were observed feeding dependent young, but a nest has yet to be found. **Sexes**: *alike.* **Immatures**: *not known.* **Voice**: *thin peeps and twitters.*

16 PAINTED HONEYEATER

Grantiella picta 16 cm

An attractive honeyeater, it lives more on mistletoe berries than on the nectar of the flowers of that plant. It is nomadic in pursuit of its favourite food. It may also take insects on the wing. At times it is high and inconspicuous in a tree, but is easily located by its distinctive calling. Singly, in pairs or colonies. Breeds October-March. Nests so frail as to be net-like, 3-20 m high. Endemic. **Female**: *smaller, duller.* **Immatures**: *like females, underparts more spotted.* **Voice**: *double whistle with emphasis on second note and loud clear calls of 'georgie-georgie'.*

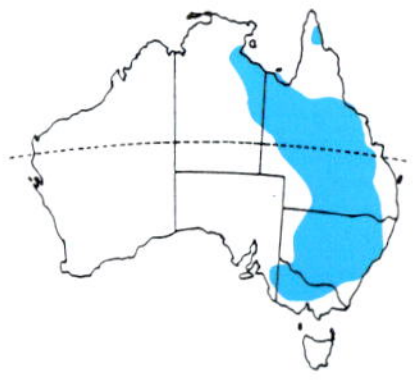

17 BROWN HONEYEATER

Lichmera indistincta 15 cm

Widespread in tropical rainforest, mangrove edges, heath and arid scrub, this honeyeater has one of the most beautiful songs of all Australian birds. Singly, in pairs and sometimes large flocks in blossom feeding on insects and nectar. It is much attracted by the red-flowering calistemon. Southern birds migrate northwards to winter. Breeds June-January. Nests up to 7 m. Also in New Guinea. **Female**: *smaller, crown olive-brown like back.* **Immatures**: *lack yellow behind eye.* **Voice**: *a glorious melodious song, very much like the Reed Warbler's.*

18 WHITE-STREAKED (Brush-throated) HONEYEATER

Trichodere cockerelli 16 cm

Distinguished by its unusual ruffled throat feathers, this honeyeater is found only north of Cooktown on Cape York Peninsula. It lives in eucalypt woodland, rainforest and watercourse vegetation. Singly, in pairs and small groups, it forages energetically in the upper foliage for nectar and insects. Locally nomadic, it sometimes gathers in large feeding flocks. Breeds January-May. Nests low down in bush or small tree often over water. Endemic. **Sexes**: *similar.* **Immatures**: *duller brown upperparts and yellow on throat.* **Voice**: *rarely quiet, with whistling and scolding.*

19 STRIPED HONEYEATER

Plectorhyncha lanceolata 23 cm

The name *lanceolata* (lance-shaped) relates to the bird's long spiky feathers on the throat and upper breast. Its habitat ranges from inland mallee scrub and native pines to dense stands of casuarinas and paperbarks towards the coast. It is active and fearless, and forages in pairs and small parties mainly for insects and fruit, sometimes raiding orchards, less often for nectar. Breeds August-January. Nests in deep cup hung by rim from twigs, 1-10 m from ground, often close to the Grey Butcherbird's nest. Endemic. **Sexes**: *alike.* **Immatures**: *less distinct markings.* **Voice**: *loud, rolling melodious song, more tuneful than most other honeyeaters.*

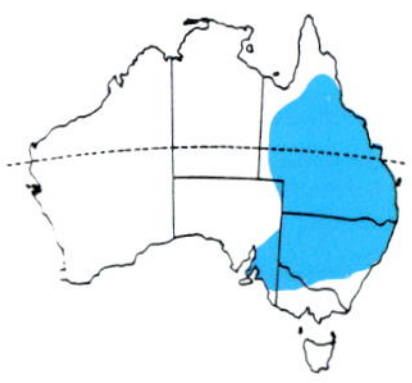

20 RUFOUS-BANDED HONEYEATER

Conopophila albogularis 13 cm

A common garden bird in Darwin and recognised by its white throat and reddish breast. It forages actively and acrobatically in tropical blossom and foliage for a wide variety of insects and nectar. It is partly nomadic in search of blossoming trees. Seen in pairs or loose companies, or in mixed feeding parties. Breeds September-April. Nest low over water or to 6 m above ground. Also in New Guinea. **Sexes**: *alike.* **Immatures**: *breast band absent or indistinct.* **Voice**: *pretty melodious song with a note of 'swee-wit' or 'swee-swee'.*

21 RUFOUS-THROATED HONEYEATER

Conopophila rufogularis 14 cm

A noisy and conspicuous wanderer in small to large groups in tropical eucalypt forest and woodland following the blossoming trees and shrubs to feed on nectar. It also takes insects from the ground and on the wing. It is never far from water, and often bathes. It is aggressive, and there is much squabbling and fighting within groups, accompanied by display threats of fluffing up neck feathers and raising wings. Breeds mainly September-April. Nests to 7 m above ground. Endemic. **Sexes**: *alike.* **Immatures**: *throat white, belly cream.* **Voice**: *constant chattering notes and sharp 'zit-zit' calls.*

22 GREY HONEYEATER

Conopophila whitei 12 cm

Not discovered until 1903, this small drab honeyeater is found only in the remoteness of the inland areas, and has been observed infrequently. In pairs or small parties, it frequents acacia scrub, mulga and spinifex, sometimes in company with thornbills. It feeds on insects and mistletoe berries. Probably nomadic. Thought to breed August-November or in response to rainfall. Nests close to the ground. Endemic. **Sexes**: *alike.* **Immatures**: *cheeks and throat washed yellow; buff around eye.* **Voice**: *succession of rapid high-pitched sibilant notes.*

23, 24 PIED HONEYEATER

Certhionyx variegatus 18 cm

The habitat of the Pied Honeyeater ranges from spinifex and mallee scrub of the inland to eucalypt woodland nearer the coast. An unobtrusive and quiet bird, it is seen in pairs and small parties, and feeds on blossom, native fruits and occasionally insects. The male performs a territorial flight, rising high while singing, then diving vertically to cover with tail fanned and wings closed. Nomadic in the southern part of the range in autumn. Breeding variable, mainly September-February. Nests in a low bush. Endemic. **Recognition:** *(cf male Black Honeyeater): the Pied Honeyeater is much larger and the male lacks black on its belly.* **Immatures:** *like females.* **Voice:** *whistling 'te-tiee-tee-tee'.*

Male left, female right

13

14

15

16

17

18

19

20

21

22

23

24

25 BANDED HONEYEATER

Cissomela pectoralis 14 cm

This attractive black and white bird is highly nomadic and wanders in pairs and small parties through the wooded and shrubby country of the tropical north following the flowers and blossoms of such trees as eucalypts, paperbarks, melaleucas and grevilleas. Very large flocks can descend on areas where an exceptional effusion of blossom has occurred. Breeds October-April. Flimsy, almost see-through nest, to 3 m high. Endemic. **Sexes**: *similar.* **Immatures**: *crown brown, underparts buff, breast band brown, mantle scalloped golden-brown.* **Voice**: *brisk sharp calls and cheerful song*

26, 27 BLACK HONEYEATER

Certhionyx niger 12 cm

The tiny and spritely Black Honeyeater is highly nomadic wherever flowering shrubs are blossoming, with a special liking for red emu-bush (*Eremophila*). In pairs and flocks, sometimes large, it darts and hovers among the blossoms probing for nectar with its long bill. It also takes insects. Southern birds migrate northwards to winter. During the breeding season the males perform steep song flights. Breeds mainly October-November. Nests low in a bush. Endemic. **Immatures:** *like females; greyer.* **Voice:** *insect-like 'peeeee'; sparrow-like chirping.*

Male left, female right

28 LEWIN'S HONEYEATER

Meliphaga lewinii 21 cm

An inhabitant of rainforest, wet eucalyptus forest, brigalow, heath, gardens and orchards, from the coast to the higher altitudes. Feeds on native fruits and nectar, hanging acrobatically from branches, also spirals up tree trunks and hovers over foliage for insects. It fearlessly drives off competitors, and is inquisitive. Breeds August-January. Nests are suspended 2-6 m above ground. Endemic. **Sexes**: *alike.* Immatures: *not streaked, underparts greenish-yellow.* **Voice**: *far-carrying rapid volleys of notes, sometimes likened to machine-gun fire.*

29 LESSER LEWIN'S [Yellow-spotted] HONEYEATER

Meliphaga notata 18 cm

The range of this species partly overlaps that of the similar looking Lewin's Honeyeater, but is identifiable by its smaller size, rounder yellow ear patch and different calls. It frequents coastal rainforest, mangroves and gardens, and flits around shrubs and middle level foliage foraging for insects, native fruits, berries and nectar. Breeds September-February. Nests smaller than Lewin's, 1-3 m above ground. Endemic. **Sexes**: *alike* **Immatures**: *paler, underparts washed ochre.* **Voice**: *sharp repeated 'queak' and melodious 'chit'.*

30 GRACEFUL HONEYEATER

Meliphaga gracilis 16 cm

The Graceful Honeyeater, also found in New Guinea, is the smallest of the Australian trio of look-alike honeyeaters with yellow facial spots. It lives inconspicuously in coastal forest, mangrove, lantana thickets and gardens, which it shares with the Lesser Lewin's Honeyeater. It is slightly smaller, its bill is longer and thinner and it tends to forage higher in the tree canopy, and is particularly acrobatic. Breeds October-February. Pretty suspended cup nest above 2 m from ground. **Sexes**: *alike.* **Immatures**: *alike.* **Voice**: *utters sharp 'click' sounds continuously.*

31 WHITE-LINED HONEYEATER

Meliphaga albilineata 19 cm

This infrequently seen bird of the Kimberley region and Arnhem Land lives in pockets of monsoon forest in rocky gorges and sandstone escarpments. Singly, in pairs and in loose flocks it eats native fruits, seeds and insects as well as nectar from flowering trees where it forages high in the canopy. When feeding it will drive off smaller birds, and it is inquisitive. Breeds perhaps August-February. Nests 1-5 m above ground. Endemic. **Sexes**: *alike.* **Immatures**: *alike.* **Voice**: *up and down whistles: (Arnhem Land) drawn out; (Kimberley) rapidly repeated.*

32 WHITE-EARED HONEYEATER

Meliphaga leucotis 21 cm

This species becomes very tame during the breeding season and a female will fearlessly land on one's head to pluck a hair or onto a sweater to pull a strand for lining its nest (see illustration). In pairs and small parties, it frequents coastal and alpine wooded habitats. It feeds on manna, nectar and insects prised from bark. The western race *novaenorciae* (named after the famous Benedictine monastery just north of Perth) is smaller and darker. Breeds July-April. Nests below 3 m. Endemic. **Sexes**: *alike.* **Immatures**: *duller, crown green, earpatch buff.* **Voice**: *ringing notes and 'chop, chop'.*

33 YELLOW-THROATED HONEYEATER

Meliphaga flavicollis 21 cm

One of Tasmania's endemic honeyeaters. Its habits are similar to those of the White-eared Honeyeater, including plucking human hair for its nest. It is widespread and abundant, and frequents a wide range of habitat. Solitary and seen in family parties only during breeding season. Takes insects from foliage, probes bark, and hawks for flying insects; also feeds on nectar and manna from foliage. Breeds July-January. Nests in low bush or grass clump, sometimes in foliage to 10 m. Endemic. **Sexes**: *alike.* **Immatures**: *paler.* **Voice**: *various warbling notes, rapid 'tonk, tonk' uttered from topmost twig of low tree.*

34 BRIDLED HONEYEATER

Meliphaga frenata 20 cm

Restricted to the Cooktown, Atherton and Mount Spec areas, this is a bird with a notable yellow and black bill, and mainly lives in the highlands; some descend to lower altitudes in winter. Singly, in pairs or small flocks it frequents high forest and swampy woodland foliage. It is active and aggressive and feeds on insects, nectar and native fruits. Breeds September-January. Nests in cup of stems, ferns and vine tendrils hung in vines or trees up to 8m from ground. Endemic. **Sexes**: *alike.* **Immatures**: *more extensive yellow throat patches.* **Voice**: *clear melodious whistling and harsh rattling notes.*

35 EUNGELLA HONEYEATER

Meliphaga hindwoodi 19 cm

This bird was known for twenty years before it was named as a new species in 1983. In rainforests it follows flowering trees and mistletoe. Movement to lower altitudes have been reported. In non-flowering periods it occurs in singles and pairs only, but when flowers are prolific large numbers assemble. Usually quiet, but noisy when in numbers at food source. Breeds possibly August-December. Only nest recorded was cup-shaped in a low bush 2.5 m high. Endemic. **Sexes**: *alike. Females slightly smaller.* **Immatures**: *crown has rufous coloration* **Voice**: *very loud calls, or single note.*

36 SINGING HONEYEATER

Meliphaga virescens 19 cm

Loud duet singing by pairs from prominent perches may account for the name 'singing', as several other honeyeaters have more tuneful voices. This, the most widespread of honeyeaters, lives in pairs and small flocks in open habitat ranging from arid scrub to coastal woodland. It feeds on nectar, native fruits and seeds. Northern birds are smaller and paler and are less streaked on breast. Breeds July-February or depending on rainfall. Nests low to 5 m above ground. Endemic. **Female**: *smaller* **Immatures**: *face band paler.* **Voice**: *resonant trilling notes.*

25

26

27

28

29

30

31

32

33

34

35

36

13

37 VARIED HONEYEATER

Meliphaga versicolor 20 cm

Its bold varied notes are a familiar sound in mangroves and trees lining the streets and waterfronts of Cairns and other coastal towns of north Queensland, and on the islands of the Barrier Reef. Quarrelsome, inquisitive and active, in pairs and parties it flits among leaves and blossoms in search of nectar and insects. Breeds April-December. Nests in mangroves or beachside trees 2-5 m above ground. Also in New Guinea. **Sexes**: *alike.* **Immatures**: *back browner, not streaked.* **Voice**: *varied resonant and melodious notes.*

38 MANGROVE HONEYEATER

Meliphaga fasciogularis 20 cm

Sometimes treated as a race of the Varied Honeyeater and commonly mistaken for it, this mangrove-dweller has scalloping rather than streaking on the breast. They intergrade between Bowen and Cardwell. A flocking bird of coast and nearby islands, it also visits woodland and gardens near the mangroves to seek out flowering trees. Breeds August-December, also April-May in the north. Usually nests in mangrove branches low over water. Endemic. **Sexes**: *alike.* **Immatures**: *upperparts brown, underparts paler.* **Voice**: *varied resonant and melodious notes.*

39 PURPLE-GAPED HONEYEATER

Meliphaga cratitia 16 cm

Distinguished by a thin streak of purple or lilac skin (or wattle) extending from mouth to ear. The widely separate populations are similar, but Kangaroo Island birds have yellow facial streaks. It is shy and usually solitary. In mallee and heath, open woodland in WA, sometimes in gardens, it follows the blossom for nectar and may congregate around flowering trees in large numbers. It also eats insects from foliage, bark and by hawking. Aggressively territorial. Breeds August-December. Nests close to the ground. Endemic. **Sexes**: *alike.* **Immatures**: *gape yellow.* **Voice**: *harsh chattering notes.*

40 GREY-HEADED HONEYEATER

Meliphaga keartlandi 16 cm

This honeyeater lives in Australia's arid areas, and frequents low eucalypts, mulga scrub, wooded hillsides, stony gorges and desert dunes. In pairs and small flocks it is constantly on the move visiting florescent eucalypts and wattles for nectar, sometimes in competition with the White-plumed Honeyeater, which it resembles somewhat in behaviour and voice. It also eats insects and the honeydew and manna exuded from insect-damaged foliage. Breeds July-November. Nests in low shrub. Endemic. **Sexes**: *alike.* **Immatures**: *duller.* **Voice**: *petulant 'chip-chip'; cheerful babbling 'chee-toyt'.*

41 YELLOW HONEYEATER

Meliphaga flava 19 cm

The attractive Yellow Honeyeater frequents open forest, riverside vegetation and mangroves. It is docile, friendly and inquisitive and is a regular visitor to gardens and orchards and is one of the most frequently seen honeyeaters in Townsville. It is jerkily active in search of nectar and insects, usually in pairs, sometimes in company with other honeyeaters. It is sedentary or locally nomadic. Breeds October-March. Nests in bark cup bound with cobweb, up to 19 m above ground. Endemic. **Sexes**: *alike.* **Immatures**: *duller, bill paler.* **Voice**: *an agreeable and cheerful whistles and notes, with a brisk 't-wheet'.*

42 GREY-FRONTED HONEYEATER

Meliphaga plumula 16 cm

Also called Yellow-fronted Honeyeater, it ranges extensively through all but the most arid desert areas of the centre, and lives in dense mallee, mulga scrub, eucalypt woodland and spinifex in some areas. Singly and in pairs it feeds among foliage and blossoms for nectar. Large flocks sometimes gather to feed when eucalypt blossoms are profuse. Breeds August-January. Nests in small trees within a metre of the ground. Endemic. **Sexes**: *alike.* **Immatures**: *duller.* **Voice**: *clear penetrating 'it-wirt, wirt, wirt', and single sharp notes; sometimes canary-like notes.*

43 YELLOW-TINTED (Pale Yellow) HONEYEATER

Meliphaga flavescens 17 cm

Complements the range of the White-plumed Honeyeater in northern Australia and is common in open forest and scrublands along watercourses. It feeds in foliage and blossom on insects and nectar and gathers in quite large flocks at sources of food, water and bathing places. Breeds July-November. Nests suspended 2-12 m high, with normally one egg. Also in New Guinea. **Sexes**: *alike.* **Immatures**: *bill brownish-yellow.* **Voice**: *in part rather like White-plumed Honeyeater but slower and higher pitch; also 'chy-uck-oo-wee' variable in pitch.*

44 YELLOW-PLUMED HONEYEATER

Meliphaga ornata 16 cm

A honeyeater of the wetter mallee in southern Australia and tall eucalypt woodland in the west of its range. It is similar in habits to the Grey-fronted Honeyeater and feeds on insects and nectar in pairs and small groups. It is aggressive and quarrelsome when feeding in blossoming trees, with loud warning cries. The male rises in steep display and song flights during the breeding season. Breeds August-December. Nests suspended from low drooping branch. Endemic. **Sexes**: *alike.* **Immatures**: *duller.* **Voice**: *calls well before sunrise. Loud animated 'chick-o-wee'; harsh sharp flight and alarm calls.*

45 WHITE-PLUMED HONEYEATER

Meliphaga pencillata 17 cm

Also known as Greenie, it is one of the most widespread of the honeyeaters. Plumage varies from greenish in the south-east to pale yellow in Central Australia. It frequents large eucalypts along watercourses, close and open woodland, and is familiar in parks and gardens. Singly, in pairs or flocks, it feeds in trees on insects, nectar and particularly manna. Fearless and aggressive. Breeds June-January. Nests low to 15 m, often host to cuckoos. Endemic. **Sexes**: *alike.* **Immatures**: *duller, brownish-yellow bill.* **Voice**: *musical 'chick-o-wee' and constant 'chip' chattering.*

46 FUSCOUS HONEYEATER

Meliphaga fusca 15 cm

Common in open forests, this noisy and gregarious honeyeater, usually in flocks of 6-12, feeds on nectar and insects in the foliage, and is frequently quarrelsome. It will visit feeding tables and gardens and can become tame. Sedentary in the north, nomadic in the south. Breeds July-March. Nests to 20 m high. Endemic. Breeding pairs have black eye-ring and black bills; out of breeding season eye-ring and base of bill is yellow. Plumage is yellower in the north. **Sexes**: *alike.* **Immatures**: *bill brown, yellowish at base.* **Voice**: *a bright 'kitty-lin-tofftoff-toff' and calls like Yellow-tinted Honeyeater.*

47, 48 YELLOW-TUFTED HONEYEATER

Meliphaga melanops 19-21 cm

One of the most glamorous honeyeaters, it lives in loose colonies of 10-100 birds in eucalypt woodland and open forest in south-east Australia, usually along creeks. There are two races: *melanops* the smaller and paler and shorter-tailed which straddles both slopes of the Great Dividing Range; and *cassidix* (the Helmeted Honeyeater) which is distinguished by the raised yellow helmet on its crown, and is restricted to a tiny area near Yellingbo in Victoria. The species eats nectar, insects, honeydew, and manna. Insects are also taken from leaves, from bark and on the wing. It raises the tufts at the sides of its head when disturbed. Birds of the high country tend to migrate northwards and to lower altitudes in winter. The population of the Helmeted Honeyeater, which is the faunal emblem of Victoria and one of Australia's rarest birds, was once found over a large area east of Melbourne. It has declined to only 100 to 150 birds within a few hectares in the Yellingbo State Wildlife Reserve. A sustained effort is being made by the Bird Observers' Club and the Victorian Fisheries and Wildlife Division to save it from extinction. Breeds June-March. Nests below 4 m. Endemic. **Sexes:** *alike.* **Immatures:** *the yellow patches are greener.* **Voice:** *very varied repertoire ranging from sweet whistling to harsh nasal notes.*

Yellow-tufted Honeyeater left, Helmeted Honeyeater right

37

38

39

40

41

42

43

44

45

46

47

48

13

49 MACLEAY'S HONEYEATER

Xanthotis macleayana 20 cm

Restricted to a small area of rainforest, woodland and mangroves, this honeyeater often frequents orchards and gardens, where its distinctive calls are familiar. It is deliberate in its movements and, usually alone or in pairs, it forages mostly in the foliage in the middle to upper layers beneath the canopy. It eats mainly insects and spiders, with some nectar and occasionally native and cultivated fruits, and is attracted to the blossom of Umbrella trees. Breeds October-December. Nests above 2 m from ground. Endemic. **Sexes**: *alike*. **Immatures**: *duller*. **Voice**: *some calls like Yellow-faced Honeyeater, and a five-note 'to-wit-too-weee-twit'.*

50 TAWNY-BREASTED HONEYEATER

Xanthotis flaviventer 20 cm

A race of species widespread in New Guinea, and a bird of tropical rainforest and riverside galleries of vine scrub. Singly and in pairs, it forages deliberately and unobtrusively in the middle to upper strata of foliage. It eats insects — often extracted from dead clumps of leaves and from crevices of trunks — and to a lesser extent fruit and nectar. Breeds November-February. Nests in a bark cup-shaped nest from 4 m above ground. **Sexes**: *alike*. **Immatures**: *duller above, paler below*. **Voice**: *rich whistling, and a call of 'you-go-bang' with emphasis on either first or last syllable.*

51 WHITE-GAPED HONEYEATER

Meliphaga unicolor 20 cm

Common across tropical Australia in riverine thickets, mangroves, swampy woodland, parks and gardens. It is noisy and aggressive, and hops along branches briskly, often with tail partly cocked (which is unique among honeyeaters), under the canopy or low in shrubs foraging for insects and spiders, sometimes even seeking them in houses. It also eats nectar, fruit and seeds. Breeds September-May. Nests in tree fork over water. Endemic. **Sexes**: *alike*. **Immatures**: *yellow wash on breast, gape yellow*. **Voice**: *a great variety of pipings, sparrow-like chirping, and notes of 'whit-whit' or 'whit-o-wee'.*

52 WHITE-FRONTED HONEYEATER

Phylidonyris albifrons 18 cm

Highly nomadic, with its movements largely governed by the flowering of eucalypts, mistletoe, grevilleas, hakeas, and eremophilas. It is usually in pairs and small parties, but in large flocks during effusions of blossom. It darts briskly among the foliage and flies in swift undulations from one bush to another. Breeds August-February, or in response to favourable conditions. Nests low on a branch or in the fold of bark against the trunk. Endemic. **Female**: *browner*. **Immatures**: *upperparts brown, throat whitish*. **Voice**: *spirited melodious song including 'quak-peter-peet' phrases and various metallic calls.*

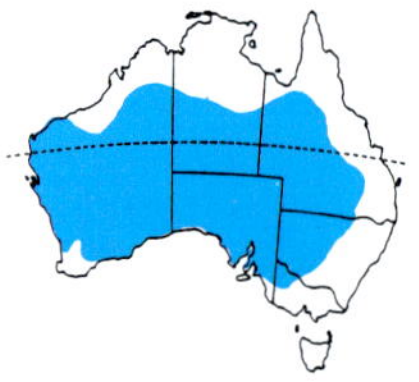

53 TAWNY-CROWNED HONEYEATER

Phylidonyris melanops 16 cm

The attractive Tawny-crowned Honeyeater lives in coastal heath, mallee scrub and sometimes in open forest country on the coast. It sips nectar from low flowering shrubs, often foraging from the ground. It also takes flying insects, both by hawking over the vegetation and by sallying forth from a vantage point. It performs a spectacular high song flight during the breeding season. Breeds July-March. Nests on the ground or to 3 m. Endemic. **Sexes**: *alike*. **Immatures**: *no tawny crown; yellow throat patch*. **Voice**: *gentle airy, flute-like song with a ventriloquial quality.*

54 YELLOW-FACED HONEYEATER

Meliphaga chrysops 17 cm

Common in east and south-eastern Australia and familiarly known as Chick-up from its cheerful call. Singly, in pairs and flocks, it frequents forest, woodland and mangrove, and is familiar in surburban gardens. It feeds on nectar in spring and manna in summer and autumn, also insects and honeydew. In autumn huge flocks set off daily from the south to winter in the north, returning to their previous territories in spring. Breeds July-March, especially October-January. Endemic. **Sexes**: *alike*. **Immatures**: *duller, rump coloured rufous*. **Voice**: *'chick-up' or 'quitch-up' amongst other whistles.*

55 CRESCENT HONEYEATER

Phylidonyris pyrrhoptera 16 cm

Most abundant in Tasmania where it is known as the Tasmanian Honeyeater. It frequents moist forest, dense heath and visits surburban gardens and is seen alone, in pairs or in loose companies. It feeds on nectar, honeydew and insects taken on the ground or from tree bark. In winter it migrates from highlands to low and coastal country. Breeds July-January or March-April. Nests to 2 m from ground in bush or fern. Endemic. **Female**: *duller*. **Immatures**: *no crescent*. **Voice**: *calls of 'egypt-egypt' amongst other notes.*

Male

56 NEW HOLLAND (White-bearded) HONEYEATER

Phylidonyris novaehollandiae 18 cm

Also known as the White-eyed Honeyeater. The white beard, black cheek or white eye instantly differentiates it from the similar looking but dark-eyed White-cheeked Honeyeater. It is common on coastal heath, closely shrubbed woodland and gardens. In pairs and small flocks, it feeds principally on nectar from some 100 species of shrubs, but especially banksias in NSW. Breeds March-May and August-October. Nests to 3 m from ground. Endemic. **Female**: *smaller*. **Immatures**: *streaking brown and grey rather than black and white*. **Voice**: *chattering and single high notes.*

57 WHITE-CHEEKED HONEYEATER

Phylidonyris nigra 18 cm

The large white cheek patch, black eye, and different voice distinguish this handsome honeyeater from the look-alike New Holland. It inhabits heath lands and open timber and forages in pairs, loose colonies and flocks on flowering shrubs for nectar and insects. Breeds March-May, or August-October. Nests in low vegetation. Endemic. **Female**: *smaller*. **Immatures**: *browner and greyer, lores yellowish*. **Voice**: *brisk 'chippity-chip', often in concert.*

58 BLUE-FACED HONEYEATER

Entomyzon cyanotis 32 cm

A conspicuous facial patch, shading from vivid blue to pale turquoise characterises this large honeyeater. It lives in a variety of habitat throughout its exclusive range including fringes of rainforest and open woodland; it also occurs in New Guinea. Insects are its main diet, also nectar from blossoms, and native fruits. It also raids orchards and takes syrup from burnt sugar canes. Breeds June-January. Often uses discarded nests of other species such as babblers. **Sexes**: *alike*. **Immatures**: *eye skin green-yellow*. **Voice**: *loud monotonous call of 'queet', alarm notes like the Noisy Miner.*

59 REGENT HONEYEATER

Xanthomyza phrygia 22 cm

This strikingly-plumaged black and creamy-yellow honeyeater is a nomad and may suddenly turn up anywhere in its habitat of woodland and open forest when there are flushes of blossom of flowering trees such as eucalypts and banksias, often in large feeding flocks. It feeds aggressively and quickly in the outer foliage and it also takes insects and native fruits. It may not return to a locality for many years. Its range and numbers are declining. Breeds August-January. Nests in tree-fork, up to 10 m. Endemic. **Sexes**: *similar*. **Immatures**: *black parts grey or brownish*. **Voice**: *a rich bell-like note and clicking sounds.*

60 SPINY-CHEEKED HONEYEATER

Acanthagenys rufogularis 26 cm

An attractive large endemic honeyeater characterised by its distinctive malar feathers. It is common in most areas of coastal woodland and across the open scrubby country of the arid and semi-arid inland. It feeds variously on insects both gleaned from foliage and flying, nectar from flowering trees, manna, native and grapes from vineyards. It is locally nomadic and flocks up to 100 can gather when there are flushes of blossom. Breeds variably, mainly June-January. Deep pendant nest up to 9 m above ground. **Sexes**: *alike*. **Immatures**: *duller; yellow cheek spines*. **Voice**: *liquid bubbling trills.*

49

50

51

52

53

54

55

56

57

58

59

60

13

61 BELL MINER
Manorina melanophrys 18 cm

Popularly known as the Bellbird this pugnacious species monopolises pockets of woodland from which it drives out other small birds, and often stays for many years. Its famous sharp 'pings' ring out all day and can sometimes be heard alongside a busy road. It eats mainly sap-sucking scale insects and their sugary lerps (scales). Breeds July-February and nests in trees normally 4-5 m up. Endemic. **Female**: *smaller.* **Immatures**: *eye patch greenish-yellow.* **Voice**: '*... sweeter than singing the notes of the bellbirds are running and ringing.*' (Henry Kendall).

62 NOISY MINER
Manorina melanocephala 28 cm

The conspicuous and noisy Micky is among the best known of the honeyeaters. It lives in open woodland in loose colonies of perhaps several hundred birds which erupt into an uproar of alarm calls when danger threatens, such as from a hawk overhead or the approach of a goanna, and mass mobbing follows. Its other name 'Soldierbird' is well suited. It feeds on insects, their honeydew and nectar. Breeds mainly June-December. Colonies of cup nests up to 20 m from the ground. Endemic. **Sexes**: *alike.* **Immatures**: *washed brown on back and wings.* **Voice**: *raucous alarm calls and rich musical notes.*

63 YELLOW-THROATED MINER
Manorina flavigula 26 cm

Race: *melanotis* (Black-eared Miner)

In a very restricted habitat of mallee and riverine woodland where three southern state boundaries meet, this race — which is distinguished by a large black ear patch and greyish rump — is being hybridised out of existence through interbreeding with the nominate Yellow-throated Miner and extensive clearance of its habitat. It is shy and little known. Breeds August-January. Nests in communities of several breeding groups with females outnumbering males. Endemic. **Sexes**: *alike.* **Immatures**: *alike.* **Voice**: *petulant and grating, rather like the Noisy Miner.*

64 YELLOW-THROATED (White-rumped) MINER
Manorina flavigula 28 cm

Common in patches in woodland, mallee and arid scrublands. A dark race *obscura*, the Dusky Honeyeater, is confined to the south-west, and race *melanotis*, the Black-eared Honeyeater, to a small area in the south-east (see 63). Highly nomadic, but in other respects its habits including feeding are similar to the Noisy Miner. The yellow throat is faint but the white rump is prominent. Breeds July-December. Nests in colonies. Endemic. **Female**: *smaller.* **Immatures**: *upperparts browner, wing coverts edged buff.* **Voice**: *variety of strident calls, rather similar to the Noisy Miner.*

65 LITTLE (Western) WATTLEBIRD
Anthochaera lunulata 32 cm

Compared with the Brush Wattlebird in eastern Australia, the crown is plain, the whitish feathers at the sides of the throat are more prominent, the bill is more slender and the eye is red. Often in loose companies, it wanders in search of blossoming trees and shrubs in woodland scrub and heath. In summer, it is partly migratory from inland to coast. Breeds July-October. Nest is often in a banksia. Only one egg is laid (the eastern relative lays two). Endemic. **Sexes**: *similar, females sometimes smaller.* **Immatures**: *browner.* **Voice**: *cackles and chuckles, often by pairs in duet.*

66 BRUSH (Little) WATTLEBIRD
Anthochaera chrysoptera 31 cm

The smallest of the wattlebirds, and without visible wattles. Birds in Tasmania are darker and larger. Solitary or in parties, it frequents woodland, heath, parks and gardens and feeds mostly on the nectar of flowering banksias, grevilleas and eucalypts. It also eats insects, manna and honeydew. It is active, continuously vocal, and aggressively drives off other honeyeaters. Breeds mainly July-December. Nests in trees up to 15 m above ground. Endemic. **Sexes**: *similar, females sometimes smaller.* **Immatures**: *browner.* **Voice**: *includes repeated braying phrases 'ke-kay-o', and one-syllable squawks such as 'yoiks'.*

67 RED WATTLEBIRD
Anthochaera carunculata 35 cm

Also called the Gillbird because of its red wattles which lengthen and deepen in colour with age. It is the largest of the mainland honeyeaters and is as well known in southern city gardens as in forest and woodland. It is nomadic, sometimes in loose flocks up to 100, following the blossoms for their nectar; also eats fruit, berries, insects, manna and honeydew. Breeds mainly July-December. Some migration from highlands towards the coast in winter. Nests 2-20 m above ground. Endemic. **Sexes**: *alike.* **Immatures**: *no wattles.* **Voice**: *loud harsh notes including characteristic 'tchock', sometimes in duet.*

68 YELLOW WATTLEBIRD
Anthochaera paradoxa 45 cm

Confined to Tasmania and King Island, it is the world's largest honeyeater and is characterised by dangling yellow-orange wattles. It lives in all but the rainforests of the west. It is strongly nomadic following the seasonal flowering flushes of blossom for its nectar. Also feeds on spiders and insects, particularly those attacking eucalypt foliage, and on cidergum sap. It migrates from highland to coast in winter. Breeds July-January. Nests to 30 m above ground. Endemic. **Sexes**: *alike.* **Immatures**: *no wattles.* **Voice**: *extraordinary discordant gurgling and regurgitation noises.*

69 LITTLE FRIARBIRD
Philemon citreogularis 29 cm

The smallest of the Australian friarbirds and the only one without a bill knob. The yellow throat to which the specific name *citreogularis* refers applies only to immature birds. In pairs and small flocks and in mixed parties, it frequents open woodland, orchards and vineyards to feed on nectar, insects and fruit. It also feeds on burnt-off sugarcane. Southern birds migrate northwards to winter. Breeds August-February. Nests 2-11 m above ground. Also lives in New Guinea and adjacent islands. **Sexes**: *alike.* **Immatures**: *yellow throat, back scalloped with white.* **Voice**: *loud monotonous 'ar-coo'.*

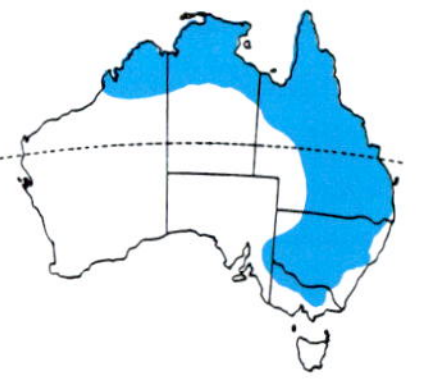

70 SILVER-CROWNED FRIARBIRD
Philemon argenticeps 31 cm

Like the Helmeted Friarbird, but its bill-knob is more prominent and the plumage on head and neck is silvery white. It lives in tropical forest and woodland near water, and shares the nomadic habits of other friarbirds, often in mixed flocks, following the blossoms of eucalypts and paperbarks, to feed on nectar and the insects it attracts. Breeds September—March. Solitary bowl nest suspended 2—11 m above ground, often host to Koel. Endemic. **Sexes:** *alike.* **Immatures:** *no bill-knob; white markings on back.* **Voice:** *a variety of cacophonous calls, especially prominent in the dawn chorus.*

71 HELMETED FRIARBIRD
Philemon buceroides 36 cm

The largest of the friarbirds. It often congregates in feeding parties of 10 to 30 with exceptionally noisy squawking and flapping in woodland, mangrove or gardens around flowering trees and shrubs — probing blossom for nectar. It also hawks for insects and eats more fruit than other friarbirds. Breeds September-April. Nests up to 14 m from ground. Also in New Guinea and Indonesia. **Sexes**: *similar.* **Immatures**: *some lack knob in bill; wings washed greenish-yellow.* **Voice**: *constantly repeated call of 'poor devil', 'poor devil' and a medley of cacophonous clankings.*

72 NOISY FRIARBIRD (Leatherhead)
Philemon corniculatus 36 cm

This raucous and pugnacious bird is the barest headed of all the friarbirds. Singly, in pairs or flocks, it frequents eucalypt forest and open woodlands, and feeds on nectar, insects, native and cultivated fruits. Southern birds migrate northwards in winter, and to lower altitudes from the high country. It breeds August-January and nests in trees up to 15 m above ground. The northern birds are smaller. Also in New Guinea. **Sexes**: *alike.* **Immatures**: *no bill-knob; white scallops on back and yellow on throat.* **Voice**: *main call of 'four-o-clock' amidst an extraordinary jumble of sounds.*

61

62

63

64

65

66

67

68

69

70

71

72

14

Finches and Sparrows, Other Small Brightly-coloured Birds

Opposite: Gouldian Finches

In every corner of the globe where grass and other seed-bearing plants grow, there will be found small to medium-sized birds with conical bills known collectively as finches and sparrows. There are over 900 species and the problem of sorting this vast complex of birds into natural groups that reflect their ancestry and relationships one with the other has long perplexed ornithologists. As at present conceived, the species fall into three main families, but the divisions within them as adopted below are not universally agreed upon.

TRUE FINCHES Family Fringillidae

This very large family comprising about two-thirds of the world's finches is confined entirely to the northern hemisphere, except for species introduced elsewhere. It divides into three subfamilies:

Fringillinae This nominate subfamily of only three species is typified by the Chaffinch, one of the most abundant and familiar birds in Europe, and widespread in eastern Asia. It is not represented in Australia.

Carduelinae The Goldfinch and Greenfinch, both successfully introduced into Australia, typify this subfamily of about 120 species distributed almost equally between the Old World and the New World. The group is mostly arboreal, but is partial to thistle seeds; this is reflected in the name of the subfamily which is derived from the Latin *Cardus* for thistle. The Canary, which has been the most familiar of all cage birds since the sixteenth century, is also a cardueline finch. Wild Canaries still survive on the Canary Islands. 25, 26

Emberizinae The largest group of fringillines, of some 550 species worldwide, includes the buntings of the Old World and the American sparrows. They range eastwards including to Japan, but do not reach Australia. Darwin's famous finches of the Galapagos Islands, a small group of fourteen species, are also counted among the emberizids. Although the bills of these finches have evolved specifically for husking and cracking seed, Darwin discovered that one was adept at using a twig or a thorn as a tool to dislodge insects from crevices.

WEAVERS AND WAXBILLS Family Ploceidae

This Old World family comprises about 230 species of small, compact seed-eating birds with strong pointed bills. It ranges from Europe and Africa through southern Asia to Australia and islands of the western Pacific. It divides into two main groups: the weaver finches and the waxbills.

Weaver finches Weaver finches are named after the intricate, beautifully constructed nests that many of them weave, and Africa is their stronghold. The weavers are represented by three subfamilies which comprise three buffalo weavers (Bubalornithinae), nine parasitic widow weavers (Viduinae) and about ninety-three typical weavers (Ploceinae). No representatives of this group occur naturally in Australia, but there have been two unsuccessful introductions of typical weavers: the White-winged Whydah was introduced to Sydney in 1931, and a breeding population of the Red Bishop was established near Adelaide in 1926. The former has not been seen since the early 1970s and the latter was considered extinct in 1976. (The male Red Bishop illustrated was photographed at Taronga Zoo while nest building.) 27

Waxbills The waxbills, which comprise the subfamily Estrildinae of 125 species, get their strange name from the nominate genus *Estrilda* of Africa, some of which have bills the colour of red sealing wax. They build flimsy and untidy nests compared with those of the weavers, and the birds are smaller. Waxbills range over the tropical and subtropical areas of Africa and Asia and are the only group to reach Australia and the islands of the western Pacific. Twenty species are represented in Australia, fourteen of them endemic. These include five distinctive groups: five mannikins (two introduced) of the widespread genus *Lonchura*; four firetails of the genus *Emblema*; one with a red brow of the genus *Aegintha*; two mainly crimson finches of the genus *Neochmia*; five white/black rumped finches of the genus *Poephila*; the solitary Plum-headed Finch of the genus *Aidemosyne*; and the strikingly beautiful parrot-finches — the Gouldian and Blue-faced Finches — of the long-tailed genus *Erythura*. Several of the Australian finches have red bills like the African waxbills. 1-24

OLD WORLD SPARROWS Family Passeridae

There are thirty-seven species of Old World sparrows native to Europe, Asia and Africa. They are typified by the ubiquitous House Sparrow which is perhaps the most widely distributed and familiar small bird in the world. It and the Tree Sparrow were introduced into Australia progressively from 1863, and odd birds occasionally arrive on ships from Asia. Both are prohibited in WA and the Northern Territory, and are shot on sight. 28-30

WHITE-EYES Family Zosteropidae

Of some eighty species of White-eyes spread widely through the tropical areas of the Old World, three occur in Australia where they have traditionally been known as silvereyes. There are many plumage variations. 31-35

SUNBIRDS Family Nectariniidae

The tiny nectar-feeding sunbirds have long thin decurved bills, and are notable for the brilliantly coloured plumage of the male. About half of the 118 species of sunbirds live in Africa; the remainder are spread through tropical Asia to islands of the western Pacific. Only one has reached Australia. 36, 37

FLOWERPECKERS Family Dicaeidae

Of fifty-eight species of this family, which extends from Australia through Melanesia to India, the Mistletoebird is the sole representative in Australia, and is so named because it feeds mainly on mistletoe berries. 40-43

PARDALOTES Family Pardalotidae

The endemic pardalotes take their name from *Pardalotus*, Greek for spotted, and are found over most of the country. There are five species, and these can conveniently be grouped as four with spotted crowns, and one comprising four distinct forms with mostly-striped crowns. Most burrow into soil banks to nest; some in tree hollows. 38, 41, 42, 44-48

PHOTOGRAPHERS: H. & J. Beste 14; G. Chapman 2, 15, 31, 34, 41; V.A. Drake 35; J. Frazier 45; T. & P. Gardner 39; C.L. Gill 3; E. & D. Hosking 30; R. & D. Keller 36, 37, 47; W.J. Labbett 43; J. McCann 8; D. Milledge (NPI) 46; F. Park (ANT) 44; J. Purnell 5, 11, 32; T.A. Waite 1, 25, 26, 48; G. Weber 42; Authors p. 139; 4, 6, 7, 9, 10, 12, 13, 16-24, 27-29, 33, 38, 40

1 BEAUTIFUL FIRETAIL

Emblema bellum 12 cm

Tasmania's only native finch, it is now rare on the mainland and confined to the coastal strip there. It lives in wooded habitats, especially tea-trees near towns, in sedentary pairs or small flocks. It has a strong arrow-like flight, and forages on the ground for seeds, hopping swiftly like a mouse. It is difficult to flush, but when it is it rises with a loud whirr of wings. It also picks small insects from foliage. Breeds September-January. Nests in low bush or tree to 6 m above ground. Endemic. **Female**: *no black on belly.* **Juveniles**: *duller, less distinct markings, blackish bill.* **Voice**: *penetrating sorrowful 'weee'.*

2 RED-EARED FIRETAIL

Emblema oculatum 12 cm

Restricted to south-west Western Australia, its range having contracted from former days. It frequents dense scrub, heavy forest and coastal paperbark swamp. It is solitary, never in flocks, and sedentary. With exceptional dexterity it weaves through low foliage foraging for seeds and insects. Breeds September-January. Builds large bottle-shaped nest with entrance tunnel up to 40 cm long very high in tree up to 15 m. Also builds an untunnelled roosting nest. Endemic. **Sexes**: *similar, but breeding female paler ear patch.* **Juveniles**: *duller, no ear patch, bill black.* **Voice**: *floating 'oowee' call, repeated faint 'quirk', harsh alarm notes.*

3 DIAMOND FIRETAIL

Emblema guttatum 11 cm

Sometimes called the Diamond Sparrow. It prefers to be near watercourses in forest, mallee or sparse woodland, but will range further. It has been considerably displaced by land development. Feeds almost entirely on the ground, hopping rapidly, sometimes jumping up to reach grass seeds. Sedentary, often nesting in the same tree year after year. Usually seen in groups or flocks. Breeds August-January. Nests in bush, mistletoe or trees to 25 m, occasionally under the roof of buildings. Endemic. **Female**: *smaller, narrow breast band.* **Juveniles**: *browner, breast band indistinct or absent.* **Voice**: *call of 'twoo-hee', low rasping song.*

4 PAINTED FIRETAIL

Emblema pictum 10 cm

A finch of the arid centre and north-west with a patchy distribution. It lives among spinifex and low scrub in rocky gullies and hills near water, and sometimes appears in station gardens. It is seen in pairs and small groups, rarely in large flocks and forages on the ground for fallen seeds. Breeding is related to rainfall. Nests in dense clump of grass, in ground depression, or on platform of earth and stones. Endemic. **Juveniles**: *much duller, no red on rump.* **Voice**: *loud harsh call 'trut'; song 'che-che-che-weree-ooee'.*

Female left, male right

5 RED-BROWED FINCH

Aegintha temporalis 12 cm

This finch has a red rump like the firetails. On Cape York, birds have a brighter olive-yellow back, and white throat and abdomen; in South Australia it is browner. The Western Australian population is descended from aviary escapees. Common in wooded habitats and near habitation, it is trusting and confiding. Sedentary or nomadic; large flocks form out of breeding season. Breeds September-November (south), January-April (north). Nests in thorny bushes or shrubs. Endemic. **Sexes**: *similar.* **Juveniles**: *browner, bill blackish, no red eyebrow.* **Voice**: *incessantly-repeated call on the move of 'ssitt-ssitt' or 'ssee-ssee-ssee'.*

6 STAR FINCH

Neochmia ruficauda 12 cm

The Star Finch lives in tall rank grass around water interspersed with bushes or low trees usually in small flocks, but these can sometimes be quite large. Plumage colour varies considerably; the northern populations are brighter and the intensity of red and the size of the white spots changes with age. It eats grass seeds, including spinifex, often taken from the plant head; also insects. Breeds December-August. Nests to 3 m in a bush or tree. Endemic. **Juveniles**: *dull olive-brown above; grey crown, cheeks and breast.* **Voice**: *high-pitched contact call 'tlit' or 'psilt' in flock; single contact call of 'sseet'.*

Male left, female right

7, 8 CRIMSON FINCH

Neochmia phaeton 14 cm

The range of the Crimson Finch spans New Guinea and Australia. There are three isolated Australian populations and two forms: black-bellied in the northern and eastern populations, and white-bellied on Cape York Peninsula. In pairs and small family parties, it lives in the long grasses of eucalypt woodland, sometimes in pineapple groves and the margins of canefields — always near water. It feeds on the seeds of pandanus, grasses and insects. It flits somewhat like a butterfly between grass clumps and constantly flicks its long tail upwards. It has adapted to human settlement, appearing at homesteads or wherever water is available in wells and troughs. It is quarrelsome and will sometimes drive away bigger birds from its nest. Breeds September-May. Nests in a low bush, buildings or even rooms. **Juveniles:** *(black-bellied form) dull brown upper-parts, paler below, bills black.* **Voice:** *call — loud penetrating 'chee-chee'; song — soft rasping sounds ending with three melodious notes.*

Black-bellied form
Male above, female below

(The photograph opposite is included for its rarity)
White-bellied form
Male above, female below

9 ZEBRA FINCH

Poephila guttata 10 cm

Popular with bird fanciers all over the world, it is the most widely-distributed and common finch in Australia and lives in a wide variety of open country habitats, always near water. Sedentary, sometimes in very large flocks. It eats grass seeds and insects. Breeds variably in colonies. Nests in bush, ground hollow or building. Also in Timor. The most popular cage and aviary bird of all Australian finches, and many colour mutations have been bred. **Juveniles**: *like females; bill black.* **Voice**: *calls of 'tiah' like a toy trumpet; aggressive 'wsst', elusive song with low ventriloquial trills.*

Male left, female right

10, 11 DOUBLE-BARRED FINCH

Poephila bichenovii 11 cm

Double-barred Finches are black-rumped in the west and white-rumped in the east, and intergrade in a broad area south of the Gulf of Carpentaria. This owl-faced finch lives in sizeable parties, and frequents woodland with grassy understorey, low scrub, and is often seen on the margins of canefields, near homesteads, and in parks and gardens where it is relatively tame. It is a frequent visitor to cleared areas and is believed to have benefitted from human settlement. Its habitat is always near water, where it gathers in large numbers. It drinks by sucking. It spends much time on the ground picking up seeds or jumping up to reach them from seeding grasses. Highly sociable in habits, it indulges in mutual preening and roosts communally. Northern birds breed from June—November, southern birds from January—March. Nests in bush or low tree, often thorny bushes like *Bursaria*. Endemic. **Females:** *may have narrow barring.* **Juveniles:** *duller; barring indistinct; upperparts greyish-olive.* **Voice:** *soft kitten-like mews and 'tiaat-tiaat' calls.*

Black-rumped form left, white-rumped form right

12 PLUM-HEADED FINCH

Aidemosyne modesta 11 cm

In pairs or small to very large flocks, this finch is highly nomadic, and appears to be migratory in the southern parts of its range. In the north it has become rare. It frequents rank grass and reed vegetation and feeds on seeds on or near the ground. It often settles in gardens and is remarkably tame. Breeds August-March (north), September-January (south). Nests within 1 m of ground. Endemic. **Female**: *less claret-red on forehead; chin and throat whitish, less barring on underparts.* **Juveniles**: *no red patches, generally olive brown.* **Voice**: *noisy when in flocks; call of 'tlip' almost inaudible song of chirps and trills.*

Male

1

2

3

4

5

6

7

8

9

10

11

12

14

13, 14 MASKED FINCH

Poephila personata 12 cm

The Masked Finch occurs in two isolated populations: race *personata* in northern Australia, and race *leucotis* (the White-eared Finch) on Cape York Peninsula. Masked Finches are amongst the most sociable of Australian grass-finches. They form strong pair bonds which persist even in the non-breeding season when they flock. Members of breeding colonies hold daily social gatherings at places outside the nesting area and indulge in a ritual of greeting with much chattering and mutual preening. The species frequents lightly timbered grasslands, and feeds on seeding grasses, termites, flies and foliage insects. It spends much time on the ground, and is never far from water. Unlike other grass-finches, it drinks only morning and evening, which enables it to forage for food far afield during the day. Flocks of hundreds may gather at water, sometimes in much greater numbers in dry conditions. It is particularly adept at drinking by sucking and can do so while hanging downwards from a branch to reach water or from the edge of a water trough. Breeds March-June. Nests in bush, tree or grass clump. **Sexes**: *alike, but female smaller.* **Juveniles**: *duller, no black mask, bill blackish.* **Voice**: *calls of 'twaat-twaat'; hissing aggressive call; loud song.*

Race *personata* left, race *leucotis* right (White-eared Finch)

15 LONG-TAILED FINCH

Poephila acuticauda 15 cm

In pairs and flocks of 30 or more, often with Masked Finches, this finch frequents savannah usually near watercourses. Flocks may be 1,000 or more during drought. It eats seeds and ground and catches flying insects. Bobs its head on landing. The bill changes colour from orange in the west to red-dish in the east. Pair bonding is very strong. Breeds January-May. Nests in loose colonies in grass or low in trees. Endemic. **Female**: *throat patch much smaller.* **Juveniles**: *duller; bill and legs black.* **Voice**: *low 'tet', far-carrying floating 'weet' or 'tee-wheet'; plaintive flute-like song.*

16, 17 BLACK-THROATED (Parson) FINCH

Poephila cincta 10 cm

The highly social Black-throated Finch is black-rumped in the north of its range (A) and white-rumped to the south (B). The northern population is common but the southern has declined greatly and is possibly extinct in southern-most parts including NSW. There is a broad band of integration where the two forms meet. Between Cairns and Normanton, the birds are rich fawn on the back and breast and paler grey on the head, and are known as Diggles Finches. The Black-throated Finch is the eastern counterpart of the closely-related Long-tailed Finch. The habits are similar including head bobbing and drinking by sucking, and the latter's long tail is the main difference between them. It frequents open timber with grassy understorey and pandanus clumps near the coast, especially in the vicinity of water. It is shy of settled areas. It eats seeds, often pulling the seed head to the ground. Breeds (north) after the monsoon from January to May, (south) August-January. Nests high in eucalypts or in pandanus palms near the coast. Endemic. **Females:** *smaller throat patch.* **Juveniles:** *paler and duller.* **Voice:** *similar to Long-tailed Finch, but hoarser calls and more flute-like song.*

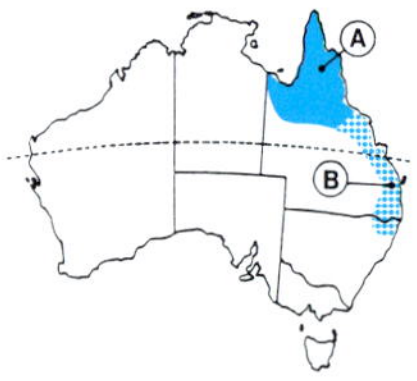

White-rumped form left, black-rumped form right

18 BLUE-FACED FINCH

Erythrura trichroa 12 cm

This brilliant bird, one of the so-called parrot-finches, is widespread in New Guinea, the Pacific Islands and Indonesia. Believed to be a relatively recent arrival in Australia, it is confined to coastal rainforest to 1000 m of Cape York Peninsula where the first breeding record was obtained only in 1944. Singly or in small groups, it forages in thick veg-etation for seeds and insects. Breeding recorded in March and November. **Female**: *blue face patch duller and smaller.* **Juven-iles**: *no blue face patch, dull green plumage.* **Voice**: *high-pitched call note; trilling followed by a whistle.*

19 YELLOW-RUMPED MANNIKIN

Lonchura flaviprymna 10 cm

The Yellow-rumped Mannikin is similar in many respects to the Chestnut-breasted Mannikin and the two feed and roost together, and nest in the same colonies, sometimes interbreeding. It is a ground feeder on seeds in grassland with scattered trees and shrubs near water. It seems to move southwards and away from the coast during the wet season. Breeds January-March. Nests in bush, sugar cane, bamboo, maize stalks, or green grass. Endemic. **Sexes**: *similar.* **Juveniles**: *back and throat dark brown, breast pale grey, underparts grey-buff.* **Voice**: *like Chestnut-breasted Mannikin.*

20, 21 GOULDIAN FINCH

Erythura gouldiae 14 cm

This brilliant bird is named after John Gould's wife Elizabeth. About three-quarters of Gouldian Finches in a given population have black heads, a quarter red heads and, on rare occasions, yellow-ochre heads. The Gouldian Finch lives in open timbered grassland always near water, and in watercourse vegetation. Gregarious and highly social in habits, it is seen in small parties or flocks, which become quite large during the non-breeding season. It has only one partner, probably for life. It climbs grasses and reeds to reach the seed heads, and rarely feeds on the ground, except after fires. It also eats insects especially during the breeding season. Once widespread across the north of the continent, it has totally disappeared from some areas, particularly on Cape York Peninsula and country south of the Gulf of Carpentaria. Its numbers have also fallen alarmingly elsewhere; where flocks of a thousand or more used to be common at waterholes, fewer than thirty are now seen. Excessive trapping and over burning of grasslands are blamed for much of the decline. Breeds December-April. Nests uniquely in tree hollows or termite mounds. Endemic. **Female:** *almost as colourful in some birds, duller in others.* **Juveniles:** *ash-grey; underparts ashy-brown.* **Voice:** *generally rather silent; calls of 'ssitt'. Song has high-pitched, almost inaudible whistling, hissing and 'cluck'.*

Red-headed left, black-headed right (female left, male right)

22 CHESTNUT-BREASTED MANNIKIN

Lonchura castaneothorax 10 cm

In Australia, this mannikin, widespread in New Guinea, lives in two populations. Birds of the north-west are nomadic; those of the east coast are sedentary. Usually seen in tightly-knit parties and breeding colonies, or very large flocks, together with Yellow-rumped Mannikins. It frequents swamps, rice fields, crops etc., and climbs seeding heads to feed. Also catches flying termites. Breeds most months of the year. Nests in grass and crop stalks or bushes within 1 m of ground. **Female**: *similar.* **Juveniles**: *dark-olive above, underparts whitish-buff.* **Voice**: *tink-ling, bell-like 'teet'.*

23 PICTORELLA MANNIKIN

Lonchura pectoralis 11 cm

Often seen in flocks flying over grassland. It is very vocal when gathering at water, drinking by taking rapid sips, then flying off. In the breeding season it is seen in groups of a dozen or less. A bird of open grassy plains with bushes, it is a mostly terrestrial insect and seed eater. Nomadic in dry conditions and moves away from the coast in the Wet. Breeds January-April (west), April-May (east). Nests low in bush or tussock. Endemic. **Juveniles**: *underparts brownish-grey, paler below.* **Voice**: *call of sparrow-like 'chip', long drawn-out 'teet', almost non-existent song.*

Female front, male rear

24 NUTMEG MANNIKIN

Lonchura punctulata 11 cm

This native of south-eastern Asia, also known in Australia as the Spice Finch, has colonised success-fully as a result of introductions and escaped caged birds. It has become the 'house sparrow' of the north, and displaces the less vigorous finches in the bush. It eats seeds and scavenges on rubbish tips and on roads for carrion. Often seen flying in very large wheeling flocks. Breeds spring and summer (south), all months (north). Nests 1-6 m high. **Sexes**: *alike.* **Juveniles**: *upperparts light brown, underparts yellowish brown, unscalloped.* **Voice**: *high-pitched 'chip' 'kit-teeee'. Almost inaud-ible song with whistling and bill-rattling.*

13

14

15

16

17

18

19

20

21

22

23

24

14

25 GREENFINCH

Carduelis chloris 15 cm

Introduced from the northern hemisphere in the 1860's, it has spread patchily in south-eastern Australia and Tasmania. Out of the breeding season, it flocks with the Goldfinch. In the winter, small flocks move from city parks and gardens into farmland and woodland. It is a seed eater mainly in vegetation associated with introduced pines. Breeds October-January. Open cup nest 2-12 m high in tree. **Female**: *duller and less yellow.* **Juveniles**: *brown, streaked darker.* **Voice**: *call is a hard dry 'ti-ti-ti-ti'; nasal 'dzhwee' by male during season; song is mixture of these two calls with musical warblings and twitterings.*

Male

26 GOLDFINCH

Carduelis cardeulis 13 cm

A bird of the northern hemisphere, introduced to the three south-eastern states of Australia between 1860-1880. Escapees have become established around Brisbane, Perth and Albany. It lives in grassy settled areas often associated with weed infestations especially thistle, feeding mainly on seed. Seen in pairs and groups; out of season it forms large nomadic flocks. Breeds September-March. Open cup nest 2-12 m high. **Sexes**: *alike.* **Juveniles**: *duller head and face, body finely striped.* **Voice**: *liquid 'tswitt-witt-witt' with pretty tinkling quality; song with harsh notes and canary-like twitterings.*

27 RED BISHOP

Euplectes orix 11 cm

Native to Africa, this member of the large family of weaver finches was introduced into Australia some fifty years ago as caged birds. A colony established itself from escaped or released birds on the Murray River, south of Adelaide, but has apparently died out. The breeding male is one of the brightest-coloured of all the weavers and has one of the most interesting courtship displays before his harem of 2 or more females. (The bird opposite was photographed in the aviaries of Taronga Park Zoo, Sydney and illustrates the neat construction of a weaver bird's nest).

Male (breeding)

28, 29 HOUSE SPARROW

Passer domesticus 15 cm

Native to Europe, Western Asia and Northern Africa, the House Sparrow is the most successful, and familiar, of city and town dwelling birds. It is more closely attached to man than any other bird and has followed European settlement all over the world, including to North America and New Zealand, as well as to Australia. Introduced from Britain progressively from 1863, it is now very common in most parts of its Australian range. It has not succeeded in crossing the Nullarbor Plain, and the authorities in Western Australia and the Northern Territory maintain constant vigilence to prevent it spreading into their States. Arrivals, sometimes by ship, are shot on sight. It is bold and noisy and lives in flocks in trees and on the ground. It feeds on seeds, scraps dropped by humans, also insects especially when feeding young. It competes with native birds for nesting sites. Breeds spring and summer and variably when conditions are favourable. Nests are untidy and are built in a variety of situations, including trees, eaves and holes in buildings. **Immatures:** *like females.* **Voice:** *a variety of unmusical chirruping, cheeping and twittering notes.*

Male left, female right

30 TREE SPARROW

Passer montanus 15 cm

Native to Eurasia and introduced from Britain around 1860, the Tree Sparrow has spread little beyond Victoria. It lives in urban and cultivated areas, parks and gardens. It feeds on seeds and insects on the ground or in foliage, in small groups. Sedentary. Breeds September—January. Nests in bush or hollow, less frequently in buildings. **Recognition:** *chestnut crown and nape; black ear patch. Slimmer than the House Sparrow and different calls.* **Sexes:** *alike.* **Immatures:** *paler.* **Voice:** *flight call of 'tek'; shrill twittering.*

31, 32, 33 SILVEREYE

Zosterops lateralis 12 cm

The Silvereye, widespread and familiar in Australia, also occurs on islands of the south-western Pacific and New Zealand. There are many plumage variations, but they can be grouped in six fairly distinct Australian races: the western race, the Tasmanian race, and four eastern races. Their distribution is continuous and they hybridise where the races meet. Silvereyes frequent low trees, bushes and gardens and feed on nectar, insects and fruit, sometimes making damaging raids on orchards. Usually they are seen in small numbers, but large flocks may gather in the non-breeding season. The Tasmanian birds migrate to the mainland to over-winter in huge flocks, often at night, and can be heard calling as they fly overhead. They disperse widely, some birds even reaching southern Queensland some 1500 km away. During the winter period, Tasmanian and mainland birds are often seen in the same areas. Silvereyes live to remarkable ages — one bird was recaptured ten years after banding. Breeds August-February, sometimes three broods. Nests in a bush 1-4 m up. **Sexes**: *female usually paler.* **Immatures**: *like adults.* **Voice**: *joyous trills, some mimicry.*

31 Western form (race *gouldi*)Perhaps the commonest small bird over much of south-west WA. *Olive green upperparts, yellow throat*

32 Tasmanian form (race *lateralis*) *Chestnut flanks, grey throat* (The area stippled on the map represents the winter range)

31

33 Eastern forms

(A) North of Mackay (race *ramsayi*): *bright yellow throat and dark yellow under tail* (illustrated)

(B) Barrier Reef form on the Capricorn-Bunker group of islands (race *chlorocephala*): *much larger; throat yellow, white under tail*

(C) South of Rockhampton to central Victoria (race *familiaris*): *throat yellow*

(D) Western Victoria to South Australia (race *halmaturina*): *throat grey; flanks buff*

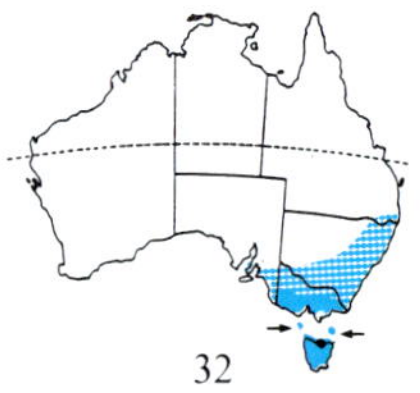

32

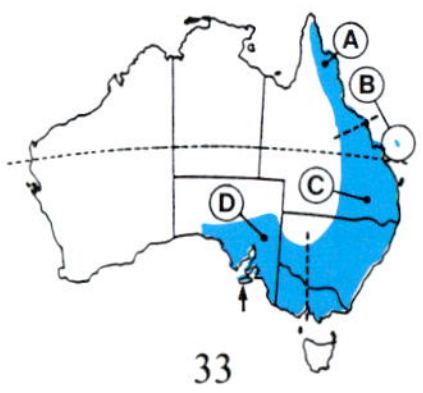

33

34 YELLOW SILVEREYE

Zosterops lutea 10 cm

In small parties to large flocks, Yellow Silvereyes frequent mangroves and coastal scrub. The plumage of populations east of Wyndham in Western Australia is brighter. Nomadic. Breeds September-March. Nest is a deep neat cup of bark or grass bound with spiderweb suspended from a horizontal twig. Endemic. **Sexes**: *alike.* **Immatures**: *like adults* **Voice**: *similar to other silvereyes, but is reported to sing more loudly once described as a 'marvellous volume of song'.*

35 PALE SILVEREYE

Zosterops citrinella 12 cm

The Pale Silvereye occurs only on the Lesser Sunda Islands of Indonesia (from Bali to Timor), on the islands of Torres Strait and the small offshore islands of northern Queensland. In pairs or parties, it frequents scrubby thickets and feeds on insects, native fruits and berries. Reported to breed in June and December. Its nest is much larger and bulkier than its mainland relatives, and is suspended from a horizontal branch or fork. **Recognition**: *a conspicuous yellow forehead.* **Sexes**: *alike.*

Juveniles

36 YELLOW-BELLIED SUNBIRD

Nectarinia jugularis 12 cm

The tiny Yellow-bellied Sunbird is a common in the coastal rainforests, mangroves, parks and gardens of tropical Queensland, and is usually seen in pairs among the blossoms of flowering trees and shrubs. Its long narrow downcurved bill, often with side serrations, enables it to probe deeply into most flower heads to sip nectar, or pierce them at their base when they are too deep. Sucking is aided by its tubular tongue. It also catches insects on the wing, and sometimes plucks spiders from their webs. The sunbird darts swiftly from flower to flower and is sometimes compared with the (*continued overleaf*)

Male

25

26

27

28

29

30

31

32

33

34

35

36

14

37 YELLOW-BELLIED SUNBIRD
Nectarinia jugularis 12 cm

hummingbirds of the Americas, but it more often clings to a petal or a leaf when feeding rather than hover like a hummingbird. It is unafraid of human beings and will enter houses. It builds a long spindle-shaped nest with a hooded side entrance suspended from a twig or a piece of rope or wire hanging from an eave or a verandah, and is oblivious to the comings and goings of people. Breeds most months but mainly August-March sometimes more than once during the year. The Australian bird is a race *(N.j.frenata)* of a species which is also found in New Guinea, South-east Asia and the Philippines. **Immatures**: *like adult females* **Voice**: *squeaky staccato notes.*

Female

38 RED-BROWED PARDALOTE
Pardalotus rubricatus 10 cm

A bird of arid country which is similar in behaviour to the Spotted Pardalote. Singly and in small parties, it is fairly readily seen in outer foliage of less leafy trees. Insects make up its diet. Breeds February—November. Nests at the end of a tunnel bored into an earth bank. Endemic. **Recognition** (cf Striated Pardalote *P.s. melanocephalus*): *speckled head and small but distinctive orange-red eyebrow spot.* **Sexes**: *alike.* **Immatures**: *upperparts yellow-green; underparts yellow.* **Voice**: *repetitive distinct five-note phrase.*

39 STRIATED PARDALOTE
Pardalotus striatus melanocephalus 11 cm

A black non-striated head, and a red spot and a broad white stripe on the wings distinguish this race (often called the Black-headed Pardalote). The rump of northern birds is bright yellow; southern birds cinnamon brown. There is interbreeding with *substriatus* in the zones where the two races meet or overlap. Their habitats and habits are similar. Breeds May-December. Endemic. **Adults and immatures**: *see nominate race.* **Voice**: *loud sharp 'chip-chip' repeated monotonously.*

40, 43 MISTLETOEBIRD
Dicaeum hirundinaceum 10 cm

The Mistletoebird belongs to the so-called flowerpecker family of some thirty-six species which extends to southern Asia. It is one of Australia's most widespread and familiar birds, and also occurs on islands in the Torres Strait and others further west to Indonesia. It is found wherever mistletoe grows, in rainforest, mangrove or desert. Its digestive system, in which its stomach is supplemented by a duct, enables it to pass large quantities of berries in a short time, thus further spreading the seeds of the berries on which the species mainly relies. It too eats the fruit of other plants too, and some insects. It flies swiftly on narrow wings and calls in flight. Breeds October-March. Suspended purse-shaped nest. **Immatures**: *like females.* **Voice**: *warbling song with mimicry; also short high-pitched note in flight.*

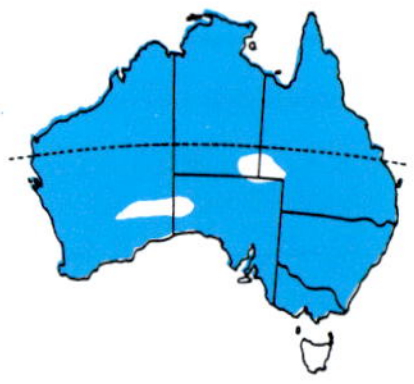

Male above, female below

41, 44 SPOTTED PARDALOTE
Pardalotus punctatus 9 cm

This spectacular little bird is often seen in tree-lined city streets, gardens, parks and golf courses as well as in forest and woodland. It feeds on insects in eucalypts and wattle. Usually sedentary, but with some movement from south to north in autumn. It sometimes gathers in flocks in winter. Breeds June-January. Nests mainly in earth bank. Endemic. **Immatures**: *upperparts yellow-green, underparts yellow.* **Voice**: *a distinctive bell-like three-syllable 'slee-p ba-bee'. Sometimes called a 'bellbird' in error.*

Male above, female below

42 STRIATED PARDALOTE
Pardalotus striatus ornatus 11 cm

An orange-red spot and narrow white stripe on the wings distinguish this race. It frequents wattle scrub and mulga and other low trees. Its habits are similar to *striatus* to which it is closely related and of which it is sometimes regarded as a hybrid. In winter, it may form travelling flocks with *substriatus* in the west, and with *melanocephalus* in the north. Breeds August-December. Nests mostly in earth banks. Endemic. **Adults and immatures**: *see nominate race* **Voice**: *'wit-e-cheu'.*

45 STRIATED PARDALOTE
Pardalotus striatus substriatus 11 cm

A red spot and broad white stripe on the wings distinguish this widespread race. In winter it gathers in travelling flocks; in Western Australia some birds from the south-west move to winter in drier northern areas. Breeds August-December. Nests in earth hole; in Western Australia in tree holes or hollows, rarely in building crevices. Endemic. **Adults and immatures**: *see nominate race* **Voice**: *'wit-e-cheu'; in WA 'wit-a-witta', 'be-quick', or 'pick-wick'.*

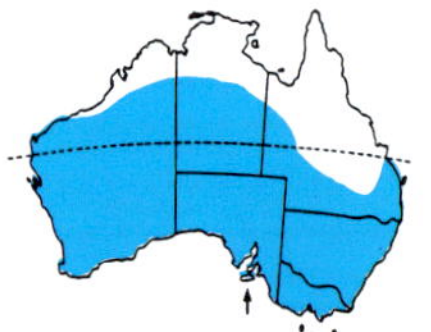

46 FORTY-SPOTTED PARDALOTE
Pardalotus quadragintus 10 cm

Endemic to Tasmania, its numbers have declined alarmingly this century. With a total population estimated at only 850 birds in 1982 it has become an endangered species. Competition for nesting sites from the Spotted and Yellow-tipped Pardalotes is believed to be one of the major reasons for its decline. It frequents the foliage of eucalypts. Breeds August–January. Nests in tree hollow to 20 m. **Recognition**: *plain crown and uniformly pale yellow face.* **Sexes**: *alike.* **Immatures**: *face, undertail and coverts paler.* **Voice**: *double-noted whistle, second note lower.*

47 YELLOW-RUMPED PARDALOTE
Pardalotus xanthopygus 10 cm

Its distribution, which spans the Nullarbor Plain, links the split populations of the Spotted Pardalote to which it is closely related. It frequents mallee and nearby taller eucalypt woodland, and feeds on insects. Usually sedentary but, in winter, feeding flocks may wander some distance. Breeds July—December. Nests in sloping tunnel dug into flat ground, with entrance extremely difficult to spot. Endemic. **Recognition**: *lemon yellow rump.* **Female**: *duller, throat off-white.* **Immatures**: *duller, crown grey.* **Voice**: *resonant repeated 'ti-i-ing'.*

Male

48 STRIATED PARDALOTE
Pardalotus striatus striatus 11 cm

This, the nominate race, was discovered and described by William Anderson at Adventure Bay, Tasmania, on 29 January 1777 during Cook's Third Voyage. (See note on page 124). It is distinguished by a yellow spot and narrow white stripe on the wings. After the breeding season, most birds are believed to migrate to the mainland to overwinter. Singly, in pairs or family parties, it feeds high in foliage on insects. Breeds August-December. Nests mostly in a hole in an earth bank. Endemic. **Female**: *duller.* **Immatures**: *grey-black unstreaked crown.* **Voice**: *'pick-it-up' or 'will-you'.*

37

38

39

40

41

42

43

44

45

46

47

48

15 Bowerbirds, Orioles and Figbirds, Birds of Paradise

Opposite: Satin Bowerbird

BOWERBIRDS AND CATBIRDS Family Ptilonorhynchidae

This family is confined to Australia and New Guinea. There are eighteen species, nine of which occur in Australia; all are endemic except the Fawn-breasted Bowerbird and the Spotted Catbird. Their sizes range from 22 cm to 40 cm and all have short wings and tail, and strong, short legs. The plumage of some of the males is strikingly different from that of the females. Most species live in wet forests, but the four species of the genus *Chlamydera* (such as the Great Bowerbird) inhabit drier vegetation.

John Gould coined the name 'bowerbird', which refers to the structures built and decorated by males to entice females to a focal site where courtship display and mating can occur – a habit described by Gould as 'incredible'. Bowerbirds exhibit skills as architects, builders and decorators unmatched by any other birds. Each prefers bower decorations of a particular type and colour. All fourteen true bowerbird species and the Tooth-billed Catbird (also known as the Tooth-billed Bowerbird) are promiscuous, and the males compete with each other for mates by displaying at or near the bower to visiting females. A male may woo ten or more females in a season, but most displays do not end in mating. The female bowerbirds, including the female Tooth-billed Catbird, build the nests, incubate the eggs and rear the young entirely on their own.

In contrast, the Green and Spotted Catbirds and the White-throated Catbird of New Guinea form monogamous breeding pairs and build no bowers or courts. The female catbird of a pair builds the nest and incubates the eggs, and both sexes feed the young. Catbirds get their name from their weird cat-like wails. 1–13

Bowerbirds can be classified into groups according to bower structure:

- **Avenue builders** There are nine species of avenue builders, five of which occur in Australia. Avenue bowers are intricate structures consisting of a stick platform on which two vertical, parallel walls of twigs are erected so they form a narrow corridor. Ornaments decorate the bower and in some species piles of display objects adorn the platform at one or both ends of the avenue. As far as is known, all avenue builders paint the inside walls of their bowers with masticated plant matter or charcoal and saliva, and often use a saturated wad of leaves or bark held in the tips of the mandibles for this purpose. Thus bowerbirds are among the very few birds in the world, along with the famous Galapagos Finch, to use a 'tool'.
- **Maypole builders** There are five maypole builders, all of which build their bowers around one or more saplings or tree ferns. The exquisite Golden Bowerbird is Australia's only member of the group. Its four New Guinea relatives are known as Gardener Bowerbirds because they decorate their bower platforms with a moss lawn and fresh leaves, flowers and fruit, and beetles and fungi. Maypole bowers range in complexity from a single cone of sticks to elaborate roofed huts.
- **Mat builder** Archbold's Bowerbird, found only in New Guinea, simply lays down a mat of dried ferns on the forest floor and decorates it with heaps of brown snail shells, beetles and blue fruit.
- **Stagemaker** The Tooth-billed Catbird prepares a stage or court by clearing leaf litter from an area on the forest floor and decorating it with a layer of large, freshly cut leaves. From singing perches above their courts, males pour out bouts of loud, varied songs.

ORIOLES AND FIGBIRDS Family Oriolidae

Orioles The Golden Oriole *Oriolus Oriolus*, Europe's only oriole, appears to have been the first to receive the name, which is derived from the Latin word *aureolus* for golden. Orioles mainly dwell in forests and woodlands, and mostly forage in foliage for native fruits and insects. They are robust birds with pointed bills and long wings, and range in size from 20 cm to 30 cm. They are accomplished vocalists. Australia has two of the world's twenty-four species. (They are unrelated to the birds called orioles in the New World.)

Figbirds Figbirds live only in New Guinea, Australia, Timor and some adjacent islands. The single species divides into five races, two of which occur in Australia. Figbirds differ from the orioles in having shorter bills and a bare skin patch around the eye. They are gregarious, whereas orioles are not. 14–18

BIRDS OF PARADISE Family Paradisaeidae

In 1522 when the King of Spain received the gift of two preserved specimens of birds of paradise brought home from the Moluccas by Magellan's surviving ship the *Victoria*, the Spaniards believed that such beauty could only come from paradise, and so the name has persisted. Several species were named after the royal heads of Europe. The name 'manucode' appears to come from a Malayan phrase, meaning 'birds of the gods', adopted by Portugese traders in the 16th century, and the strange name 'rifle-bird' arose because their plumage colour resembled the uniform of the British Army rifle regiments. Tragically, by the end of the last century, an estimated 50 000 birds of paradise skins were being exported annually to Europe for feminine adornment – until outlawed in the 1920s. Today, habitat destruction threatens some species.

During their spectacular courtship displays, the males exhibit much bright and iridescent plumage; erectile plumes, shields and capes; and magnificent tail feathers. Most species display on perches in tree-tops; others clear ground courts and display above them on vertical or horizontal perches. There are forty three species, ranging from the size of a starling to over a metre in length. Most species live in the New Guinea mainland. Only four species occur in Australia – the Trumpet Manucode and three riflebirds, two being endemic. In most bird of paradise species the males are promiscuous and the comparatively drab females alone provide all the brood-care. Between five to ten species are monogamous and both parents care for the young. 19–24

PHOTOGRAPHERS: H. & J. Beste 7, 10, 15, 19, 21; C.A.C. Cameron 6; N. Chaffer 9; T. & P. Gardner 4, 23; P.N. Green p. 149; 1, 2, 3, 13; R. & D. Keller 11; W.S. Peckover 20, 24; J. Purnell 17; K. & B. Richards 18; B.A. & A.G. Wells 5; E.E. Zillmann 8; Authors: 12, 14, 16, 22.

15

1, 2 SATIN BOWERBIRD

Ptilonorynchus violaceous 27–33 cm

The elegant violet-eyed Satin Bowerbird has two main populations comprising race *violaceous* in the south and the smaller race *minor* of the Atherton Tablelands. During the breeding season, the species lives in rainforest, wet eucalypt forest and adjacent woodland, and moves out, often in large flocks, to gardens and orchards in winter. It eats fruits throughout the year but feeds more on leaves and flowers in winter and insects in summer. Each breeding seasons, an adult male builds and maintains an avenue bower of two vertical, parallel walls of twigs, usually running north-south, on a mat base perhaps a metre or more in diameter. It is decorated with a great diversity of both natural and man-made objects and more than 300 items have been counted at a single bower. Blue objects predominate, but they are often yellow and brown and are mostly placed on the northern platform. A male will raid and flatten a neighbour's bower and steal ornaments. During the breeding season the bower is attended daily and the inside avenue walls are painted with charcoal or vegetable matter and saliva. A female visitor triggers a striking visual and vocal display by the male. It holds an ornament in its bill and performs a routine of strutting, bowing and flinging open its wings as it dashes back and forth across the avenue entrance uttering rythmic buzzing and rattling notes and quiet mimicry. It mates in or near the avenue, and may mate with five or more females in a season. Immature males can become sexually active before attaining full adult plumage in the seventh year. There may be more than 20 bowers in an area and these are dispersed evenly far apart. The same bower territories tend to be occupied year after year by the same or a different male. Breeds September–February. Nests up to 35 m in a tree as close as 24 m from the nearest bower. Endemic. **Immatures**: *like adult female until the fourth-year male undergoes changes in throat and bill colour.* **Voice**: *drawn out whistle, harsh churrs, ringing advertising calls, buzzing and rattling.*

Male left, female right

3, 4 REGENT BOWERBIRD

Sericulus chrysocephalus 24–28 cm

The handsome Regent Bowerbird inhabits wet forests and scrub, particularly rainforest. It forages from ground level to the canopy and feeds primarily on fruits, but also on leaves, insects and spiders. It may be found in the same rainforest as Satin Bowerbirds and Green Catbirds and sometimes all three species can be seen feeding simultaneously in a fig tree. In the non-breeding season, it visits gardens and orchards and forages singly and in small groups. Mostly solitary and sedentary in the breeding season. The striking plumage of the adult male is attained at 4–6 years although females may breed when 3 years old. The adult male builds an avenue bower similar to that of the Satin Bower-bird, but it is smaller. Many bowers are concealed by a thick vine layer directly above the bower, and hence they are seldom found. This has led to speculation that this species is discarding its bower-building behaviour. But, as the social organisation of the male Regent and Satin Bowerbird is similar, it is highly likely that the Regent Bowerbird regularly builds bowers. Male Regents sparsely decorate their bowers with brown small shells, blue, red or black berries, fresh leaves and sometimes flowers and man-made objects. Unlike the Satin Bowerbird, ornaments are placed mostly within the avenue. The male paints the inside avenue walls. An adult male at its bower holds an ornament in its bill and performs strutting, head-bobbing, bowing and wing-opening to a visiting female or immature male. Quiet chattering, twittering and mimicry, only audible a metre or so from the bower, occur throughout display. Like the Satin Bowerbird, immature males display on platforms and rudimentary bowers to females and immature males. Breeds October–January. Nests up to 10 m. Endemic. **Sexes**: *distinctly different.* **Immatures**: *like adult female.* **Voice**: *metallic te-aar, chattering and mimicry. No loud advertising calls.*

Male left, female right

5 SPOTTED BOWERBIRD

Chlamydera maculata 25–31 cm

The western race *guttata* is blacker on the throat and upperparts and the rufous-brown spotting is brighter and more pronounced than in its eastern relative. It also has a band of lustrous lilac-pink plumes on the nape which it exposes when displaying. Its range approaches that of *maculata*, but apparently does not overlap. It frequents vegetation in the rocky ranges of the dry inland particularly where the native fig occurs, and visits homestead gardens. The diet is predominantly native figs but includes other fruits, seeds, nectar, insects and spiders. Like *maculata*, it occurs singly and in small groups and forms winter foraging flocks. Also similar are the bower ornaments, display and the painting of the inside avenue walls. Breeds August–February. Rainfall regulates the start of breeding. Endemic. **Female**: *nape patch duller and smaller.* **Immatures**: *paler body plumage than adults.* **Voice**: *like the eastern race.*

6 SPOTTED BOWERBIRD

Chlamydera maculata 25–31 cm

The eastern race *maculata*, more common in the north, inhabits dry, open woodland, inland riverine vegetation, and visits homestead gardens. The male builds an evenue bower of twigs and grasses and places hundreds of display objects on the platform and in the avenue. Ornaments (preferably white, green, amber and mauve) include bones, pebbles, shells, fruits and man-made glass and metallic objects. In display at the bower the male turns its hindneck towards the female to expose a fan of iridescent lilac-pink plumes. Mates in the avenue or on the bower platform. Breeds October–January. Nests from 2–15 m in tree branches or mistletoe clumps. **Female**: *nape plume smaller or absent.* **Immatures**: *lack plume.* **Voice**: *renowned for mimicry of natural and mechanical sounds; ringing metallic advertising calls; harsh grindings, hissing and clicking.*

1

2

3

4

5

6

15

7 GREAT BOWERBIRD

Chlamydera nuchalis 32–37 cm

The Great Bowerbird, widespread within its range, is Australia's largest bowerbird. It inhabits rainforest, eucalypt forest, riverine vegetation, dry woodland, mangroves and urban gardens. An adult male builds and maintains one large avenue bower. Bower ornaments and ornament colour are similar to those of the Spotted Bowerbird, but the Great Bowerbird also collects red objects. The male's bower display is similar to that of the Spotted Bowerbird. It has a brilliant patch of lustrous rose-lilac feathers on its nape which, normally hidden, is suddenly opened into a beautiful fan when displaying. Occurs singly and in small groups. Feeds on fruits, insects and spiders. Breeds August–February. Female nests to 10 m. Endemic. **Sexes**: *alike, but female smaller, sometimes paler; lilac neck patch reduced or absent.* **Immatures**: *paler than adults, lack neck patch.* **Voice**: *like Spotted Bowerbird.*

8 FAWN-BREASTED BOWERBIRD

Chlamydera cerviniventris 25–30 cm

The Australian range of the Fawn-breasted Bowerbird, which has an extensive coastal distribution in New Guinea, is restricted to Cape York Peninsula. It lives in coastal woodlands, forest edge, riverine vegetation and mangroves. The drab coloured adult male builds the most elaborate, ornately decorated of the two-walled avenue bowers. One end of the platform and the tops of both avenue walls are decorated profusely with many fresh green fruits, particularly bunches of green berries. The male paints the inside walls with vegetable matter. The twigs are green when freshly painted and turn a dull grey and red-brown when dry. Occurs singly and in small groups. Mainly breeds September–December. Nests to 10 m. **Sexes**: *alike.* **Immatures**: *like adults, but forehead more streaked greyish.* **Voice**: *like other Chlamydera.*

9, 10 GOLDEN BOWERBIRD

Prionodura newtoniana 23–35 cm

The brilliant plumage of the adult male Golden Bowerbird contrasts sharply with that of the female. The species lives in tropical rainforest above 900 m and has a restricted distribution in north-east Queensland from near Cooktown to the Seaview Range, north of Townsville. It is the smallest bowerbird, yet the male builds a massive 1–3 m high maypole bower comprising two pyramids of sticks built around two saplings about a metre apart. These structures are fused together with fungus and bridged by a horizontal perch on which the male displays to visiting females. The male festoons the display perch and nearby parts of the bower with grey-green lichen, fruits and dried cream flowers with black seeds attached. Bowers are built on sloping hillsides and are dispersed evenly throughout suitable habitat. An adult male maintains and defends one bower, but a single bower may be attended by several female-plumaged males. This bowerbird occurs singly and in small groups, and feeds predominantly on fruits, but also takes beetles and cicadas. Breeds September–February. Female nests in tree crevices 0.1–15 m above ground. Endemic. **Immatures**: *first year plumage like adult female, then plumage of immature male becomes washed with yellow before assuming the adult male plumage.* **Voice**: *croaks, rattles, mimicry.*

Male left, female right

11 SPOTTED CATBIRD

Ailuroedus melanotis 26–30 cm

Sedentary in tropical upland rainforest in north-east Queensland, New Guinea and adjacent islands. Males do not build bowers or courts. Monogamous breeding pairs hold year-round territories and defend them in the breeding season by song and chasing. Noted for their strange wailing notes uttered throughout the day particularly at the onset of breeding. Females build the nest and incubate the eggs and both sexes feed the young. They occur in small foraging flocks in the nonbreeding season and singly or in pairs at any time of the year. Feeds chiefly on fruits, but also leaves, flowers, seeds, insects and nestling birds. Forages from ground level to the canopy. Breeds September–March. Nests to 15 m. **Sexes**: *alike.* **Immatures:** *head greener and less mottled; upper parts duller.* **Voice**: *repertoire includes loud cat-like wailing notes; baby cries; single or double high pitched squeak or 'chip'.*

12 GREEN CATBIRD

Ailuroedus crassirostris 28–33 cm

Monogamous breeding pairs occupy year-round territories in rainforest. Males do not build bowers or courts. Quiet in the non-breeding season but can be located by their 'chip' call. In the nesting season, neighbouring territorial males engage in singing bouts throughout the day. Male and female members of a pair have different songs and sing a duet against neighbouring pairs. Females build the nest and incubate the eggs and both sexes feed the young. The diet is primarily fruits, especially figs, and includes flowers, leaves, seeds, frogs, insects, millipedes and nestling birds. Forages from ground level to the canopy but mostly in the middle and upper vegetation levels. Breeds September–February. Nests to 18 m. Endemic. **Sexes**: *alike.* **Immatures**: *duller than adults and less distinctly streaked.* **Voice**: *like Spotted Catbird; song dialects vary throughout its range.*

7

8

9

10

11

12

15

13 TOOTH-BILLED CATBIRD (Stagemaker)

Ailuroedus dentirostris 24–27 cm

The Tooth-billed Catbird lives in tropical rainforest between 600 and 1400 m. Around September, the male clears a display court on the forest floor and decorates it with fresh leaves placed with the under-surfaces uppermost. Throughout the breeding season, males perch above their courts and the forest rings with their loud calls as neighbouring males sing and counter-sing against each other to advertise their presence and attract mates. Fruits predominate in the diet, which includes insects and leaves. The species gets its name from its bill serrations which are used primarily for tearing and eating leaves but also for snipping off leaves for the display court. Breeds November–February. Female nests up to 27 m. Endemic. **Sexes:** *alike.* **Immatures:** *like adults, but browner above and paler below.* **Voice:** *loud and varied repertoire including mimicry; harsh 'chuck'.*

15 YELLOW ORIOLE

Oriolus flavocinctus 25–30 cm

The Yellow Oriole ranges from southern New Guinea and Aru Islands to coastal northern Australia, and inhabits wet tropical forests and woodlands, mangroves, parks and gardens. It is noted for its melodious, incessant song delivered for up to 2 hours at a time from perches within tree foliage. The larger range of the Olive-backed Oriole extensively overlaps that of the Yellow Oriole. Where both species occur, the Olive-backed Oriole apparently inhabits drier vegetation. Like the Olive-backed Oriole, it feeds on fruit and insects and mostly occurs singly and in pairs, but small groups visit fruiting trees. Breeds August–January. Nests to 15 m. **Sexes**: *similar, but female paler.* **Immatures**: *upperparts paler than adults; body more heavily streaked; bill dark.* **Voice**: *loud, repetitive, liquid bubbling notes, harsh scolding.*

17 OLIVE-BACKED ORIOLE

Oriolus sagittatus 25–28 cm

Ranging from southern New Guinea and adjacent islands, the Olive-backed Oriole is widespread in northern and eastern Australia. Northern populations are sedentary and south-eastern populations migrate north in the non-breeding season. It lives in eucalypt forest and woodland and adjacent rainforest, visits parks and gardens, and feeds on fruit and insects. Monogamous breeding pairs hold nesting territories. Both sexes feed the young and vigorously attack and chase potential avian predators from the nest area. Breeds September–January. Suspends nest in fork of outer foliage up to 20 m high. **Sexes**: *slightly different; male greener and female greyer.* **Immatures**: *duller than adults, greyish upperparts, wing and tail feathers edged rufous, bill and eye blackish.* **Voice**: *loud, clear, rollicking song resembling 'olly-olly-ole'; mimicry.*

14, 16, 18 FIGBIRD

Sphecotheres viridis 27–29 cm

Figbirds are arboreal and frequent rainforest edge, eucalypt forest, woodland and urban parks and gardens. They feed on many kinds of native and cultivated fruits but, as their name suggests, they specialise in eating fig fruits. Locally nomadic when searching for fruiting trees. Unlike orioles, figbirds forage in large, noisy flocks of 100 or more in the non-breeding season. In the breeding season they form monogamous pairs, and both sexes feed the young. The facial skin patch of the male sometimes flushes red. **Immatures**: *like females.* **Voice**: *chattering and chirping, squeaks and short, sharp yelps.*

(A) Race *flaviventris* (YELLOW FIGBIRD)

The distinctive Yellow Figbird ranges from north-east and north-west Australia to the Kai Islands west of the Aru Islands. It frequents tropical woodlands and gardens. The yellow race of figbird hybridises and intergrades with the green southern race where their ranges meet around Townsville. Breeds September–March and nests to 20 m high.

Male above, female below

(B) Race *vieilloti* (SOUTHERN FIGBIRD)

Breeds October–February. Nests to 20 m. Female: (not illustrated – but similar to illustration of female race *flaviventris*).

Male

13

14

15

16

17

18

15

19, 21 VICTORIA'S RIFLEBIRD

Ptiloris victoriae 23–25 cm

Named in honour of Queen Victoria in 1850 by John Gould. This riflebird, restricted to northeast Queensland and adjacent offshore islands, lives in tropical rainforest and frequents woodland and gallery forest. It forages like treecreepers up tree trunks to the canopy land uses its stout decurved bill to search and probe for insects and spiders in arboreal epiphytic ferns and orchids. It also feeds on fruits. The species is promiscuous and, in the breeding season, solitary adult males deliver advertising calls and display their plumage to attract females for mating. Display sites include the tops of tree stumps and high exposed tree perches. Each adult male may have more than one display site within its territory. In response to a visiting female, a displaying male adopts an upright posture with tail raised, opens and fans its wings so the wing tips extend above the head and moves its head from side to side. In a display between an adult and immature male, the two males face each other and almost touch. Each bird in turn extends its opened wings forward, throws back its head and opens its bill wide. The tail is then spread and the two birds sway from side to side with their wing tips meeting out from the breast. In flight, the wings of the male make an audible, rustling sound. Breeds October–January. Female nests in tree or palm foliage to 5 m. Sometimes the nest rim is decorated with a sloughed snake-skin. Endemic. **Immatures**: *like adult females until immature male develops black feathers and wing rustling.* **Voice**: *single rasping 'yaas'; female less vocal.*

Male above, female below

20, 22 MAGNIFICENT RIFLEBIRD

Ptiloris magnificus ♂ 28–33 cm ♀ 26–28 cm

In Australia, the Magnificent Riflebird is restricted to the tropical rainforests of Cape York Peninsula. It also occurs in the Aru Islands and has an extensive distribution in lowland New Guinea. Like a treecreeper, it forages singly up tree trunks and along branches in the middle and upper vegetation levels in search of insects and spiders. it also feeds on fruits. The species is promiscuous and, in the breeding season, solitary adult males call from high, exposed branches to advertise their presence and attract mates. When a female visits the display perch, the male spreads open both wings, stretches the head back and swings it from side to side to expose the metallic iridescent throat and breast shield. A displaying male may relax its wings then snap them back to produce a rustling sound in unison with its swaying head. No vocal sounds accompany the display. Unlike the Paradise Riflebird, the greenish-yellow interior of the mouth is apparently not exposed in display. Breeds October–February. Nests from 2–12 m in palms or dense tree foliage. Sometimes the outer nest rim is decorated with a sloughed snake-skin. **Sexes**: *distinctly different.* **Immatures**: *like adult female.* **Voice**: *loud reverberating wee-hoo whistle by adult male; female whistles are more subdued.*

Male above, female below

23 PARADISE RIFLEBIRD

Ptiloris paradiseus 25–30 cm

The Paradise Riflebird lives in rainforest and disperses to adjacent eucalypt woodland and forest in the non-breeding season. It mostly forages singly in trees and feeds on fruit and insects, and can be located by showers of debris from above. Its wings rustle in flight. In the breeding season, solitary adult males deliver advertising calls from high display perches. On the approach of a female, the promiscuous male opens and fans its wings, cocks the tail, points the bill up and exposes the lime-green gape. The male encloses the female with its outstretched rustling wings, claps its wing tips together and performs mating. Breeds October–January. Nests up to 25 m. Endemic. **Male**: *similar in coloration to the male Victoria's and Magnificent Riflebirds.* **Immatures**: *like females.* **Voice**: *double, sometimes single, harsh, loud explosive 'yaas'.*

Female

24 TRUMPET MANUCODE

Manucodia keraudrenii 27–32 cm

The loud trumpet-like blast of the male, which gives the species its name, is produced by modified coiled windpipes. It lives in rainforests in Cape York Peninsula and has an extensive range in southern New Guinea and adjacent islands. Like Green Catbirds, monogamous breeding pairs hold a territory and specialise in feeding on figs. The female of a pair builds the nest and incubates the eggs and both sexes feed the young. Adults feed on fruit and insects and usually forage singly, in pairs or in mixed species flocks in the middle to upper levels of the forest. Some birds may migrate between northeast Australia and southern New Guinea. Breeds October–January. Nests to 20 m. **Sexes**: *similar.* **Immatures**: *like adults, but less glossy.* **Voice**: *drawn out, for carrying trumpet-like call.*

19

20

21

22

23

24

16

Mostly Black-and-white Birds

In this Section are grouped the Australian representatives of six passerine families, simply because their plumage for the most part is black and white. Individual pied birds (for example the Willie Wagtail) have been kept with their families and will be found in the Sections given in the last paragraph.

MAGPIES, CURRAWONGS, BUTCHERBIRDS   Family Cracticidae

All ten species of this family are confined to the Australian Region with eight in Australia, five of them endemic. They are among the most familiar and widely distributed of Australia's birds, and are noted for their loud distinctive voices, some quite beautiful. They are also aggressive and some prey on other birds.

Magpies The name 'magpie' is one of several instances where the popular name of an English bird was bestowed upon a superficially similar but entirely unrelated new species discovered in Australia during the early days of colonization. Australian magpies broadly divide into three populations on the basis of plumage pattern, but the differences are not always clear cut where populations meet and hybridise. 1-3

Currawongs The familiar currawongs, of which there are three endemic species, are the largest members of the family and are conspicuous from their large powerful bills, prominent yellow eyes and loud ringing calls. They are often seen in city parks and gardens, especially in winter when they descend from the hills. They are sometimes known as bell-magpies, amongst a host of vernacular names. 4-7

Butcherbirds Butcherbirds are smaller than magpies, and the four species in Australia, three endemic, are amongst the finest of songbirds, and sing, often in duet, almost the year round. They are aggressive hunters of small birds, rodents, reptiles, and large insects which, if too large to eat whole, are impaled, perhaps on a thorn, in a 'larder' to be dismembered with their strong hooked beaks and eaten later.

MAGPIELARKS   Family Grallinidae

There are only two members of this family, one, the Peewee, lives in Australia, and the other, the Torrentlark, in New Guinea. They are neither magpies nor larks, only resembling the former in their pied plumage. They build mud nests and are sometimes called Mudlarks. The Peewee has benefitted enormously from spreading land cultivation, and is found almost everywhere on the mainland. 13, 14

DRONGOS   Family Dicruridae

There are twenty species of drongos found mostly in tropical Africa and Asia. The term 'drongo' comes from the native name for the Crested Drongo of Madagascar. Most drongos are black with bright metallic sheens of green to purple, and have rather long, deeply forked tails. When nesting drongos will vigorously tear bark from trees to get at grubs and insects, and indulge in spectacular aerial acrobatics, often accompanied by shrill stuttering shrieks and whistles. One species only is represented in Australia, the Spangled Drongo, whose range also extends to New Guinea and southern Asia. 15

STARLINGS AND MYNAHS   Family Sturnidae

Members of this family, of which there are 111 species, are distributed worldwide, excepting in South America, and are particularly well represented in the Ethiopian and Oriental Regions. Only three occur in Australia — one native (Shining Starling) and two introduced (Common Starling and Indian Mynah). Some starlings are brightly coloured but most are dark black with metallic sheens. Their name 'starling' connotes the starlets that appear at the tips of the fresh autumn feathers of the Common Starling after moulting. These disappear with wear through the winter, leaving the bird a shining glossy black by the time breeding starts in spring. Common Starlings often gather in huge wheeling flocks towards dusk before coming in to roost for the night — sometimes crowding the roofs and ledges of edifices in thousands, even in the heart of the largest cities. 16-18

MUDNESTERS   Family Corcoracidae (Part)

This small Australian family of two consists of the White-winged Chough and the Apostlebird. (The latter is included in Section 4.) Although different in appearance, they are of similar build and behaviour. They are highly sociable and keep together in tightly-knit family units of up to a dozen or more birds, and do everything together, including foraging for food, roosting, nest-building and rearing young. Their pudding-bowl shaped nests are built of mud and fibre which is cemented to the fork of a tree. Occasionally more than one female may lay in a nest which can contain up to ten eggs. 19

RAVENS AND CROWS   Family Corvidae

Ravens and crows of the genus *Corvus* are found in all continents except South America, and belong to an almost worldwide family of about 105 species, which also comprises jackdaws, rooks and jays. No group of birds has been more successful in coexisting with man. The five endemic Australian species — three ravens and two crows — are the only corvids with white eyes. Three ravens are now recognised on the grounds of their distinct differences in anatomy, behaviour, voice and ecology. While corvids as a group are readily distinguishable by their all-black glossy plumage, they are not easily separated one from another. The problem for the observer can be simplified by taking note of the following: distribution, call, behavioural idiosyncrasies, and physical features, in that order. As to the latter, crows are smaller than ravens, and ravens have longer and more conspicuous throat feathers. If feathers are ruffled in wind, crows reveal white bases to their feathers and ravens grey. 20–24

OTHER BLACK-AND-WHITE BIRDS: see Sections 2, 6, 7, 8, 10, 13 and Part II.

Opposite: Pied Currawong

PHOTOGRAPHERS: H. & J. Beste 18; G. Chapman 4, 20, 21, 24; J. Dart 1; T. & P. Gardner 2, 8, 10, 23; C.L. Gill 7; P.N. Green 3; H.J. Pollock p. 159, 15; K. Vagg (BOC) 19; T.A. Waite 22; G. Weber 5; E.J. Whitbourn (NPI) 12; Authors 6, 9, 11, 13, 14, 16, 17

16

1, 2, 3 AUSTRALIAN MAGPIE

Gymnorhina tibicen 38-44 cm

The beautiful carolling of this familiar bird — heard particularly at dawn and dusk — is one of the most nostalgically evocative of all Australian sounds. It lives in all types of woodland in the vicinity of watercourses, and is equally at home in parks and suburban gardens, except inland where it is wary of humans. However, although it can become tame, it will not hesitate to swoop down viciously on any bird or person deemed to be a threat to the nest. Adult magpies live in well defined groups with dominant polygamous male; usually only one female successfully raises a brood, building the nest and hatching the chicks alone. Birds not yet of breeding age form large flocks with other unattached birds. Young birds are fed by their parents until they are able to fly, but even then will run after any adult to beg with incessant plaintive piping. Food is provided by insects and invertebrates taken on the ground or by probing into any likely crevices. Breeds July-March. Nests usually in three up to 20 m high, sometimes even on a pole. One race occurs in New Guinea. The three main Australian races are illustrated, but many hybrid forms occur where populations overlap (such as south of Canberra) — ranging from fully black-backed to fully white-backed. Conversely, the change from black-backed to white-backed can be abrupt — with fully black and fully white backed individuals in adjacent territories. **Adults and immatures**: *(see below)*. **Voice**: *likened to 'quardle oodle ardle wardle doodle', some mimicry; females in north-west are less vocal.*

Race: *dorsalis* (Western form) **Male**: *white-backed like race hypoleuca*. **Immatures**: *like dull female — black parts greyer.*

Female

Race: *hypoleuca* (White-backed form) Tasmanian magpies are smaller. **Female**: *white back indistinctly mottled.* **Immatures**: *like dull female — black parts greyer.*

Male

Race: *tibicen* (Black-backed form) The nominate race and the most widespread. **Female**: *nape mottled dusky grey* **Immatures**: *like dull female — black parts greyer.*

Male

4, 7 GREY CURRAWONG

Strepera versicolor 50 cm

The Grey Currawong shows great variation in plumage colour throughout southern Australia, ranging from grey in the east, dusky grey in central and Western Australia, through a darker form (the Black-winged Currawong which lacks a white wing patch) in South Australia and the Murray mallee, to the darkest of all, the race *arguta* in Tasmania, once called the Clinking Currawong. It frequents forests, woodlands, dense scrub and mallee, gardens and orchards. In the east it is plentiful in alpine country. It is unobtrusive and shy living mainly in pairs, its presence usually betrayed only by the loud ringing double-note call. Unlike the Pied Currawong, flocks are rarely encountered except in south-western Australia and Tasmania. It feeds on native and cultivated fruits and berries, insects taken on the ground or from under tree bark prised up with the powerful beak. It also raids nests for eggs and nestlings. Breeds July-November. Nests to 15 m high. Endemic. **Female**: *smaller.* **Immatures**: *duller and browner.* **Voice**: *clinking, likened to 'an anvil being struck'; brassy 'tew-tew' or 'dink-dink'; also a feline mew.*

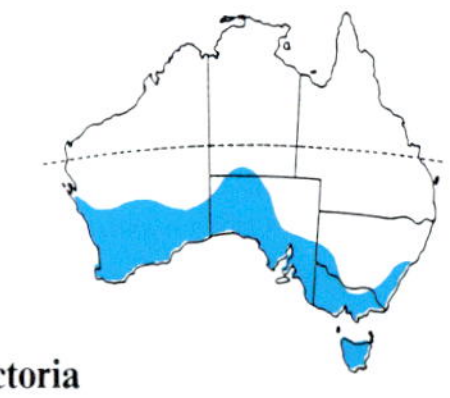

Top: photographed in Western Australia

Lower: photographed in Victoria

5 BLACK CURRAWONG

Strepera fuliginosa 47 cm

Confined to Tasmania and islands of the Bass Strait, and widespread. It frequents the wetter forest and higher tablelands in ones and twos in summer and forms large flocks in winter. Some birds descend to lower altitudes during this period, and become nomadic. Breeds August-December. Nests to 20 m. Endemic. **Recognition** *(cf Grey Currawong): blacker; white tail and wing tips almost absent.* **Sexes**: *alike.* **Immatures**: *like adults except for yellow gape.* **Voice**: *loud ringing kar-week week-kar.*

6 PIED CURRAWONG

Strepera graculina 46 cm

The bold and aggressive Pied Currawong, is one of eastern Australia's most conspicuous birds. It lives mainly in wooded country and is often seen and heard in suburban parks and gardens. High country populations tend to descend to lower altitudes in winter, and may gather in large flocks. It feeds on insects, nestlings and eggs, carrion and berries. Nests to 20 m. Endemic but extends to Lord Howe Island. **Sexes**: *alike.* **Immatures**: *grey-brown.* **Voice**: *ringing 'cur-ra-wong' and 'wolf-whistles'.*

8 BLACK BUTCHERBIRD

Cracticus quoyi 32-45 cm

There are two forms of this the largest of the butcherbirds. More northerly populations are larger, have a narrower bill and are wholly black; those south of Cooktown are smaller with a broader bill, and go through a rufous phase from immaturity. Sedentary, it frequents all levels of forest and mangroves and feeds on small animals, eggs, crustaceans and fruit. Breeds September-February. Nests to 10 m or in lower mangrove tree. One race in New Guinea. **Female**: *smaller.* **Immatures**: *generally — dull black washed brown; north-eastern Queensland only — mostly rufous.* **Voice**: *beautiful bubbling or yodelling 'ah-oo-ah'.*

9 PIED BUTCHERBIRD

Cracticus nigrogularis 32-35 cm

In pairs, small parties and family groups, the Pied Butcherbird lives in open woodland or timbered pastoral land and is widespread across its range except in desert areas. In the south-west it is slowly extending its range southwards as forests are cleared. Northerly populations are smaller than those in the south. It feeds on nestlings, insects, mice, snakes and lizards. Breeds August-December. Nests high. Endemic. **Sexes**: *alike.* **Immatures**: *brown-grey where adult plumage is black.* **Voice**: *outstandingly beautiful song of loud mellifluous notes, sometimes by moonlight; some mimicry.*

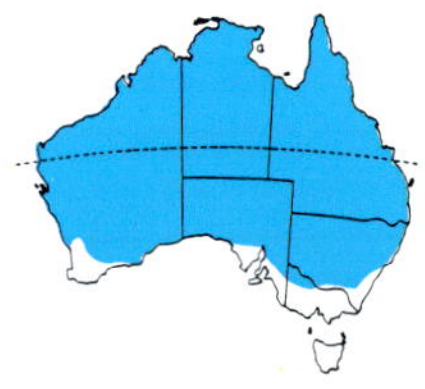

10 SILVER-BACKED BUTCHERBIRD

Cracticus torquatus argenteus 28 cm

Two silver-backed races of the Grey Butcherbird occur in north-west Australia — *argenteus* (illustrated) in the Northern Territory, and *latens*, a slightly different form in the Kimberley. They closely resemble the Grey Butcherbird in behaviour. Breeds July-January. Nests 1-7 m from the ground. Endemic. **Female**: *brown wash on crown.* **Immatures**: *browner, white parts duller.* **Voice**: *melodious rollicking whistling, often in duet; mimicry.*

11 GREY BUTCHERBIRD

Cracticus torquatus 28-32 cm

Its pre-eminently lovely song is one of the hallmarks of this familiar bird. It is equally well-known for impaling nestlings and other prey to be eaten later. In any timbered areas it may be seen singly, in pairs or family parties. Being sedentary, it may become relatively tame in gardens but is fearless in defence of the nest. Tasmanian birds are larger. Breeds July-January. Nests 1-7 m high. Endemic. **Female**: *brown wash on crown.* **Immatures**: *browner, white parts duller.* **Voice**: *loud melodious rollicking whistling, often in duet; some mimicry.*

Male

12 BLACK-BACKED BUTCHERBIRD

Cracticus mentalis 25 cm

This Cape York-dwelling bird is a tropical variant of the other butcherbirds. In forest or woodland, and around any settlement, it is seen singly or in pairs. It forages in trees and shrubs for insects, lizards, small birds and nestlings, and eats some seeds. Breeds October—December. Nests high in tree fork. Also in New Guinea. **Recognition**: (cf Pied Butcherbird): *the differences are mainly its white instead of black throat and smaller size.* **Sexes**: *alike.* **Immatures**: *duller* **Voice**: *euphonious carolling, softer than Grey Butcherbird; some mimicry.*

1

2

3

4

5

6

7

8

9

10

11

12

16

13, 14 PEEWEE (Australian Magpie-lark)

Grallina cyanoleuca 26-30 cm

All over Australia, except in Tasmania where it only occurs as a vagrant, the widespread Peewee lives mostly in open timbered country, always near water for the mud needed to construct the nest, and is frequently seen in wooded surburbia and city parks. It is bravely combative and noisy in defence of its territory and chases off much larger birds. While readily accepting the presence of humans, it may unexpectedly attack them in the breeding season. It sometimes also attacks its own image in reflective surfaces like windows, or even a car windscreen. It feeds on small invertebrates and insects, also aquatic creatures. When foraging on the ground, its head bobs pigeon-like as it walks. Established pairs are generally sedentary, but non-breeding birds form nomadic flocks that can reach several thousand. Breeds all months in response to rain. Builds mud nest plastered to a tree, pole or building. Endemic, but recently recorded in New Guinea. **Female**: *similar, but throat white encircled with black band, no white eyebrow.* **Immatures**: *white eybrow and throat.* **Voice**: *mainly strident 'pee-o-whit' or 'pee-o-whee', and a liquid 'orch-id'.*

Male left, female right

15 SPANGLED DRONGO

Dicrurus hottentottus 30 cm

Singly, in pairs or parties, this bird, distinguished by its forked 'fishtail', frequents forest fringes, open woodland, mangrove, parks and gardens. It feeds mainly on insects snapped up in the air, or gleaned from foliage; also nectar. Eastern birds are partly migratory after breeding, some moving northwards as far as New Guinea, others southwards. Breeds September-March. Nests on horizontal fork to 15 m high. Also in SE Asia and islands to the north and north-east of Australia. **Female**: *smaller.* **Immatures**: *head dull black spangled blue, spotted white below, brown eye.* **Voice**: *harsh twanging and tearing sounds, chattering, creaky whistle.*

16 INDIAN [Common] MYNAH

Acridotheres tristis 24 cm

Not to be confused with the native Noisy Miner, this bird, a native of India, was originally introduced to Melbourne in 1860 and subsequently released in Sydney and parts of Queensland, including the canefields to combat insects, but without success. It has become one of Australia's worst avian pests, and many bemoan the way its aggressive territorialism has driven off much-loved native birds. It lives in pairs and family groups and roosts communally with great hullaballoo in park and street trees, under bridges and elsewhere. Any source of food is exploited. Breeds October-March. **Sexes**: *alike.* **Immatures**: *duller.* **Voice**: *strident rollicking chatter; accomplished mimic.*

17 COMMON STARLING

Sturnus vulgaris 21 cm

A native of Europe, the Common Starling, introduced into Victoria in the mid-19th century, has since spread extensively through urban areas, cleared land and native bush. Although it usefully preys on insect pests, it displaces native birds, damages fruit crops, and becomes a nuisance by nesting in house roofs. Declared a pest in Western and South Australia and vagrants are shot on sight. It feeds on fruit and insects. It gathers in very large flocks in autumn and winter and thousands may settle for the night in communal roosts. **Sexes**: *alike.* **Immatures**: *grey-brown upperparts, paler below.* **Voice**: *whistling, wheezy rasping, rattling notes; mimicry.*

18 SHINING [Metallic] STARLING

Aplonis metallica 24 cm

This highly sociable bird arrives from New Guinea to breed in north-eastern Queensland in August and departs in March. Huge breeding colonies take over large rainforest trees in which to roost and nest and hundreds of their large untidy nests may festoon a single tree. Fruit is the main diet and dense feeding flocks may crowd a fruiting tree, with birds dashing continuously between the feeding and nesting areas. The noise set up by a colony can be deafening. Also in Indonesia and the Solomon Islands. **Sexes**: *alike.* **Immatures**: *duller, white underparts streaked black.* **Voice**: *incessant medley of brassy chattering and whistling.*

19 WHITE-WINGED CHOUGH

Corcorax melanorhamphos 47 cm

The highly sociable White-winged Chough lives in forest, woodland and mallee, always near water. It forages mainly on the ground, often at roadsides and is usually seen in parties of up to 20 methodically turning over litter in search of insects, small animals and seeds. Flocks of up to 100 are not uncommon in winter. Breeds June-March. Builds mud nest in tree up to 10 m. Endemic. **Recognition** *(cf Currawongs): white wing window visible mostly in flight; smaller down-curved bill; red eye and longer tail.* **Sexes**: *alike.* **Immatures**: *smaller, tail shorter, eye brown.* **Voice**: *plaintive whistling and mournful piping, strident alarm call.*

20 LITTLE CROW

Corvus bennetti 48 cm

Familiar in country towns and outback stations. Highly nomadic when conditions are bad inland and may travel great distances in large flocks after breeding. Mainly eats insects, also grain and carrion. Performs spectacular aerobatics in flight. Breeds in spring or after rain in desert regions. Nests in loose colonies in scrubby trees and on poles. Endemic. **Recognition**: *flips wings up and tail down in time with each 'nark'; no throat hackles.* **Sexes**: *alike.* **Immatures**: *underparts sooty brown, eye brown.* **Voice**: *nasal 'nark-nark-nark' and warbling calls.*

21 TORRESIAN (Australian) CROW

Corvus orru 50 cm

This mainly tropical corvid lives a mostly sedentary existence in farmland, woodland and along timbered inland water courses. It feeds on insects, crops and carrion. Immatures form flocks until 2 years old. Breeding season varies across range. Nests high in eucalypts and among pylons. Also in New Guinea. **Recognition**: *(cf Australian Raven) voices differ; repeatedly shuffles the wings on landing; does not flip them when calling; throat hackles inconspicuous.* **Immatures**: *brown tinge to feathers, brown eye.* **Voice**: *repeated nasal, high-pitched, clipped notes 'uk-uk-uk-uk' or 'ok-ok-ok-ok'.*

22 FOREST RAVEN

Corvus tasmanicus 52 cm

The only corvid in Tasmania. Small isolated populations also live on the mainland where they frequent dense forest within the ranges of the Australian and Little Ravens. Widespread in Tasmania where it frequents all types of habitats from dense forest in the high country to coastal woodlands. Nomadic flocks form in the non breeding winter months. Breeds August-November. Nests high in tree. Endemic. **Recognition**: *short and indistinct throat hackles; shorter tail and larger bill than other ravens.* **Sexes**: *alike.* **Immatures**: *underparts sooty brown, eye brown.* **Voice**: *loud very deep 'kor-kor-kor-r-r'.*

23 LITTLE RAVEN

Corvus mellori 50 cm

The gregarious Little Raven lives in open alpine woodlands, inland plains and well watered pastoral areas. Often in nomadic flocks, it feeds on large insects, carrion and spilt grain. Inland birds migrate in large flocks south-east during summer heat, and alpine birds descend to lower levels to winter. Breeds August-December; later in alpine regions. Nests in trees, also on poles or in low bushes inland. Endemic. **Recognition**: *shuffles or flicks wings above back when calling; throat hackles inconspicuous.* **Sexes**: *alike.* **Immatures**: *underparts sooty brown, eye brown.* **Voice**: *clipped 'kar-kar-kar'.*

24 AUSTRALIAN RAVEN

Corvus coronoides 56 cm

Largest of the corvids, and common in most wooded habitats and well watered pastoral areas. Often wary and difficult to approach. Usually in pairs or small flocks of immatures. Feeds on large carrion, including dead or dying lambs, also insects and grain. Breeds July-September. Nests high in tall tree. Endemic. **Recognition**: *calls with head lowered and tail nearly horizontal — seldom flicks wings; long conspicuous hackles on bulging throat; bare skin on sides of chin.* **Female**: *slightly smaller.* **Immatures**: *brown wash on underparts, eye brown.* **Voice**: *drawn out 'ahh-ahh-ahh-r-r-r' including in flight.*

13

14

15

16

17

18

19

20

21

22

23

24

PART TWO
AQUATIC BIRDS
BIRDS OF THE INLAND WATERS, SHORE AND OCEAN

17

Long-legged Waterbirds

The species included in this Section, although belonging to three different orders and five families, have one feature in common — they are all long-legged and most wade deep into marshes, river margins and tidal estuaries in search of food.

HERONS, EGRETS AND BITTERNS Family Ardeidae

The sixty-four species of this world wide family of lanky birds are characterised by their long dagger-like bills, long necks, and by the way they fly — with necks neatly tucked back on the body in an S-shape and legs trailing. They feed on fish, amphibians, crustacea and other aquatic creatures, and adopt a variety of hunting techniques, from waiting motionless for a moment to make a strike to active trawling and clever strategems for luring prey. Another feature of these birds is their highly-developed powder-down feathers which are never shed and continually fray at the ends to produce a talcum-like powder for dressing the other feathers. Removing mud and grime picked up from hunting grounds is further assisted by combing with a serrated nail on the middle toe. The family divides into four fairly distinctive groups or tribes: Day (or Typical) Herons and Egrets; Night Herons; Bitterns and Tiger Herons. The first three groups are represented in Australia with fourteen species:

Day (or Typical) Herons and Egrets The name 'heron' or 'egret' has no taxonomic significance, but there is a tendency to call the white birds 'egrets' and the variously plumaged birds 'herons'. The Cattle Egret, however, adopts orange-buff markings when breeding, and the plumage of the Reef Heron is either white or grey. The bills, legs and face of egrets undergo distinctive colour changes when breeding and are an important aid to identification. At other times, clues are pretty well limited to size alone. Egrets are famous for their beautiful plumes, or aigrettes, developed when nesting. Until the early part of this century, vast numbers of them were slaughtered to satisfy the vogue for 'osprey feathers' in women's hats. They were mostly killed during the breeding season, and countless orphaned broods perished too. The practice was worldwide, and several species of egrets had all but been wiped out before public outrage outlawed the trade. Of thirty-five species around the world of this tribe, ten occur in Australia. Their sizes range from 48 cm for the Pied Heron to 1.5 m for the Great-billed Heron. 1-15

Night Herons Night Herons are mostly active from dusk to dawn and are rarely seen during the day. Of nine species worldwide, one only occurs in Australia. 16-17

Bitterns Bitterns live a secluded existence in reedy marshes where the camouflage of their streaked plumage gives them protection. When threatened, they freeze, with body, neck and bill pointing skyward — and they even sway in rhythm with moving reeds. Three of the twelve species breed in Australia; the Yellow Bittern is a rare non-breeding visitor. 18-20

IBISES AND SPOONBILLS Family Plataleidae (or Threskiornithidae)

Ibises have long slender downcurved bills and spoonbills have straight bills with broadened spatulate tips. Despite these marked differences, they have much in common, including bare skin heads and faces and the absence of a voice box. They also fly with necks extended forward and legs trailing, sometimes in huge wheeling flocks at great heights. There are about thirty-two species distributed in the tropical and temperate regions of the world. Twenty-six of them are ibises, with three in Australia; and six are spoonbills, two of which are common in Australia. In times of grasshopper plagues, pasture-lands are sometimes covered with hordes of ibises. The Sacred Ibis is related to its namesake in Africa which was deified and mummified in antiquity by the Egyptians — but now a rare bird in that country. Its generic name *Threskiornis* embodies the Greek word *threskos* meaning sacred. 21, 22, 25, 27, 29

CRANES Family Gruidae

Cranes are famous for their spectacular dancing displays, in which pairs and groups leap up and down in gay abandon with deep bowings to each other. They are distributed widely except in South America and Antarctica and live mostly in wetlands. They fly in V-formation with head stretched forward and legs trailing, sometimes at great heights. Of fourteen species, two occur in Australia: the Brolga, which is endemic, and the Sarus Crane, which was first recorded in Australia only in 1966. 23, 24

STONE-CURLEWS Family Burhinidae

This family of nine species, found throughout the tropical and temperate parts of the world, is represented in Australia by two species which live in very different habitats: the Bush Stone-curlew in lightly-forested country, and the Beach Stone-curlew on the beaches, estuaries and around the coastal islands. The inelegant name 'thick-knee' has the merit of being descriptive. 28, 30

STORKS Family Ciconiidae

The Jabiru, or Black-necked stork, is Australia's only representative of the widespread stork family of seventeen species. The name Jabiru, familiar to most Australians, has a long history of usage and dates from the listing of the species in John Latham's *Index of Ornithology* of 1790 as the 'New Holland Jabiru'. Jabiru is the generic name for the world's largest stork, which is native to South America. Its origin is given in Oliver Goldsmith's *History of the Earth and Animated Nature* (1774) — a work which also influenced the adoption of the name Emu (see Section 1) — and is described thus: 'Jabiru Native of Brazil. The skin of the neck is wrinkled, and so flaccid that it depends like the dewlap of a cow. This circumstance has given rise to the name of Jabiru, which is in the language of the Guaianis signifies anything inflated by the wind.' 26, 31

Previous page: Sooty Terns
Opposite: Royal Spoonbill

PHOTOGRAPHERS: H. & J. Beste 3, 11, 12, 18; G. Chapman 28; J. Frazier 31; T. & P. Gardner 13, 14, 15, 16, 17, 25, 29; R.A. Graves 20; C.A. Henley 7; K. Ireland 26; P. Jordan 30; J. McCann 23; A.R. McGill 5; P.D. Munchenberg 19; L.F. Schick 4; E.C. Slater 8; F. Spolc p. 167; G. Weber 6; Authors 1, 2, 9, 10, 21, 22, 24, 27.

17

1, 2 REEF HERON [Eastern Reef Egret]

Egretta sacra 61-66 cm

Reef Herons, widespread throughout south-eastern Asia, New Guinea and islands of the south-west Pacific, are essentially coastal birds and are found all around the coastline and offshore islands of Australia — though rarely in Tasmania. The Australian birds can be either grey or white. White birds exceed grey in the north and vice versa in the south — with only grey birds in southern Australia. Both forms interbreed and grey and white nestlings often hatch in the same brood. The Reef Heron is usually seen alone patrolling tidelines and reefs in search of small fish and other marine creatures. It breeds all months, but mostly September-January, and nests in loose colonies in trees or on rock ledges. **Recognition**: *the white form is distinguishable from other white egrets by its shorter and thicker legs and crouching hunting gait.* **Adults and immatures**: *similar. Immatures of the grey form are brown.* **Voice**: *loud croaks.*

Grey form left; white form right

3 LITTLE EGRET

Egretta garzetta 56 cm

Usually seen singly in shallow water actively 'puddling' to dislodge prey from the bottom. Breeds all months depending on the food supply, and nests in colonies in trees often with other water birds. Also occurs in New Guinea, New Zealand, Europe and Africa, and throughout Asia. **Recognition**: *(Breeding) two ribbon-like nuptial plumes on head and others on back and breast; black upper mandible, legs and feet; yellow face, eyes and soles. (Non-breeding) similar except for absence of plumes.* **Adults and immatures**: *alike.* **Voice**: *croaks and growling sounds.*

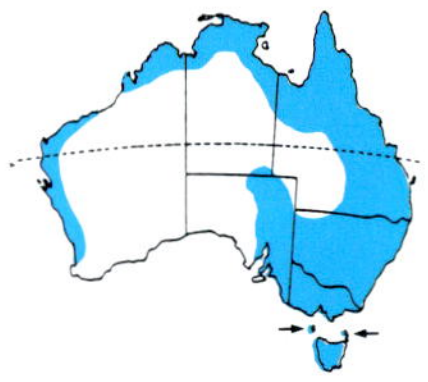

4, 5 CATTLE EGRET

Ardeola ibis 53 cm

In 1933, a small number of Cattle Egrets, native to Africa and Asia, were introduced into north-western Australia. Coincidentally, there was occurring one of the most remarkable events in the story of ornithology when Cattle Egrets began spreading from their original homes across the world to colonise new territories including America. Their invasion of Australia was first reported in 1948, since when the species has spread rapidly and continues to do so. The birds are usually seen in flocks in the vicinity of grazing cattle and other farm animals, feeding on the insects they disturb. They sometimes perch on the animal's back. Breeds all months depending on food availability. Nests in colonies in trees with other water birds. **Recognition**: *bill and neck shorter and thicker than other herons. (Breeding) orange-buff head, breast and back with nuptial plumes especially on back; reddish-orange bill, face and eyes; dark legs. (Non-breeding) yellow bill, face and eyes; dark legs lighter above knee.* **Adults and immatures**: *alike.* **Voice**: *low croaks and guttural sounds.*

Breeding left; non-breeding right

6 PLUMED [Intermediate] EGRET

Egretta intermedia 63 cm

The spectacular nuptial plumes of this bird fall in a cascade of delicately-fringed aigrettes to form a mantle over its back which sweeps well beyond the tail. Solitary or in groups, its behaviour and general appearance is akin to that of the Great and Little Egrets. Breeds throughout the year when food is plentiful. Nests in colonies often with other water birds. Also in New Guinea, Africa and parts of Asia. **Recognition** *(Breeding) bill orange-red; eyes yellow, facial skin blue-green, upper legs red, feet black. (Non-breeding) bill orange-yellow, eyes yellow, facial skin yellow-green, legs blackish.* **Immatures**: *as winter adult.* **Voice**: *croaks and growling sounds.*

Breeding

7 GREAT-BILLED HERON

Ardea sumatrana 1.5 m

This giant heron is wary and secretive and spends most of the day skulking in mangrove swamps bordering tidal rivers and estuaries. It flies out from its deep seclusion on the receding tide, especially at dusk, to feed on the mudflats and in shallow waters. Its prey includes quite large fish. Sedentary. Breeds in all months. Nests in a tree over water. Also in New Guinea and South-east Asia. **Sexes**: *alike.* **Immatures**: *rusty brown.* **Voice**: *harsh croaks and said to 'roar like an angry bull'.*

8 WHITE-NECKED (Pacific) HERON

Ardea pacifica 90 cm

Australia is the stronghold of this stately heron, but it is occasionally seen in New Guinea and New Zealand. It is fairly widely distributed inland, less so towards the coast, and lives in mainly shallow fresh water areas and wet paddocks. It seldom frequents salt or brackish waters. Usually alone, it stalks through wet grass or wades in shallow waters hunting for frogs, insects, crustacea and fish. It is strongly nomadic and large eruptions may occur in an area after good rains. Breeds all months. Nests in tree usually in or over water, often in small colonies. **Sexes**: *alike out of breeding season.* **Immatures**: *grey neck spotted with black.* **Voice**: *deep croaks.*

9 GREAT EGRET

Egretta alba 91 cm

This elegant bird, the largest of the white egrets in Australia, is distributed worldwide in warm and temperate areas and is probably known by more English names than any other bird, including Large Egret (Australia), Great White Egret (Europe), Common Egret (America), Great Egret (Japan and Australia) and White Heron (New Zealand). It is usually seen stalking alone in shallow inland and coastal waters ready to strike with its long bill to seize a fish or other aquatic prey. Breeds at anytime according to food supply but mainly October-March. Nests in colonies in tree-tops usually with other water birds. Recognition: *(Breeding) nuptial plumes on back extend well beyond tail; bill black, eyes yellow, face greenish, legs dark below paler above. (Non-breeding) bill, face and eyes yellow, legs blackish.* Adults and immatures: *similar.* **Voice**: *deep croaking and cawing notes at breeding colony.*

Non breeding

1

2

3

4

5

6

7

8

9

17

10 WHITE-FACED HERON

Ardea novaehollandiae 68 cm

Familiar in estuaries, inland lakes, swamps and wet grasslands across the continent, it is probably Australia's best known heron. It waits and watches in shallow water for prey, sometimes puddling with its feet to disturb aquatic animals, and also stalks for reptiles, young birds, frogs, rodents and insects. It often feeds in associated flocks. Sedentary or nomadic in search of food. Breeds all months. Nests high in a tree, sometimes well away from water. Also in New Guinea, New Zealand and New Caledonia. **Sexes**: *alike.* **Immatures**: *lack white on head and cinnamon on breast.* **Voice**: *harsh croak.*

11, 12 PIED HERON

Ardea picata 48 cm

This dainty heron of the tropics, frequents salt and freshwater swamps and nearby grasslands usually in loose feeding flocks of 30 or more. It feeds on fish and small invertebrates and sometimes scavenges at garbage dumps in search of food. Large flocks of several hundreds may gather around feeding areas in times of drought. It is aggressive towards other species to the point of stealing their prey or forcing them to disgorge. Sedentary, migratory or nomadic. Breeds September-November. It nests in large colonies in trees. Also in New Guinea and Indonesia. **Sexes**: *alike.* **Immatures**: *head white without a crest.* **Voice**: *loud 'ohrk', soft coo.*

Breeding left, eclipse right

13, 14, 15 MANGROVE [Striated] HERON

Butorides striatus 43-51 cm

The Mangrove Heron is found extensively throughout the tropical and subtropical regions of the world and belongs to a group of some thirty races sometimes referred to as the 'green herons'. In Australia, there are two forms: a greenish grey-backed form, which is commonly seen all along the northern and eastern coasts; and a rufous form, which occurs in north-western Australia, and sporadically on the Queensland coast and elsewhere. It is characterised by short legs and stumpy body, and skulks looking more like a bittern than a heron. It mostly frequents mangrove swamps, tidal estuaries and mudflats. It roosts when the tide is in among the mangroves, and moves out onto the mudflats as it recedes. It usually hunts alone crouching low on the mud, sometimes waiting and watching absolutely still and at other times stalking stealthily along, poised to jab at its prey. It feeds on crustaceans, fish, molluscs, insects and particularly mud-skippers. Very wary, it freezes if disturbed or flies off low. Breeds September-December. Nests in dense foliage over water. **Sexes**: *alike.* **Immatures**: *(see illustration).* **Voice**: *squawks, explosive 'hoo'.*

Rufous form left; grey form centre; immature right

16, 17 NANKEEN [Rufous] NIGHT HERON

Nycticorax caledonicus 56-65 cm

The strange term 'nankeen', traditional also for the Nankeen Kestrel, was popularly used in earlier days for a buff-coloured cotton cloth imported from Nanking in China. This elegant bird, which sports usually two long head plumes throughout the year, is essentially nocturnal in habits. It is widespread across the continent in a varied habitat which includes swamps, lagoons, river margins and mangrove estuaries. It roosts communally by day, sometimes in large colonies, in leafy trees usually near water and takes wing at dusk in slow silent flight to the feeding grounds. It feeds mainly on fish, but also takes a variety of other aquatic prey, insects, and even mice. Its range extends through New Guinea to south-eastern Asia, and includes islands of the western Pacific. Breeds September-March/April. Nests in a tree or shrub up to 25 m high above water; also on ground on treeless islands. **Sexes**: *alike.* **Immatures**: *(see illustration).* **Voice**: *loud gutteral 'kwak'.*

Adult left; immature right

18 BROWN [Australasian] BITTERN

Botaurus poiciloptilus 66-76 cm

This heavy bird is confined to Australia and New Zealand and is the biggest of the Australian bitterns. It lives in dense reed beds fringing rivers and swamps. It is most active at dusk and during the night and hunts alone. If approached, it silently fades away and is rarely seen. Its food includes yabbies, eels, fish and small mammals and insects. Seems to be nomadic. Breeds October-January. Nests sometimes colonially in reeds. **Recognition**: *cf illustration of immature Nankeen Night Heron if in doubt.* **Sexes**: *alike.* **Immatures**: *paler.* **Voice**: *legend has it that its weird loud booming at night is the call of the fabled bunyip.*

19 BLACK BITTERN

Dupetor flavicollis 54-66 cm

The Black Bittern lives in thick vegetation along rivers, creeks and mangroves, sometimes in rainforest near water, and is more inclined to show itself in the open than other bitterns. By day, it roosts in a tree, at night it forages singly or in pairs for fish, frogs or crustaceans. Probably sedentary. Breeds September-January. Nests semi-colonially in trees over water. Also in New Guinea to India, China and the Philippines. **Female**: *upperparts browner, less streaked below.* **Immatures**: *duller, breast brown streaked yellow.* **Voice**: *long drawn out croaking, soft 'coo-oorh'.*

20 LITTLE BITTERN

Ixobrychus minutus 30 cm

The extremely secretive and elusive Little Bittern lives in thickly vegetated swamps, lakes and rivers and is believed to be nocturnal. It freezes at the slightest provocation, rarely takes flight, and steals away if approached. Usually solitary, it feeds on small fish, insects and frogs. Probably migratory. Breeds October-December. Nests on a platform of rushes and reeds in swamp vegetation or on tree branch over water. Widely distributed except in the Americas where a closely-related species occurs. **Female**: *crown and back brown, underparts streaked brown.* **Immatures**: *browner, heavily streaked.* **Voice**: *deep croaking monotonously repeated.*

21 GLOSSY IBIS

Plegadis falcinellus 52 cm

This, the smallest of the ibises in Australia, frequents mostly inland fresh shallow waters including wet grasslands, swamps and lagoons. It is gregarious and generally feeds in flocks on a variety of insects, molluscs, crustaceans; sometimes rice seeds. As it is strongly nomadic, its distribution is patchy and variable. Breeds variably in response to rainfall. Nests in colonies in low vegetation standing in water, mostly with other waterbirds. Almost worldwide distribution. **Adults**: *glossier, brighter in colour, head less streaked, bill more curved.* **Female**: *bill shorter.* **Voice**: *croaks and grunts.*

Immature

10

11

12

13

14

15

16

17

18

19

20

21

17

22 SACRED (White) IBIS

Threskiornis aethiopica 70 cm

Related to the famed Sacred Ibis of Egyptian antiquity, as distinct from the White Ibis of America. Common and locally abundant — flocks of over 10,000 have been recorded in Victoria. It lives in swamps, margins of streams and lakes, adjoining grasslands, and city parks where it can become tame. It feeds variously on aquatic creatures, insects such as grasshoppers and garbage. Breeds August-November in south, February-May in north. Nests in mixed colonies over water or in swamp vegetation. Range extends to Africa. **Sexes**: *similar but bill of female shorter.* **Immatures**: *head and neck feathered.* **Voice**: *grunts.*

25 STRAW-NECKED IBIS

Threskiornis spinicollis 70 cm

The most numerous of the ibises in Australia — over 200000 breeding birds have been recorded in South Australia. In pairs and flocks it frequents swamps and the margins of lakes and streams, also dry pastures, and feeds on frogs, small reptiles, mammals and, usefully, locusts and grasshoppers. It is highly nomadic, and odd birds stray as far as New Guinea; otherwise it is an endemic species. Breeds variably, depending on rain. Nests colonially in swamp vegetation and trees. **Female**: *bill shorter.* **Immatures**: *lacks neck plumes, feathered to crown, neck shorter.* **Voice**: *grunts and croaks.*

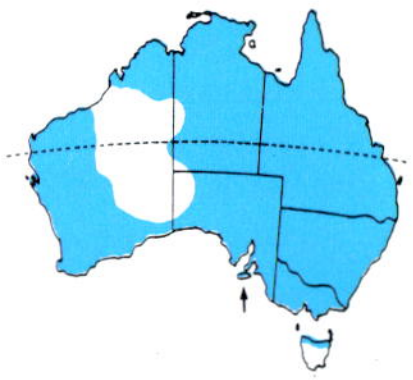

27 ROYAL SPOONBILL

Platalea regia 75 cm

The Royal Spoonbill is distinguished by its black bill and legs, and a spectacular crest of head plumes in the breeding season. It is frequently seen wading singly or in small parties, sometimes in flocks in shallow coastal and inland waters, and feeding by sweeping its long spoon-shaped bill from side to side. It is partial to shrimps and also takes fish and molluscs. Sometimes flies at great heights. Breeds October-May. Nests in colonies in tall trees over water. Range extends to Indonesia and New Zealand. **Female**: *bill shorter.* **Immatures**: *bill shorter, flight feathers tipped black.* **Voice**: *low grunting and croaking; bill clapping.*

29 YELLOW-BILLED SPOONBILL

Platalea flavipes 89 cm

When breeding, this spoonbill grows long nuptial breast plumes instead of a crest. It is the more abundant of the two spoonbills, and more often seen inland and in small areas of water, such as farm dams. It feeds on fish, aquatic insects and molluscs. It sweeps methodically with its flat bill and stirs the bottom to dislodge prey. It is highly nomadic with an erratic distribution. Breeds September-January or according to conditions. Nests in colonies or in tree in or near water. Endemic. **Sexes**: *male's bill longer than female's.* **Immatures**: *face not outlined black.* **Voice**: *feeble grunts, bill claps.*

23 BROLGA

Grus rubicundus 1.25 m

The elegant Brolga, one of the world's largest cranes, is famed for its graceful mutual dancing display rituals performed during courtship and socialising, when the birds bow, advance and retreat, trumpet and fling objects into the air. It lives near swamps, shallow lakes, wet grasslands and pastures, and has a varied diet of vegetable matter, insects, cereal grains and sedge tubers. Sometimes seen in very large flocks in the north of its range during the drier months, and it is nomadic. Breeds October-April. Nests on mound in shallow water. Also in New Guinea. **Sexes**: *similar.* **Immatures**: *facial skin paler pink.* **Voice**: *various trumpetings and hoarse croaks.*

28 BEACH STONE-CURLEW [Thick-knee]

Esacus magnirostris 53-58 cm

The Beach Stone-curlew, alone and in pairs, frequents quiet sandy beaches and mudflats of northern and eastern Australia and around coastal islands. It is sedentary and sightings are usually far between. During high tide it shelters in mangroves or foreshore trees, emerging as the water recedes to feed on crabs, crustaceans and insects. It is active at night. Breeds October-February. Nests in a scrape in the sand. Also in New Guinea and SE Asia, and Pacific islands. **Sexes**: *alike.* **Immatures**: *similar but duller.* **Voice**: *banshee wail higher pitched than Bush Stone-curlew.*

30 BUSH STONE-CURLEW [Thick-knee]

Burhinus magnirostris 50-58 cm

The Bush Stone-curlew lives in open woodland, edges of forest, and along inland watercourses. It is still abundant in parts of the tropical north but is approaching extinction in pastoral areas in the south where once it was common. It is wary, and may crouch near ground if approached rather than fly away. It is mostly active nocturnally and emerges at dusk sometimes in groups to fly perhaps some distance to a feeding area to forage for insects, spiders and fruits. Sedentary. Breeds August-January. Nests in a scrape on the ground. Endemic. **Sexes**: *alike.* **Immatures**: *paler.* **Voice**: *loud wailing 'wee-loo', often at night.*

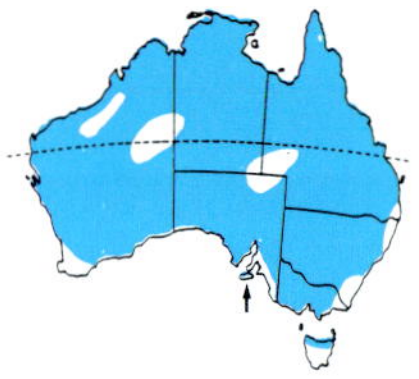

24 SARUS CRANE

Grus antigone 1.3-1.4 m

The existence of the Sarus Crane in Australia was not known until about 20 of them were spotted on the Normanton River in North Queensland in 1966. It is distinguished from the Brolga by an extension of the bare red facial skin downwards to form a distinct collar on the neck, and by its pink legs instead of grey. Its habits are similar. It disperses after breeding and forms large, loose feeding flocks. Breeds mainly January-March. Like the Brolga it nests on a platform of vegetation just above water level. Also in India and South-East Asia. **Sexes**: *alike, female smaller.* **Immatures**: *like Brolga.* **Voice**: *wild trumpeting usually in duet.*

26, 31 JABIRU [Black-necked Stork]

Xenorhynchus asiaticus 1.29 m-1.70 m

It is not without interest that John Latham named the Australian stork the New Holland Jabiru (see Introduction) for it is about the same size as the large South American Jabiru and has a similar hefty black bill — whereas the more familiar White Stork of continental Europe is smaller and has a thin yellow bill. The stately Jabiru is better known in the north than in the south but it is nowhere common. It mostly frequents freshwater lagoons, estuary mudflats, mangrove swamps, swampy ground and flooded pastures, and quietly paces through the shallows probing for fish, amphibians, and crustaceans. Except when breeding, it is usually seen alone. It is wary and flies off if approached. It labours to become airborne and its flight is ponderous. Sometimes soars to great heights. It is partly nomadic. Breeds March-June. Nests at or near the top of a tree or a large bush up to 25 m. Its range extends through New Guinea to south-eastern Asia and India. **Sexes**: *female — eye yellow, male — eye black.* **Immatures**: *like adults but black is brown and white is dusky.* **Voice**: *almost mute, claps bill at nest.*

Male above, female below.

22

23

24

25

26

27

28

29

30

31

18 Waterfowl

The waterfowl family Anatidae embraces Swans, Geese and Ducks and consists of 148 species. In Australia, the driest of continents, an exceptionally small proportion of the land provides habitats favourable to waterfowl and only twenty-three species are represented. Of these, two are introduced and ten are endemic. Most are nomadic, with their movements largely influenced by Australia's erratic rainfall patterns. After breeding, nearly all waterbirds moult their flight feathers all at once and become flightless for several weeks. The groupings which follow accord with the divisions adopted by H.J. Frith in his *Waterfowl in Australia* (1967), and consist of three subfamilies and thirteen tribes.

ANSERINAE (subfamily)

The five tribes of this subfamily are somewhat goose-like in appearance.

Swans, tribe Cygnini Swans have been extolled from time immemorial for their grace and beauty, and have been immortalised in mythology — such as in the legend of Leda and the swan. They are the largest of the waterfowl, with seven species worldwide. Five of them are pure white and live in the higher latitudes of the Northern Hemisphere. The other two, the Black-necked Swan of South America and Australia's Black Swan, are confined to the Southern Hemisphere. It was during an exploratory expedition to the west coast of Australia of three Dutch ships led by Willem De Vlamingh in January 1697 that the Black Swan was discovered on what became known as the Swan River. Three live specimens were taken to Batavia. There is an ancient belief that swans sing only when they are dying. According to this legend, it is only then that the Mute Swan becomes vocal and when mortally wounded glides to water singing a strange and haunting 'swan-song'. 1-2

Geese, tribe Anserini Except in farmyards, the fourteen species of the large spectacular geese, which live mainly in the Northern Hemisphere, are unrepresented in Australia. (The Cape Barren Goose, Magpie Goose and Pygmy-geese, though exhibiting certain goose-like characteristics, are not true geese and are included in other groups.)

Whistling Ducks, tribe Dendrocygnini When Whistling Ducks are around, the air vibrates with a chorus of their shrill whistling — and the whistle of their wings in flight. Two of the eight species of these aptly-named birds occur in Australia. The name Tree Duck, sometimes used for this tribe, is misleading for the Australian birds, since they tend to perch less in trees than do the perching ducks (tribe Cairinini). 4, 7

Cape Barren Goose, tribe Cereopsini This sole representative of its tribe has no close living relatives and is endemic to Australia. Relatively few in number, it is protected as a rare species. 3

Freckled Duck, tribe Stictonettini This rare and unique endemic genus and species also constitutes a tribe unto itself. 10

ANSERANATINAE (subfamily)

Australia's Magpie Goose is so different from all other waterfowl that it constitutes a whole subfamily of its own. 6

ANATINAE (subfamily)

This large subfamily embraces all other waterfowl and divides into seven tribes. Five, described below, are represented in Australia, though some sparsely. Absent are the Somateriini (Eider Ducks, famous for eiderdown, which live in the arctic regions) and the Mergini (Scoters, Goldeneyes and Mergansers, the fish-eating sea ducks of the Northern Hemisphere).

Perching Ducks, tribe Cairinini These birds spend more time in trees, including nesting, than any other ducks; and are equipped with long, sharp strong claws to facilitate perching. Included in the tribe is the spectacular Mandarin Duck of China and the famous Muscovy Duck of Central and South America. The Australian members are the endemic Wood Duck and two species of Pygmy-geese. 5, 8, 9

Shelducks, tribe Tadornini Shelducks are large heavy birds and somewhat goose-like in appearance. Of fifteen species of the tribe scattered throughout the Old World, two are represented in Australia, one endemic. 11-12

Dabbling Ducks, tribe Anatini The dabbling, or dipping, ducks comprise a tribe of thirty-five species and is by far the largest. It is typified by the Mallard of the genus *Anas* which is arguably the world's most familiar duck. Dabbling ducks mostly frequent fresh shallow water and are non-diving surface feeders. Their bills are equipped with fine lamellae for filtering mud and water. The typical Dabbling Ducks — the Grey Teal, Chestnut Teal and Black Duck — feed by dredging the mud at the water's edge and upending in the shallows to reach the bottom for crustaceans, yabbies and shrimps; or by stripping the seeds and flowers from vegetation at the water's edge. They also filter the surface water for insects and seeds. The ducklings especially do this. Almost all have a patch of iridescent bright feathers (the speculum) in the wing. The sexes are usually dimorphic and the male (drake) somewhat larger. There is an annual eclipse phase with the males when they moult out their bright feathers and assume the appearance of the females. Eight dabbling ducks occur in Australia, including the endemic Pink-eared Duck of the genus *Malacorhynchus*. Of these, the Black Duck is the best known. It is however a different species from the American Black Duck and the African species of the same name. (The Garganey, which is a rare non-breeding visitor is not illustrated.) 13-17, 20

Pochards, tribe Aythini Pochards are diving ducks and are most at home in deep fresh water. They are characterised by a rather short heavy body and large legs placed far apart which makes them strong underwater swimmers. Sexes are dissimilar. Of fourteen species, the endemic Hardhead, or White-eyed Duck, is Australia's only representative, and is one of only a small number of ducks in which the males have white eyes. 18, 21

Stiff-tailed Ducks, tribe Oxyurini These expert divers are characterised by the thick and hard shafts of the tail feathers, like those of the cormorants. Of nine species worldwide, the Musk Duck and the Blue-billed Duck occur in Australia; both are endemic. 19, 22

Opposite: Black Swan

PHOTOGRAPHERS: G. Chapman 5, 16, 22; J. Frazier 6; P. Fullagar 10; T. & P. Gardner 3, 4, 19; P.N. Green 2, 15; E. & D. Hosking 14; K. Ireland 12; P. Roberts 1; L. Robinson p. 175; L.F. Schick 20; E.C. Slater 11; B.A. & A.G. Wells 7, 21; Authors 8, 9, 13, 17, 18

18

1 BLACK SWAN

Cygnus atratus 1.17-1.42 m

Native to Australia, the Black Swan is most often seen on extensive shallow stretches of permanent water where it can reach to the bottom to feed, and occasionally on brackish or even salt water. It flies with neck outstretched in V-formations. After breeding, enormous flocks, sometimes of many thousands, gather to moult. Breeds variably from February to September. Nest is a heap of vegetable matter in water. Introductions to New Zealand have multiplied dramatically; some have also reached New Guinea. **Sexes**: *males, larger; females, bill and iris lighter.* **Cygnets**: *grey-brown.* **Voice**: *high-pitched 'creaky-gate' bugles, including in flight.*

2 MUTE SWAN

Cygnus olor 1.27-1.56 m

The graceful Mute Swan, bigger than the Black Swan, was introduced progressively into Australia from the late 1800s for release on ornamental lakes, but not always successfully. Small breeding colonies survive, such as at Northam, WA and Lake Leake, Tasmania. It can be very tame, although extremely aggressive when breeding. It eats aquatic plants, frogs, fish and small birds. It is still found in the wild state in eastern Europe and Asia. Breeds mostly in late winter or spring. Nests on floating vegetation. **Sexes**: *alike.* **Cygnets**: *brown-grey.* **Voice**: *mostly silent, completely so in flight, but hisses and grunts savagely when breeding.*

3 CAPE BARREN GOOSE

Cereopsis novaehollandiae 1 m

The Cape Barren Goose breeds on small islands off the southern coast of Australia vegetated with tussock grassland and heath, from the Furneaux Group in the east to the Recherche Archipelago in the west. After breeding many birds disperse to adjacent mainland and northern Tasmania. It grazes on grasses and herbage. Pairs probably mate for life and are exceptionally belligerent in defence of their breeding territories. Breeds May-August. Nests in trees, on rocky ledges or on the ground. It was hunted to near extinction in earlier days. Endemic. **Sexes**: *similar.* **Immatures**: *paler.* **Voice**: *honking and grunting.*

4 WATER [Wandering] WHISTLING-DUCK

Dendrocygna arcuata 60 cm

Rarely seen away from water, and frequents deep tropical lagoons and swamps, where it feeds almost exclusively, diving deep for aquatic vegetation. It is usually seen in closely-knit flocks either on the water or roosting on the bank and is very watchful and wary. During the wet season it disperses inland and vagrants may appear in the South. Breeding is linked with rainfall. Nests in long grass. The range extends to New Guinea and many Pacific, Indonesian and Philippine islands. **Sexes**: *alike.* **Immatures**: *duller.* **Voice**: *high whistles and twitters.*

5 MANED (Wood) DUCK

Chenonetta jubata 48 cm

A mane of elongated black neck feathers characterise this duck, and a small pointed bill and long neck give it a rather goose-like appearance. It has claws on its feet which provide purchase for climbing and perching in trees. It lives mostly on land in country near water, and grazes on grass and plants, including at night. It gathers in large flocks after breeding. Locally nomadic. Breeds in spring or after heavy rains. Nests in tree hollow. Ducklings jump out to calling parents below soon after hatching. Endemic. **Immatures**: *paler.* **Voice**: *'mew' or 'mmwarp'.*

Male left, female right

6 MAGPIE (Pied) GOOSE

Anseranas semipalmata ♂75-92 cm ♀71-81 cm

Normally in flocks of several thousand, it frequents swamps and marshy margins of rivers and lagoons of the tropical north, and feeds on aquatic vegetation. Now almost entirely absent in the south where formerly abundant. Unlike other waterfowl, it has half-webbed feet, a prominent head knob, and a hooked bill; and it moults progressively without becoming flightless. Nomadic after breeding. Breeds during the Wet and nests on a floating platform of vegetation. Also in New Guinea. **Sexes**: *similar, but male's head knob sometimes larger (also larger in older birds.)* **Immatures**: *mottled brown-grey.* **Voice**: *resonant honking.*

7. GRASS (Plumed) WHISTLING-DUCK

Dendrocygna eytoni 61 cm

Rarely swims and never dives, and is most commonly seen in very large flocks, feeding widely on open grass plains, paddocks and lagoon edges. Flocks leave their daytime camps near a water source to feed at night on seeds, legumes, and vegetable material. On returning to the daytime camp birds land on water and immediately swim ashore to roost. They disperse inland in the wet season in the north. Breeding is related to rainfall. Nests in long grass. Endemic but vagrant in New Guinea. **Sexes**: *alike.* **Immatures**: *paler.* **Voice**: *high-pitched whistling and twittering.*

8 GREEN PYGMY-GOOSE

Nettapus pulchellus 36 cm

More often seen than the White Pygmy-Goose, this beautifully-plumaged bird lives in pairs and small flocks in the tropics amongst the blue waterlilies, on dams, lagoons and lakes. It is exceptionally aggressive, and males often fiercely attack each other. It only rarely leaves the water and walks on land with difficulty. It dives deep for water plants or seeds, or takes water-lily buds. Sedentary; sometimes nomadic. Breeds during the wet season. Nests in tree hole. Also in New Guinea and Indonesia. **Female**: *neck not solid green all round, faint white eye-stripe.* **Immatures**: *like females.* **Voice**: *whistles and trills.*

Male

9 WHITE [Cotton] PYGMY-GOOSE

Nettapus coromandelianus 38 cm

This uncommon small duck is entirely aquatic and seldom leaves water. Usually in pairs or small flocks, it frequents deep lagoons and freshwater lakes and feeds on aquatic plants and their seeds during the evening and night. It is shy and retiring and hides by day far out among the waterlilies. Sedentary. Breeds January-March. Nests in tree hollow usually over water. Widespread in south-eastern Asia, Indonesia and New Guinea. **Female**: *less green on back, no black collar.* **Immatures**: *like females.* **Voice**: *staccato cackles, quacks.*

Male

10 FRECKLED DUCK

Stictonetta naevosa 59 cm

This rare and vulnerable duck is most often seen in small family parties in permanent heavily-vegetated swamps and creeks. By day it retires to dense cover in deep water, then flies to the shallows at dusk to feed. Its diet of algae, seeds and weeds is obtained by a filtering action, most often while wading on the shore line. (See illustration.) It is sedentary but wanders sometimes in flocks, during drought. Breeding related to climatic conditions. Nests at water level. Endemic. **Sexes**: *similar, but female paler.* **Immatures**: *freckled buff.* **Voice**: *piping, whistling and quacking.*

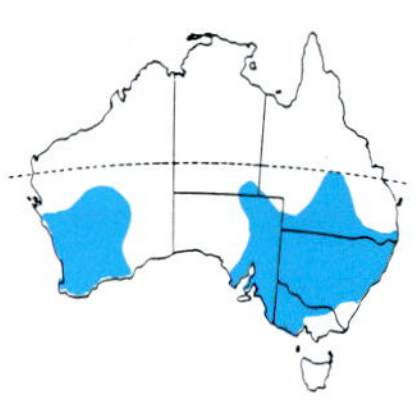

11 WHITE-HEADED [Radjah] SHELDUCK

Tadorna radjah 55 cm

Also known as the Burdekin Duck. Because of its tameness and slow flight it is an easy target for shooters and has disappeared from the east coast. Usually in pairs and small parties it frequents shallow brackish water, mudflats and mangroves near the coast. It roosts by day near cover at the water's edge and moves out to feed in the shallows for small aquatic animals, algae and sedge. Some movement further inland in wet season. Breeds April-July. Nests near water in large hollow limbs. Also in New Guinea and adjacent islands. **Sexes**: *alike.* **Immatures**: *white parts flecked brown.* **Voice**: *loud hoarse whistling.*

12 CHESTNUT-BREASTED [Australian] SHELDUCK

Tadorna tadornoides 72 cm

As this bird is most abundant at sea level and on the coast, its alternative name Mountain Duck has been described as 'a piece of ingenious folklore'. Shy and wary and keeps well out on shallow lakes and marshes. Rafts of perhaps 10 000 birds gather to moult in summer after breeding. It feeds mostly on grass and herbage. Seldom dives. It forms strong pair bonds for life. Breeds in winter. Nest in tree to 25 m high. Endemic. **Male**: *in eclipse is paler on breast and white neck ring shrinks.* **Immatures**: *duller.* **Voice**: *deep honking.*

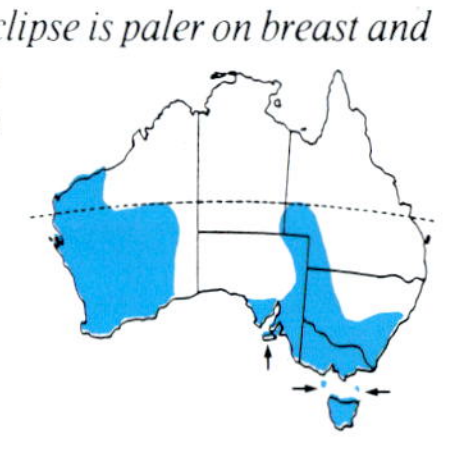

Male left, female right

1

2

3

4

5

6

7

8

9

10

11

12

18

13 PACIFIC BLACK DUCK

Anas superciliosa 50-61 cm

Widespread and probably the most numerous and familiar duck in Australia. Also in New Guinea, Indonesia and New Zealand (there called the Grey Duck). Singly, in pairs and flocks it is seen wherever there are stretches of water. Becomes tame in city parks. A typical dabbling duck, it feeds by dredging mud by up-ending, and filtering seeds from the surface. Nomadic. Breeds usually July-October south; March-May north. Nests on ground, in grass or tree hole. **Sexes**: *similar*. **Immatures**: *black above, underparts yellow*. **Voice**: *female, quacks, male loud courtship peeps*.

Female left, male right

14 MALLARD

Anas platyrhynchos 55-68 cm

Introduced from Europe, the Mallard has settled well into its Australian habitat. Too well, perhaps, for it interbreeds with the closely-related native Black Duck and, being the dominant strain, it is feared that it may overwhelm those characteristics with which the Black Duck survives the rigours of the Australian climate. Familiar in urban parks and environs. Breeds throughout the year. Nests on the ground or in tree holes. Male in eclipse is similar to female but the bill is dull green. **Immatures**: *black above, yellow wing and underparts*. **Voice**: *quacks similar to Black Duck*.

15 CHESTNUT TEAL

Anas castanea 48 cm

Usually in pairs or small groups, it mostly frequents saltwater and brackish lakes and swamps of the coastal areas, and is abundant in Tasmania. When seen together with the Grey Teal, the females are difficult to separate. Dabbles mostly for aquatic plant material and insects. Usually sedentary. Breeds June-December. Nests on the ground or in crevices. Both parents stay with the ducklings. Endemic. **Recognition**: *female is darker than Grey Teal and lacks the latter's white throat patch*. **Immatures**: *paler*. **Voice**: *female loud quack, male peep or whistle*.

Male left, female right

16 BLUE-WINGED [Australian] SHOVELER

Anas rhynchotis 45-53 cm

The shoveler's broad, shovel-like bill is fringed with larger lamellae than the other dabbling ducks and is specially adapted for sieving small aquatic organisms from the water surface. It lives in thickly-vegetated swamps, usually well out in small parties often amongst other flocking ducks — sometimes foraging communally in a V-shaped formation. Very wary. It has a habit of rising into the air with a clap of the wings. Sedentary. Breeds August-December or after flooding. Nests mainly on the ground. Male in eclipse plumage resembles female. Endemic. **Immatures**: *paler*. **Voice**: *occasional 'took-took' but mostly silent*.

Male right, female left

17 GREY TEAL

Anas gibberifrons 47 cm

The ubiquitous Grey Teal is often seen in immense flocks and is highly nomadic in adapting to Australia's erratic rainfall. It mainly lives on inland swamps and tree-lined rivers in winter, and moves to the coast as these dry up in summer. All types of water, whether salt, brackish or fresh, offer it a home. It is a typical dabbling duck and feeds on seeds of land and water plants, small crustaceans, insects and their larvae. Breeds throughout the year according to water levels. Nests on the ground, in crevices and tree holes. The species is widespread in Australasia. **Sexes**: *alike*. **Immatures**: *paler*. **Voice**: *female quacks, male peeps*.

18, 21 HARDHEAD (White-eyed Duck)

Aythya australis 42-58 cm

In pairs or flocks, sometimes very large, the Hardhead lives in pools and streams. Seldom ashore, it is a bird of deep water, a particularly strong swimmer and at home even in swiftly-flowing turbulent water. It dives deep for plant material, small animals and fish. If alarmed, it rises vertically with loudly whirring wings. Perhaps the fastest flyer of all Australian ducks. Occurs also in New Guinea, Indonesia and some Pacific Islands but breeds mainly in Australia. Breeds October-December but variable. Nests mainly in swamp vegetation, or tree hollow over water. **Sexes**: *similar, but male's eye is white*. **Immatures**: *golden-brown*. **Voice**: *soft croaks and whistles, but usually silent*.

Male above, female below

19 BLUE-BILLED DUCK

Oxyura australis 44 cm

The diving Blue-billed Duck lives entirely on water in thickly-vegetated swamps singly or in pairs; or in very large flocks on open lakes in winter. It is completely helpless on land, and labours in take-off to become airborne. It is secretive and wary; and usually keeps distant from the shore. It dives at once if alarmed and may travel far before resurfacing. Breeds September-March. Nests in dense reeds. Endemic. **Male**: *blue bill only when breeding; plumage duller in eclipse*. **Female**: *bill grey-brown*. **Immatures**: *paler*. **Voice**: *lower quacks, seldom heard*.

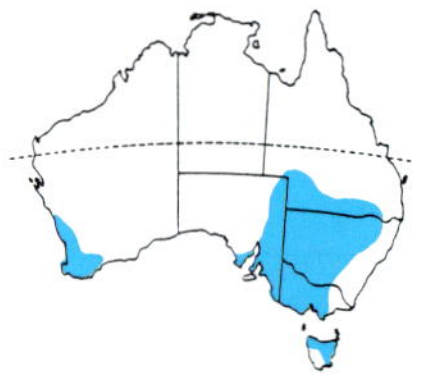

20 PINK-EARED DUCK

Malacorhynchus membranaceus 38-45 cm

This small unusually tame duck is distinguished by its remarkable bill. Usually in pairs and small groups, it frequents mainly shallow lakes and swamps. It is extremely nomadic in response to rainfall and its numbers sometimes erupt after flooding. It feeds in shallow water for small organisms and often swims in circles in pairs to stir up mud. Breeds variably according to rainfall. Nests on a log, or in a bush or tree hole to 10 m. Endemic. **Sexes**: *alike*. **Immatures**: *paler, pink ear patch less distinct*. **Voice**: *several pleasant trilling or chirruping notes*.

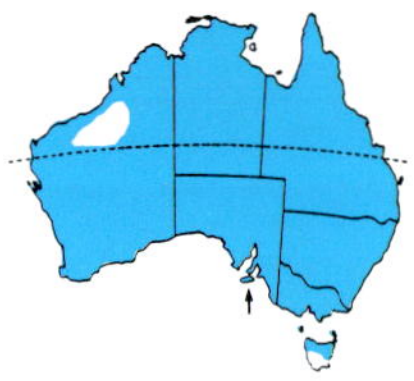

22 MUSK DUCK

Biziura lobata 60-72 cm

This grotesque bird frequents swamps and inlets singly or in companies, and dives for food and when seeking cover if approached. It is seldom seen on land. The polygamous male has a variety of lures to attract females during the breeding season: it secretes an oil which emits a powerful musky smell; it inflates its large pendulous bag below the bill during display and erects its tail feathers in a fan while kicking back 2 m jets of water. Breeds September-December but variable. Nests in dense reeds. Endemic. **Female**: *reduced bill lobe*. **Immatures**: *like females*. **Voice**: *sometimes shrill whistle or 'plonk' sound*.

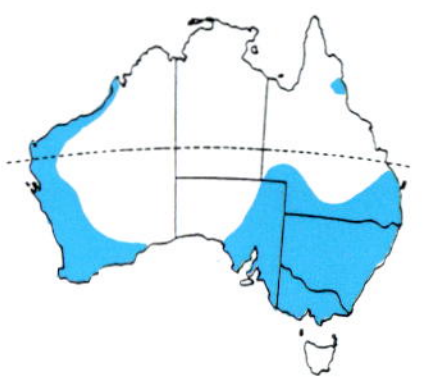

Male

13

14

15

16

17

18

19

20

21

22

19

Marsh Birds

Opposite: Lotusbird

The birds in this Section, although not forming a natural systematic unit, are linked by the common attributes of living in marshy country, feeding mainly on aquatic life, and being relatively short-legged. They represent six families of the following three Orders: Podicipediformes — grebes; Gruiformes (Part) — rails, crakes, waterhens and coots; Charadriiformes (Part) — jacanas, Painted Snipe, Japanese Snipe, stilts and avocets.

GREBES Family Podicipedidae

Grebes are entirely aquatic-living birds which propel themselves through water by means of legs placed well to the rear of their bodies. This makes walking on land awkward for them. Their webbed feet do not have the leathery connecting skin of other diving birds, but instead stiff horny flaps fringing each toe separately, usually called lobe-toed. They use their feet for steering both in the water and in the air since they have too short a tail to perform the function of a rudder. Generally they ride higher in the water than ducks, but when danger threatens they have the ability to submerge the whole body with its dense waterproof plumage, until only the eyes and bill are visible. They are not particularly swift swimmers, but are very agile when diving for prey. The chicks are carried on the backs, or tucked under the wings, of either parent. There are twenty species of grebe worldwide; three are found in Australia, and one of these is endemic. 1-3

RAILS, CRAKES, WATERHENS, COOTS Family Rallidae

The rail family is one that is not easily classified. Between 130-140 species live worldwide, fourteen of them in Australia. They are extremely shy and elusive, and are more commonly heard than seen, the hens and coots far less so than rails and crakes. When disturbed, their narrow bodies, strong legs and long toes enable them to disappear quickly through thick undergrowth to cover. The expression 'thin as a rail' derives from this characteristic. Many are poor fliers and flick their tails. Typically, they live in marshes, and as some look superficially like domestic fowls they are often referred to as waterhens. The differences between rails, crakes, hens and coots are mainly those of behaviour arising from habitat specialisation, although the coot has toe flaps (lobes) similar to those of the grebes. They feed omnivorously on a wide range of vegetable and animal matter. 4-16, 18

LOTUSBIRD Family Jacanidae

The plover-related Lotusbird is the sole breeding representative in Australia of the eight species of tropic or sub-tropic dwelling jacanas, as they are known elsewhere in the world. All have several unique characteristics, chief of which are the exceptionally long toes and claws which enable them to flit as daintily as ballet dancers across floating vegetation, such as lily pads, as though walking on water. More prosaically, they have a frontal shield like that of the coots and waterhens, and also spurs like little sharp horns on the bend of their wings to give them a formidable fighting power. They are enormously confident, allowing an observer to approach closely, but will 'freeze' in the face of danger. One of their endearing characteristics is the care taken to protect the young, which they will transport under their wings. If necessary, they will try the 'broken wing' trick to draw intruders away from the nest, or encourage the nestlings to submerge in the water, afterwards helping them to scramble back to safety by pressing down the leaf to form a ramp. The strikingly-marked eggs are also carried under the wings, and if any should fall into the water, they float until the parent can roll them back to the vegetation on which the nest is built. In his delightful book *Sharing a Dream*, Glen Threlfo describes how a nesting bird, seeming to know that one of her three eggs was missing after a storm, searched under the edges of the leaves to retrieve it. 17

STILTS AND AVOCETS Family Recurvirostridae

Stilts have very long legs in proportion to their body size. Like the avocets, they are wetland birds who wade in sedate, deliberate fashion, lifting their legs in measure tread as they search for food. The stilts use their straight bills to probe the mud deeply, while the avocet with its unusually curved bill sweeps the shallows. Having webbed feet, these birds swim freely, and fly strongly with their long legs stretched out behind. The total number of world species is variously given as seven to thirteen; three breed in Australia. 19, 22, 23

PAINTED SNIPE Family Rostratulidae

A small family of snipe-like waders of only two species. Although somewhat similar in appearance and behaviour, they live in separate geographical areas: one in the region extending from Africa to Australia, and the other in South America. Characteristically, they have rather large feet with four toes and long slightly curved bills; and the usual roles of the sexes are reversed — the larger and more brightly coloured female initiates the courting. 20

TRUE SNIPE Family Scolopacidae (Part)

The true snipe, of the genus *Gallinago*, belong to the diverse family of migratory waders included in Section 20, which comprises about a dozen species of snipe and woodcock and is scattered over most parts of the world. Illustrated is the Japanese Snipe, a regular non-breeding visitor from Japan which spends about half the year in Australia. The Pintailed Snipe is sometimes seen in the north as a rare visitor. 21

PHOTOGRAPHERS: H.& J. Beste 7; C.A.C. Cameron 22; G. Chapman 1, 9; J.A. Estbergs 2; A.L. Hertog 4; K. Ireland 17, 23; F.F. James 11; T.G. Lowe 20; J. Purnell 5, 21; K.& B. Richards 8; L. Robinson 12; L.F. Schick 18; T.A. Waite 14; C. Webster p. 181; B.A. & A.G. Wells 10, 16, 19; Authors 3, 6, 13, 15

19

1 [Great] CRESTED GREBE

Podiceps cristatus 61 cm

Widespread in many parts of the world, this is one of the most intensively studied of all birds, in part due to its unusual appearance; in part to its extraordinary and intricate courtship display of several phases culminating in a spectacular 'weed dance'. In Australia it is found in the vegetated margins of large expanses of still water, usually in pairs. Swims and dives, submerges, and patters over water when taking off in flight. Sedentary or migratory. Breeds November-January. Nests on floating vegetation. **Sexes**: *alike.* **Immatures**: *lack ear tufts and ruff.* **Voice**: *variety of discordant calls.*

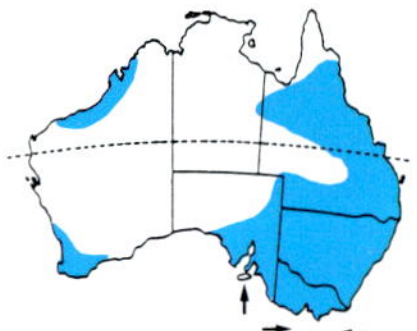

2 HOARY-HEADED GREBE

Poliocephalus poliocephalus 31 cm

The male has two plumage phases: after breeding it loses the buff breast and the white lines on the head, but can usually be identified then by the broad black bar down the nape. Usually in pairs, sometimes several of these together when breeding, feeding on aquatic plants and insects in marine or muddy estuaries. Sedentary or nomadic. Breeds November-January or after rain. Nests in clump of weeds in shallow lightly vegetated inland waters or swamps. Endemic, but has spread sparingly to New Zealand. **Sexes**: *similar.* **Immatures**: *like non-breeding adult.* **Voice**: *low notes, mostly silent.*

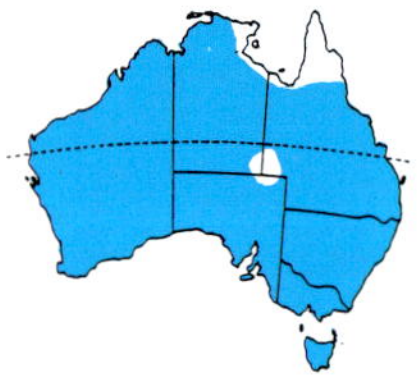

Breeding plumage

3 LITTLE [Australian] GREBE

Tachybaptus novaehollandiae 26 cm

Familiar on any vegetated stretches of water, this species is somewhat similar in appearance to the Hoary-headed Grebe with which it often associates, although in the breeding season the male Little Grebe sports rather brighter plumage and a creamy patch between eye and bill. It is less ready than the Hoary-headed Grebe to take to flight when alarmed. Undertakes nomadic flights. Mostly at night. Breeds September-January. Nests on floating vegetation. Widespread in the Australian Region extending to Indonesia. **Sexes**: *alike.* **Immatures**: *like non-breeding adult.* **Voice**: *rattling trill; noisier than the Hoary-headed Grebe.*

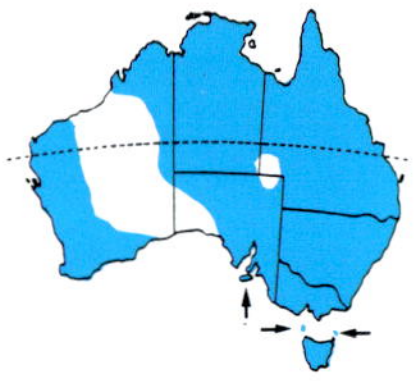

4 CHESTNUT RAIL

Eulabeornis castaneoventris 52 cm

Living as it does in remote tropical mangroves, and being wary, shy and difficult to flush, the Chestnut Rail is rarely observed. It is more confident when at low tide it emerges into the open, searching the exposed mudflats for crustaceans. Characteristically, it flicks its tail when walking. It is seen singly or in family parties. Sedentary. Breeds Ocotber-February. Energetically defends the nest which is in mangrove roots or tree to 3 m high. Also on Aru Islands. **Sexes**: *alike.* **Immatures**: *not recorded.* **Voice**: *repeated harsh screech, grunting.*

5 LEWIN'S RAIL

Rallus pectoralis 23 cm

Shy and elusive and infrequently observed in its habitat of dense reedy sections of swamps and lakes. When flushed, it prefers running to flying. Its long slender beak is used in probing for crustaceans and insects. Throughout its range there is a size variation, south-eastern birds being the smallest. The larger south-western race is possibly extinct. Solitary and sedentary. Breeds August-December. Nests in or near water. Also in New Guinea. **Sexes**: *alike.* **Immatures**: *grey faintly barred white.* **Voice**: *'cree-ek', 'jik-jik-jik'; grunting.*

6 BANDED RAIL

Rallus philippensis 31 cm

The Banded Rail is not uncommon in well-vegetated coastal areas including the margins of mangroves and inland habitats near water. Sometimes visits settled areas, forages for insects, small animals, seeds and aquatic plants. It is timid and runs away swiftly with tail flicking when danger threatens, infrequently resorting to flight. Solitary or in pairs. Birds in the west tend to be darker; those on Cape York Peninsula smaller. Some are nomadic and wander to New Guinea and islands to the north and west. Breeds mostly September – March. Nests low in grassy cover. **Sexes:** *alike.* **Immatures:** *duller.* **Voice:** *creaky squeak.*

7 RED-NECKED RAIL [Crake]

Rallus tricolor 30 cm

The Red-necked Rail lives in the obscurity of a remote habitat in the dense rainforest and watercourse vegetation of northern Queensland, and is mainly nocturnal. Especially secretive in habits, it always keeps to cover. It walks, swims underwater, flits silently, or flies in an ungainly fashion. It eats insects and freshwater animals. Solitary or in small family parties. Sedentary. Breeds November-April. Nests on or near ground. Also in islands north of Australia. **Sexes**: *alike.* **Immatures**: *duller, belly barred chestnut.* **Voice**: *shrill 'tare', sharp 'tok' mostly at night.*

8 SPOTTED [Australian] CRAKE

Porzana fluminea 20 cm

The Spotted Crake is not as timid as its relatives, and resorts less to dense cover. It lives in the vegetated margins of fresh, brackish or salt water, and feeds on aquatic plants, molluscs and insects and is seen singly, in pairs or companies. Breeds August-February. Nests in vegetation in shallow water. Endemic. **Female**: *paler.* **Immatures**: *upper parts browner.* **Voice**: *harsh 'crake'; various chatterings.*

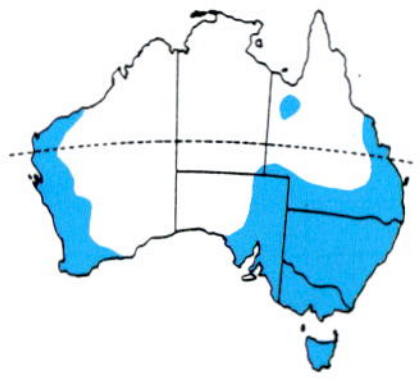

Male

9 SPOTLESS CRAKE

Porzana tabuensis 20 cm

A somewhat plain and timid bird with red legs. It frequents dense thickets around fresh water and salt swamps, and mangroves; also beaches and offshore islands. It feeds on insects, molluscs and aquatic plants. Sedentary or nomadic. Breeds September-January. Nests well hidden in vegetation in shallow water. Also in Philippines to South-West Pacific. **Sexes**: *alike.* **Immatures**: *duller, throat dull white.* **Voice**: *accelerating 'kop-kop', 'croo'.*

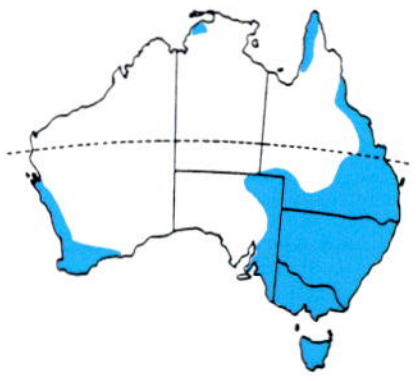

10 MARSH [Baillon's] CRAKE

Porzana pusilla 16 cm

A widespread species, known in the Old World as Baillon's Crake. Long toes enable it to walk over water plants; it also swims and dives, and may roost in nearby trees. Its diet is composed of molluscs, seeds and green matter. Shy and secretive. Nomadic or migratory. Breeds September-February. Nests in a clump in shallow water. **Sexes**: *alike.* **Immatures**: *paler, less distinct markings.* **Voice**: *squeaky 'chut', 'crak'; alarm calls.*

11 WHITE-BROWED CRAKE

Poliolimnas cinereus 18 cm

This long-toed crake runs about on water-lily pads or other floating vegetation amongst coastal swamps, lakes and mangroves of northern Australia. Besides seeds and swamp plants, it feeds on insects often taken on the wing. Walks with tail flicking and flies weakly with legs trailing. Not over-timid or shy of observers. Sedentary or nomadic. Breeds September-May. Nests in swamp vegetation or rice fields. Also in South-east Asia, Polynesia and various Pacific Islands. **Sexes** *alike.* **Immatures**: *crown brown.* **Voice**: *often in concert at dusk, repeated 'crake', chattering and churring.*

12 BUSH-HEN

Gallinula olivacea 26 cm

The mainly nocturnal Bush-hen lives in restricted habitats of wet grassy vegetation along creeks, in lantana thickets, rainforest margins, or swamps of north-eastern Australia. It calls frequently, sometimes incessantly, by day or night, especially in the wet season; were it not for this calling it would be difficult to find, for it is a reclusive bird. Seen singly, in pairs or family parties. Feeds terrestrially on insects, seeds and leaves. Sedentary or migratory. Breeds October-April. Nests low in grass. Also in New Guinea and South-east Asia. **Female**: *smaller.* **Immatures**: *duller, head black.* **Voice**: *single 'tok', drawn-out 'nee-you'.*

1

2

3

4

5

6

7

8

9

10

11

12

19

13 [Purple] SWAMPHEN
Porphyrio porphyrio 45 cm

In the south-west, the purple of the neck, shoulder edge and breast is replaced by blue; in Tasmania by purple-blue and Tasmanian birds are larger. Seen singly, in pairs or loose groups, it lives among reeds in swamps, river margins and lakes including in public parks. Flicks tail when walking and swimming, exposing white under-tail coverts. It eats green shoots. Shy, but becomes tame in parks, but if alarmed will promptly take wing. Sedentary. Breeds variably. Nests on platform in reeds. Also in Africa, Asia and extends to New Zealand. **Sexes**: *alike*. **Immatures**: *duller*. **Voice**: *mostly silent, some metallic cackling.*

14 TASMANIAN NATIVE-HEN
Gallinula mortieri 45 cm

Found only in Tasmania, this bird is flightless, and almost twice the size of its mainland relative, the Black-tailed Native-Hen. It is often seen in open grassy areas along creeks and swamps, and pastures, from sea level to 1100 m. It runs swiftly, flicking its tail when disturbed. It lives in pairs or family groups in clearly defined and jealously guarded territories, and feeds on vegetable matter. Sedentary, with little or no wandering. Breeds June-January. Nests in low grass near water. Endemic. **Sexes**: *alike*. **Immatures**: *duller, washed grey*. **Voice**: *harsh 'see-saw' uttered as a duet.*

15 [Eurasian] COOT
Fulica atra 38 cm

The Coot is distributed from Europe across the Old World to Australia and New Zealand, and its distinctive white frontal plate is as familiar in the duck ponds of rural England as in the city lakes of Australia. It frequents fresh and brackish swamps, floodwaters and artificial lakes, forming large rafts on open water. It dives for vegetable matter and feeds on land on green shoots and takes scraps of food tossed to it in public parks. Usually gregarious, it may be belligerent during breeding August-February. Sedentary or migratory. Nests low near water. **Sexes**: *alike*. **Immatures**: *paler, throat whitish*. **Voice**: *sharp shrill calls.*

16 BLACK-TAILED NATIVE-HEN
Gallinula ventralis 35 cm

In contrast with the flightless Tasmanian Native-Hen, this commonly-seen rail is a strong flier. It is highly nomadic and can appear almost anywhere quite suddenly. Singly or in pairs, but more often in large flocks, it frequents margins of swamps, lakes, and rivers and irrigated pastures. It feeds on green shoots of grasses, herbs and weeds. It walks and runs with tail cocked or fanned. Breeds variably depending on rainfall. Nests on or close to the ground near water. Endemic. **Sexes**: *alike*. **Immatures**: *duller*. **Voice**: *mostly silent; metallic cackling.*

17 LOTUSBIRD [Comb-crested Jacana]
Irediparra gallinacea 24 cm

Also known as the Lilytrotter, this appealing bird has distinctive toes and claws which help it in walking over plants growing in still waters. It used to breed as far south as Sydney until about 1976. It flies well, using the feet as a rudder, and may roost in trees. The comb colour changes from red to yellow in moments of excitement. Sometimes in open companies, sedentary or nomadic. Nests on a small platform of floating vegetation. Also in South-east Asia, although there the Pheasant-tailed Jacana mainly predominates. **Sexes**: *alike*. **Immatures**: *breast white*. **Voice**: *shrill alarm call, squeaking.*

18 DUSKY MOORHEN
Gallinula tenebrosa 38 cm

The Dusky Moorhen lives in Indonesia, New Guinea and Australia, and closely resembles its almost worldwide relative the Common Moorhen. It is distinguished by a bright red frontal plate and white-edged tail, and lives as happily on artificial city ponds and lakes as on outback permanent swamps and rivers. It looks duck-like, especially when it up-ends to feed on underwater vegetation. Gregarious out of breeding season, August-February. Nests among rushes or at foot of tree growing in water. **Sexes**: *alike*. **Immatures**: *browner*. **Voice**: *strident 'kerk', shrieks.*

19 BANDED STILT
Cladorhynchus leucocephalus 41 cm

The first breeding colony was only recorded in 1930. It lives in salt lakes, estuaries and inlets, sometimes in great flocks. It feeds mainly on brine shrimps, and rafts of several hundred sometimes crowd together in shallow water to corner prey. It is seasonally nomadic over great distances, with movement away from the interior towards the coast in summer. Breeds in immense colonies mainly May-December, on large inland salt lakes. Nests on a scrape in the ground or on floating vegetation. Endemic. **Sexes**: *alike*. **Immatures**: *head, neck, back white, no breastband*. **Voice**: *yelping and wheezing.*

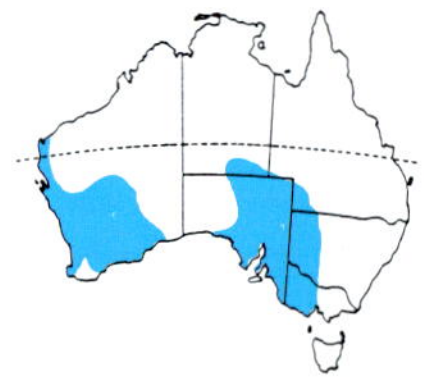

20 PAINTED SNIPE
Rostratula benghalensis ♂22 cm ♀25 cm

The Painted Snipe lives in marshy areas generally with ample cover, in pairs and small parties. The male incubates the eggs and rears the chicks and puts on an aggressive defence display. Somewhat nocturnal, secretive and solitary. Breeds October—December (south), March—May (north). Nests in mud depression, in small colonies. Also in Africa and Asia. **Recognition:** (cf Japanese Snipe): *conspicuous pale 'horse collar'; broad whitish eye ring, flight clumsy with dangling legs, shows black mark across underwings*. **Female:** *brighter*. **Immatures:** *like male but lacking black breast band*. **Voice:** clicking sounds.

Male

21 JAPANESE [Latham's] SNIPE
Gallinago hardwickii 26 cm

Nesting in the northern islands of Japan, this strong, swift-flying snipe migrates southwards, arriving in Australia during August. Over the next month others arrive in greater numbers to reach the south-eastern and Tasmanian coastal areas and disperse into wet paddocks, irrigated areas, mangroves, lake margins and swamp edges. It feeds mainly at night on worms, aquatic larvae and seeds. Extremely wary; when alarmed it will burst up in zigzagging flight before dropping back into cover some distance away. Return flights begin February April. **Sexes and immatures**: *alike*. **Voice**: *single loud 'crek' in flight.*

22 BLACK-WINGED (White-headed) STILT
Himantopus himantopus 38 cm

The word stilt is aptly descriptive of this longlegged but graceful bird. It frequents both saline and shallow fresh waters of lakes, estuaries, swamps or sewage ponds, and normally forms smaller flocks than the Banded Stilt. It gathers its food usually by wading. Nomadic according to seasonal conditions. Breeds August-December. Nests colonially on the ground adjacent to shallow water. Almost world-wide distribution. **Sexes**: *alike*. **Immatures**: *crown, nape and eye patch grey*. **Voice**: *high-pitched puppy-like yapping or yelping.*

23 RED-NECKED AVOCET
Recurvirostra novaehollandiae 46 cm

Sweeping its long upcurved bill rhythmically from side to side, the Red-necked Avocet sweeps shallow water or mud for aquatic insects, crustaceans and plant matter. Like the Banded Stilt, with whom it often associates, it will also submerge and swim readily. It is seen in small groups or large flocks, mostly in salty or brackish swamps rather than on freshwater, but occasionally appearing at tidal inlets. Sedentary or nomadic in response to rainfall. Breeds August-December. Nests colonially on the ground or in shallow water. Endemic. **Sexes**: *alike*. **Immatures**: *paler markings*. **Voice**: *yelps, yaps and wheezes.*

13

14

15

16

17

18

19

20

21

22

23

20

Migratory and Resident Waders

This Section embraces four families of mainly shore birds, usually known collectively as waders, which belong to the large and diverse worldwide order Charadriiformes of about sixteen families and over 300 species.

LAPWINGS, PLOVERS AND DOTTERELS Family Charadriidae

A family of about sixty-four species worldwide, with thirteen seen regularly in Australia and another four occasionally. They divide conveniently into two groups: lapwings in one, and plovers and dotterels in the other.

Lapwings, Genus *Vanellus* The nominate species of this genus, affectionately known as the Peewit in Britain, is the most common and widespread wader in Europe. Lapwings, of which there are about twenty-six species, are found in all temperate and tropical countries, except North America. Two species occur in Australia, one endemic. Lapwings are the largest of the plovers, and some have sharp-pointed spurs on the bend of the wing which are used in fighting. In general, they are characterised by erratic wingbeats in flight (*Vanellus* means 'winnowing fan') and their loud cries of alarm at the slightest threat of danger which alert an entire neighborhood.

True Plovers There are about thirty-eight species of plovers and dotterels worldwide, with eleven regularly in Australia and four occasional visitors. The name dotterel tends to be used for the smaller plovers and has no taxonomic significance. The majority of true plovers belong to the genus *Charadrius* and are most often seen on ocean beaches and mudflats. Some species breed in the arctic tundra and migrate across the equator to Australia for the southern summer. Six species of plover are endemic. 1-16

SANDPIPERS AND ALLIES Family Scolopacidae

This family comprises about eighty-six species worldwide. About thirty-two are regular non-breeding visitors to Australia, and several others occasionally, from their breeding grounds in the far north of the Northern Hemisphere. Most arrive from August to October in dull plumage and depart in bright nuptial plumage during the following March—April. Vast numbers reach Australia and can be seen in many places around the coast feeding on the mudflats between the tides. Concentrations of tens of thousands have been recorded on the north-west coast and in the Gulf of Carpentaria. Their thin, long bills — always as long and sometimes longer than the head — set members of this family apart from the plovers. They vary greatly in size from the sparrow-sized stints to the bulky curlews which can measure 60 cm from bill to tail, and are the largest members of the family. The species regularly reaching Australia represent eight genera. Most belong to the genus *Calidris* which are fairly small waders, and includes the stints, the knots and several sandpipers. Next come the typical sandpipers of the genus *Tringa* which vary in size from about 20 cm to 35 cm, and include the Common Sandpiper, the Tattlers and the Greenshank. Also represented are the curlews and the whimbrels of the genus *Numenius*, the godwits of the genus *Limosa* characterised by their long straight or slightly up-turned bills, and members of the genera *Limicola* (Broad-billed Sandpiper), *Arenaria* (Turnstones), *Phalaropus* (Phalaropes) and *Philomachus* (Ruff/Reeve). 17-34, 37-45

PRATINCOLES AND COURSERS Family Glareolidae

The sixteen members of this family, widespread in the warmer regions of the Old World, fall into two groups which, though greatly different in appearance and habits, are closely related anatomically. Of the seven species of pratincoles — elegant long-winged and long-tailed waders of open country — two occur in Australia; one a breeding endemic and the other a regular visitor. The former, the Australian Pratincole, is unique in that it also exhibits several features of the coursers (not represented in Australia) such as a short tail. 35-36

OYSTERCATCHERS Family Haematopodidae

Scientifically named from their red legs, this family of conspicuous black or pied coastal birds is characterised by their powerful vertically flattened bills which enable them to open oysters, mussels and clams, and prise limpets off the rocks. Two of the world's seven species are found in Australia. 46-48

Opposite: Knots

PHOTOGRAPHERS: H.& J. Beste 4, 8, 36; G. Chapman 9, 19, 48; M.A. Chappell (Animals, Animals) 6; B. Chudleigh 16; B.J. Coates 24; M.D. England (Ardea) 29; A.T. Foster 13; K.W. Fink (Ardea) 35; T.& P. Gardener 2, 25, 31, 39, 43, 46, 47; A.L. Hertog 30; E.& D. Hosking 32, 45; K. Ireland 23; R.B. Legge 11; T.G. Lowe 10; E.C. Slater 1, 5; R. Vaughan (Ardea) 44; W. Weisser (Ardea) 34; B.A. & A.G. Wells 18; E.E Zillmann 12, 22, 40; Authors p. 187; 3, 7, 14, 15, 17, 20, 21, 26, 27, 28, 33, 37, 38, 41, 42

1, 4 MASKED LAPWING
Vanellus miles

(A) Race *miles* (Masked Plover) 36 cm
(B) Race *novaehollandiae* (Spur-winged Plover) 38 cm

Native to Australia, the Masked Lapwing has introduced itself to New Guinea and eastern Indonesia (race *miles*) and progressively to New Zealand (race *novaehollandiae*) where it appeared for the first time in 1932. Both races carry sharp-pointed shoulder spurs on the wings. The rounded wing-tips and erratic wingbeats help to identify it in flight. The populations meet at about Mackay, Queensland, where some hybridize. Black 'lapels' and full black cap readily distinguish the southern birds and the very large facial wattles the northern. The Lapwing is common and widespread in open wetlands, most pastures and estuaries, and often seen on golf courses and parks. It has benefitted from land clearance. It characteristically stalks with shoulders hunched, to dart head down at insects, worms and crustaceans. Seeds also are sometimes eaten. It wanders locally including at night in pairs and small flocks — sometimes in hundreds when not breeding. It is very wary and signals danger with its calls at the slightest provocation. It retreats to cover in shallow waters to roost. Breeds June-January in the south, variably in the north. Nest is a depression in open country and it is boldly aggressive in its defence. **Sexes**: *alike.* **Immatures**: *spotted on crown and back.* **Voice**: *rapid 'keeks' also at night.*

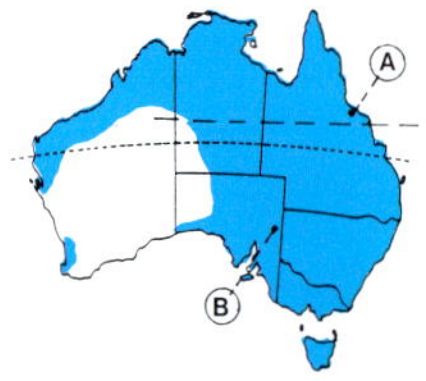

(A) above (B) below

2 GREY PLOVER
Pluvialis squatarola 29 cm

Breeds in the high Arctic regions during the northern summer. It visits Australia between August and March/April and frequents the seashore and grassy mudflats fairly well around the coast but it is uncommon. It is rarely seen inland. It is timid and wary. It is seen singly, in small parties or large congregations, and mainly keeps aloof from other species. It may assume its striking full breeding plumage of black underparts and patchy part-black upperparts before departure. **Sexes**: *alike.* **Immatures**: *paler grey-buff.* **Voice**: *poignant 'pee-o-wee'.*

3, 6 [Lesser] GOLDEN PLOVER
Pluvialis dominica 25 cm

Less shy and more gracefully erect in stance than the Grey Plover, the Golden Plover is also more commonly seen in Australia, August-April, after breeding in Alaska and Siberia. In Australia it frequents rocky coasts, beaches and sandflats, occasionally inland swamps and grassy paddocks, and even grassy airstrips and golf courses. Often in large companies, most birds are in white and buff-mottled non-breeding plumage, but some may be patchy, just losing, or beginning to attain, the handsome plumage of black dappled on the upperparts with gold spots. **Sexes and immatures**: *alike in non-breeding plumage.* **Voice**: *piping 'too-weet'.*

Non-breeding above, breeding below

5 BANDED LAPWING (Plover)
Vanellus tricolor 28 cm

The banded Lapwing often lives in short grass in inland habitats and will travel long distances during hot periods to find shade. Seen in small parties or winter flocks, it is nomadic, wary, and actively de-ensive at the nest, swooping on intruders or putting on distraction displays. It feeds on insects and tiny animals, seeds and shoots, often loosely raking the ground with one foot to flush prey. Breeds June—November. Nests in ground depression. Endemic. **Sexes:** *alike.* **Immatures:** *mottling on crown and breast.* **Voice:** *harsh 'kew-kew', often at night; chiming 'a-chee-chee'.*

7 RED-KNEED DOTTEREL
Erythrogonys cinctus 18 cm

The endemic Red-kneed Dotterel lives on the edges of swamps and vegetated lagoons in all but the most arid or most humid of regions, and is widespread throughout mainland Australia. It is commonly seen darting about in the mud, singly, in pairs or companies, probing with jerky head-bobbing for small creatures in mud or water. Sedentary or nomadic during dry conditions. Breeds September—December. Nests in ground depression close to water. **Sexes:** *alike.* **Immatures:** *upperparts brown, white below.* **Voice:** *'chet-chet', trilling.*

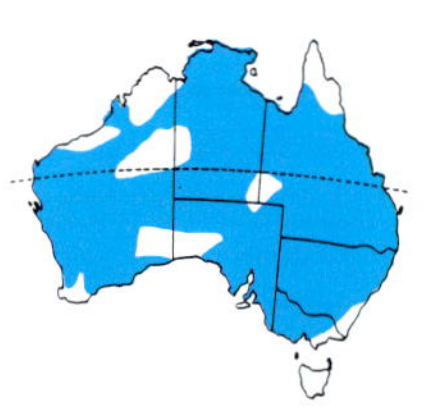

8 BLACK-FRONTED PLOVER
Charadrius melanops 15 cm

Usually in pairs or family parties, this small wader frequents dry river beds, swamps and lagoon edges and feeds on small aquatic animals and insects. It runs rapidly and bobs its head when at rest. Sedentary or nomadic in dry conditions. Breeds August-January. Nests in ground depression or among stones in a stream bed. Also in New Zealand. **Sexes**: *alike.* **Immatures**: *lacks breastband and black forehead.* **Voice**: *rapid churring.*

9 RED-CAPPED PLOVER
Charadrius ruficapillus 15 cm

Native to Australia, the engaging Red-capped Plover is distributed widely inland, as well as around the coast, and is often seen in pairs or flocks on beaches, rivers and margins of lakes. It scurries quickly with sudden pauses. If approached, it flies off a short distance to resume running. It eats insects and small crustaceans. Disperses in flocks after breeding September-January. Nests in scrape in the ground. Vagrant to New Zealand. **Female**: *lacks black markings, paler, legs greenish.* **Immatures**: *paler than adult female, upperparts mottled.* **Voice**: *'tik' or 'tink', trilling and churring.*

Breeding male

10 INLAND DOTTEREL
Peltohyas australis 20 cm

Australia's Inland Dotterel is a bird of the open plains, and is adapted to living in gibber country and semi-arid places where it may frequent the margins of lakes and streams. Though not timid, it flies well away if flushed. The scientific name *Peltohyas*, meaning shield, refers to the Y-shaped breast marking. During the heat of the day it covers it eggs with soil, and takes shelter in the shade. Becomes active after sunset. Eats insects and seeds. Some birds are nomadic. Breeds April-October. Nests in scrape on ground. Endemic. **Sexes**: *similar.* **Immatures**: *lack black markings, back dappled.* **Voice**: *'quik', alarm note 'kr-oot'.*

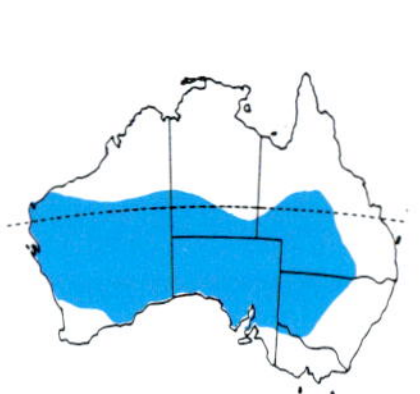

11 HOODED PLOVER
Charadrius rubricollis 20 cm

The endemic Hooded Plover is most abundant in Tasmania, coastal areas of southern Australia and around inland salt lakes in the south-west. In pairs, family parties or small flocks, it frequents open stretches of beach with grassy sand dunes. It is usally seen running along the water's edge or among jetsom. It is shy and easily disturbed. Breeds August-February. Nests in a ground depression. **Sexes**: *alike.* **Immatures**: *brown above, white below, no black markings.* **Voice**: *short piping notes and barking 'kew-kew'.*

12 ORIENTAL PLOVER
Charadrius veredus 25 cm

The Oriental Plover is a non-breeding visitor to Australia, arriving from its breeding grounds in China in September and departing in March. It is mainly seen in the north, on the seashore, swamps on inland plains, ploughed paddocks, or other little-disturbed stretches of grass. Its flight is swift and characterised by erratic zigzags. At rest it stands upright and montionless. Seen singly, in pairs and small flocks. Extremely wary. **Sexes and immatures**: *alike in non-breeding plumage; breeding male has chestnut and black breastband.* **Voice**: *sharp whistling note.*

1

2

3

4

5

6

7

8

9

10

11

12

20

13 DOUBLE-BANDED PLOVER

Charadrius bicinctus 18 cm

The Double-banded Plover is a trans-Tasman migrant arriving in February from breeding grounds in New Zealand and nearby islands. It leaves Australia in September or October by which time the non-breeding plumage of only a faintly-marked brownish breastband is giving way to a clearly-defined breastband and a black collar around the lower throat. It is seen on the seashore, mudflats and salt marshes, also inland on swamps and in ploughed paddocks, taking its food from the mud by probing. **Sexes**: *alike*. **Immatures**: *like non-breeding adults*. **Voice**: *piping whistle 'chip-chip', musical trilling.*

14 MONGOLIAN PLOVER

Charadrius mongolus 20 cm

A non-breeding visitor from breeding grounds in Central Asia arriving during September and departing in April. Commonly seen around the coast, particularly in the north and east in estuaries, mangroves, mudflats and sandy beaches, usually in small parties or flocks. Often associates with Double-banded and Large Sand Plovers. Occasionally inland. It runs swiftly and probes in wet sand or mud for food. **Sexes**: *alike; in breeding plumage breast-band is rufous-red, lores black, underparts and forehead white.* **Immatures**: *upperparts lightly scalloped.* **Voice**: *short penetrating 'drrit'.*

Non-breeding

15 LARGE SAND PLOVER

Charadrius leschenaultii 23 cm

A non-breeding visitor from northern Asia arriving in October and departing in March, and found throughout most of the coastal areas feeding on tidal sand and mudflats, singly and in flocks. It is less often seen than the Mongolian Plover, except in the west. If disturbed it usually flies up without moving far away and returns to the same spot. **Recognition** (*cf Mongolian Plover*): *larger, paler, longer legs and thicker bill.* **Sexes**: *alike in non-breeding plumage.* **Immatures**: *upper-parts scalloped lighter.* **Voice**: *trilling; louder than Mongolian Plover.*

16 RINGED PLOVER

Charadrius hiaticula 19 cm

A rare summer vagrant to Australia from its breeding grounds in northern Europe and Asia. It frequents muddy and sandy shores, sometimes with other plovers and stints. It feeds on small crustaceans, worms and insects and their larvae. **Female**: *similar, but black markings are duller.* **Voice**: *pleasant two-note call.*

17 RED-NECKED STINT

Calidris ruficollis 15 cm

One of the smallest and most-commonly seen of the waders regularly to visit Australia, migrating from north eastern Siberia or Alaska. The salmon-pink colouring about the head and neck which gives it its name occurs only in the breeding plumage, so is not usually seen here. Flying in tightly-packed wheeling flocks over a large part of the continent, it visits beaches, mudflats, estuaries and swamp margins. It feeds busily, probing wet sand or mud with a quick jabbing action, often in company with other waders. It frequently overwinters. **Sexes**: *alike.* **Immatures**: *browner.* **Voice**: *'chit-chit'.*

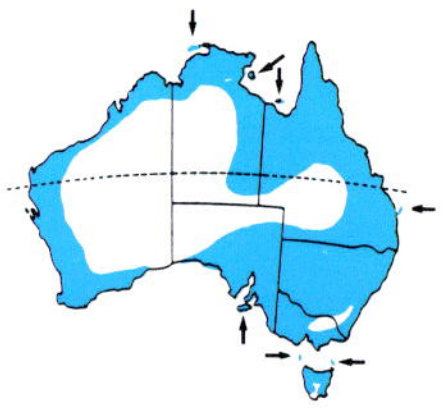

Non-breeding

18 LONG-TOED STINT

Calidris subminuta 14-15 cm

First recorded in Australia in 1886, but accepted as a regular summer visitor only in 1960 from breeding grounds in Siberia. Its distribution in Australia is patchy and it is ar less often seen than the Red-necked Stint. Frequents fresh water habitats rather than tidal flats. It feeds with darting runs and freezes if disturbed. In Australia September to April. **Recognition** (*cf Red-necked Stint*): *taller and slimmer stance, darker upper parts, legs greenish-yellow and longer.* **Sexes**: *alike in non-breeding plumage; breeding plumage darker above, crown rufous streaked darker.* **Voice**: *chirruping trilling; 'chee-chee'.*

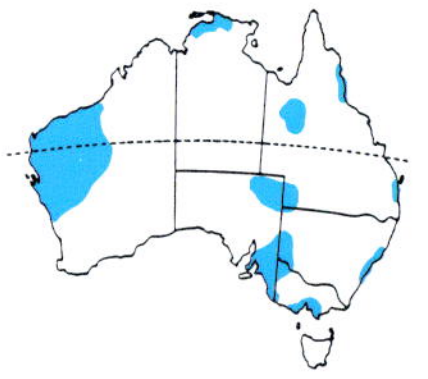

Non-breeding

19 SANDERLING

Calidris alba 18 cm

The Sanderling breeds in the Arctic Circle, and is seen as a non-breeding visitor along most Australian coasts from September to April, though not commonly. It feeds on beaches with darting movements at the water's edge, running after receding waves to pick its prey from the wet sand, sometimes taking to the air to avoid the incoming flow. Reluctant to fly far if disturbed. **Recognition:** *distinctive white under-parts contrasting with the black of the legs and feet.* **Sexes**: *alike in non-breeding plumage; male in breeding plumage is mostly cinnamon streaked black.* **Voice**: *soft twittering and chattering.*

Non-breeding

20, 21 CURLEW SANDPIPER

Calidris ferruginea 21 cm

Among the most common of the migratory waders, it reaches Australia from its northern Siberian breeding grounds north of the Arctic Circle in September and departs in April—but some overwinter here, many of which are probably immatures. It frequents a wide range of habitats, both coastal and inland, often in company with other waders. It frequently submerges its head in shallow water when probing for worms, crustaceans and insects with its long curved bill. It makes an imposing sight when massed in dense compact whirling flocks that can be numbered in thousands. **Sexes**: *alike; non-breeding males have less rufous on the head, neck and underparts than the breeding males.* **Voice**: *chirruping.*

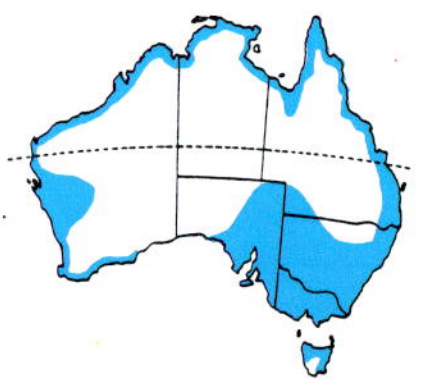

Non-breeding left, breeding right

22 BROAD-BILLED SANDPIPER

Limicola falcinellus 18 cm

Somewhat rare in Australia except on northern coasts, the Broad-billed Sandpiper is an inconspicuous non-breeding summer visitor from north Siberia, often seen in small flocks with Red-necked Stints. Mostly frequents tidal mudflats, it feeds with a deliberate up-and-down drilling movement, with the head and neck often submerging. **Recognition** (*cf Red-necked Stint*): *long sickle-shaped bill, and a branching white eye stripe.* **Sexes**: *alike in non-breeding plumage; breeding plumage has upperparts darker with tawny markings, breast heavily streaked.* **Immatures**: *slightly darker.* **Voice**: *chattering and trilling.*

23 SHARP-TAILED SANDPIPER

Calidris acuminata 23 cm

Widespread around the coasts and the wetlands inland, the Sharp-tailed Sandpiper is perhaps the most abundant migratory wader that visits Australia. Numbers are at their peak from August to April but very few stay throughout the year. Flocks of up to 1000 have been recorded in the south. It frequents a wide variety of habitats, including swamps, mudflats, irrigation areas, and mangroves, and feeds in shallow water on aquatic creatures and some plant material often with other waders. **Sexes**: *similar, but females smaller.* **Immatures**: *brighter, like breeding males.* **Voice**: *soft metallic 'pleep' and shrill piping notes.*

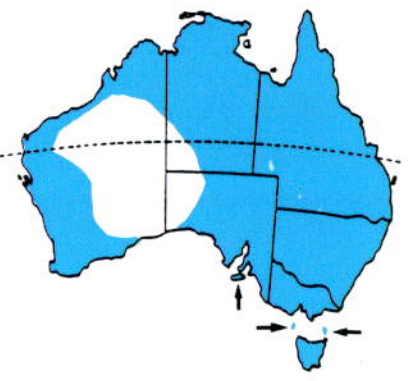

24 PECTORAL SANDPIPER

Calidris melanotos 23 cm

Breeds in Siberia, Alaska and Canada and a rare but regular summer visitor to Australia. In behaviour, habitat and appearance it is similar to its Sharp-tailed relative. **Recognition:** (*cf Sharp-tailed Sandpiper*): *abrupt change from brown breast to white belly, short tail, yellow legs and slightly longer and more de-curved bill. Different calls.* **Female**: *smaller, otherwise similar to non-breeding male; breeding male is more intensely coloured.* **Immatures**: *more buff.* **Voice**: *rasping chatter, twittering.*

13

14

15

16

17

18

19

20

21

22

23

24

20

25 GREAT KNOT
Calidris tenuirostris 28-30 cm

This, the largest of the sandpipers, breeds in Siberia, and vast numbers overwinter in northern Australia as non-breeding migrants from August to March. A concentration of over 100 000 was recorded between Broome and Port Hedland WA in September 1982. It is often seen in company with godwits, but the numbers thin out towards the south. It frequents mainly estuaries and tidal mudflats, occasionally inland lakes. Some stay all year. **Sexes**: *alike in non-breeding plumage; breeding male is darkly striped on head and neck, breast black-grey.* **Immatures**: *darker and more mottled than non-breeding adults.* **Voice**: *'knut-knut'.*

Non-breeding

26 [Red] KNOT
Calidris canutus 25 cm

Breeding in the Arctic Circle between August and April, it visits Australian shores in large numbers and frequents coastal sand flats and the margins of estuaries and rivers. Sporadically seen inland. It feeds in close-packed flocks that move in unison. The scientific name comes from its quaint King Canute-like action in running out in the wake of retreating waves. **Recognition** (*cf Great Knot*): *shorter bill, less boldly mottled, squatter posture.* **Sexes**: *alike in non-breeding plumage. Breeding males are deep chestnut red on breast and below (rarely seen in Australia).* **Immatures**: *browner.* **Voice**: *'knut-knut'.*

27 TEREK SANDPIPER
Xenus cinereus 25 cm

The very long slightly upcurved bill, bright orange part-webbed feet, and crouching stance while foraging, are distinctive features of the Terek Sandpiper. It also characteristically dashes about actively, teeters, and bobs its head. Singly and in small flocks, occasionally in large concentrations, it is seen on estuaries, mudlfats and coastal brackish swamps. It breeds in northern Eurasia and is in Australia August to April, some overwintering. **Sexes and immatures**: *alike in non-breeding plumage; breeding male is more mottled and brown.* **Voice**: *piping trill.*

Non-breeding

28 GREENSHANK
Tringa nebularia 34 cm

The distinctive Greenshank is widespread, mainly in coastal regions, but also common in some of the wetlands in the interior. Restlessly wary. Usually alone or in small parties, it is sometimes seen in large flocks. It is more common in the north. It feeds by wading in shallow water. Breeding in Eurasia, it is in Australia September to April. **Sexes**: *alike in non-breeding plumage; breeding males are darker, spotted brown on neck front and breast.* **Voice**: *staccato whistling, fluting.*

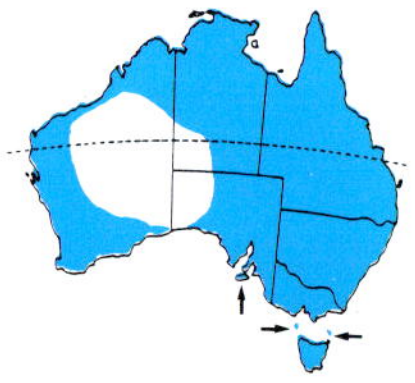

Non-breeding

29 MARSH SANDPIPER
Tringa stagnatilis 23 cm

Often called the Little Greenshank, the Marsh Sandpiper is a generally uncommon summer visitor to Australia, although sometimes seen locally in quite large numbers. Mostly solitary or in small groups, it frequents both saltmarshes and freshwater, mainly inland. It forages in shallow water, sometimes swims with head held high. It breeds in Eurasia. **Sexes**: *alike in bon-breeding plumage; breeding male is darker, spotted on throat and breast.* **Immatures**: *like non-breeding adults.* **Voice**: *subdued 'tee-oo', sharp alarm 'tchik-tchik'.*

Non-breeding

30 COMMON SANDPIPER
Tringa hypoleucos 20 cm

The Common Sandpiper, a non-breeding migrant from Europe and northern Asia, is commonly seen in the north, less so in the south, between July and May. Singly, or only in very small parties, it haunts rocky coastal streams, beaches, inlets and small inland bodies of water including stock dams. Often perches on rocks or posts. It is characterised by constant nervous up and down teetering and rapid flight close to water with down-arched, rapid flicking wingbeats. **Sexes**: *alike in non-breeding plumage; breeding male has upperparts and breast markings darker.* **Immatures**: *similar to non-breeding adults.* **Voice**: *piping 'twee-wee-wee'.*

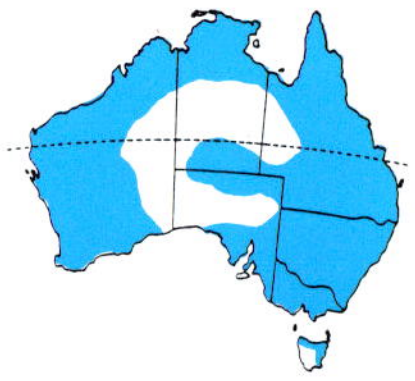

31 WANDERING TATTLER
Tringa incana 28-29 cm

The Wandering Tattler, a summer non-breeding migrant from Alaska, is comparatively rare in Australia, and has only been recorded on the east coast, chiefly on rocky areas. Its flight is swift and graceful. **Recognition** (*cf Grey-tailed Tattler*): *slightly larger, darker, longer nasal groove extending to three-quarters length of bill; also different habitat and voice.* **Sexes and immatures**: *alike in non-breeding plumage; breeding birds have strongly barred underparts.* **Voice**: *rapidly repeated 'whee-we-we' and fluting notes.*

Non-breeding

32 WOOD SANDPIPER
Tringa glareola 23 cm

A somewhat uncommon non-breeding migrant from northern Europe and Asia. It begins arriving in small numbers on northern Australian coasts in August and spreads southwards to all States. Singly or in small flocks, it frequents shallow stretches of fresh or brackish water, more usually inland, and often fringed with river red gums. Feeds on shellfish and insects. It swims well, and perches in trees. **Sexes and immatures**: *alike in non-breeding plumage; breeding plumage is more boldly marked.* **Voice**: *'chif-chif', whistling.*

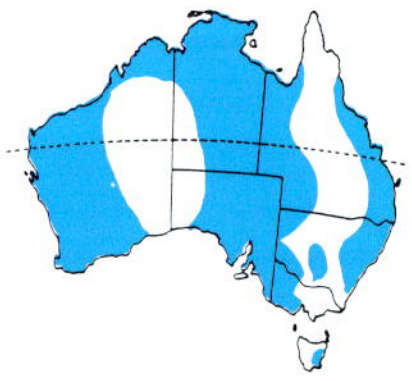

Non-breeding

33 GREY-TAILED TATTLER
Tringa brevipes 26-27 cm

Breeds in far north of eastern Siberia and one of the commonest of the migratory waders in Australia between September and April. It mostly frequents tidal mudflats and mangrove-fringed shores, singly, in small groups and larger flocks. At high tide it roosts in trees or on rocks, or on the roof of any handy building. Habitually bobs the head and flicks the rear end. **Sexes**: *alike in non-breeding plumage; male in breeding plumage is finely barred on throat, breast and flanks.* **Immatures**: *pale edges to wing feathers.* **Voice**: *variations on 'troo-eet'.*

Non-breeding

34 RED-NECKED PHALAROPE
Phalaropus lobatus 17-20 cm

This dainty member of the wader family breeds across northern Europe and Asia and winters at sea, mainly in the eastern Pacific. It was first sighted in Australia near Melbourne in 1962, and later on swamps and lakes on or close to the coast. It feeds while swimming in shallow water, picking water insects from the surface, and characteristically spins rapidly around to disturb prey from the bottom. It is very buoyant and sits high in the water. **Female**: *larger and brighter in breeding plumage than the male, a rufous band reaching from the eye round the neck.* **Immatures**: *darker.* **Voice**: *'chick'.*

Non-breeding

35 ORIENTAL PRATINCOLE
Glareola maldivarum 23 cm

A regular non-breeding migrant from eastern Asia, arriving October in flocks numbering hundreds or even thousands in north-western Australia. It spreads mainly across the north, but stragglers sometimes reach southern areas. It frequents edges of swamps and lagoons, claypans and airfields, often in mixed flocks with Australian Pratincoles. Its deeply-forked tail gives it a swallow-like appearance in flight and much of its insect food is hawked on the wing. **Sexes**: *alike in non-breeding plumage; breeding male has buff throat bordered black.* **Immatures**: *paler.* **Voice**: *'chick-chick'.*

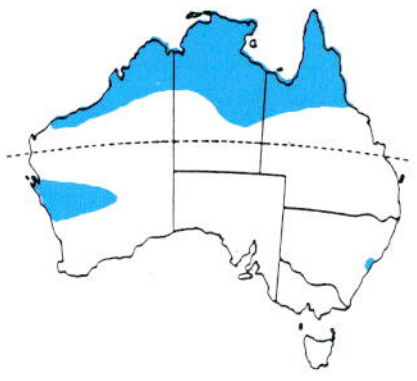

Non-breeding

36 AUSTRALIAN PRATINCOLE
Stiltia isabella 24 cm

The Australian Pratincole lives and breeds mainly in Australia, but many migrate to overwinter in New Guinea and Indonesia. It frequents dry open coastal and inland plains in the vicinity of water. It is partly nomadic and able to cope with conditions of extreme heat. It characteristically bobs up and down and runs rapidly. It flies gracefully and frequent zig-zagging. Its tail is shorter and less deeply forked than that of the Oriental Pratincole. Breeds October-January. Nests among stones on the gound, in colonies. **Sexes**: *alike.* **Immatures**: *browner and mottled.* **Voice**: *whistled 'weeteet'.*

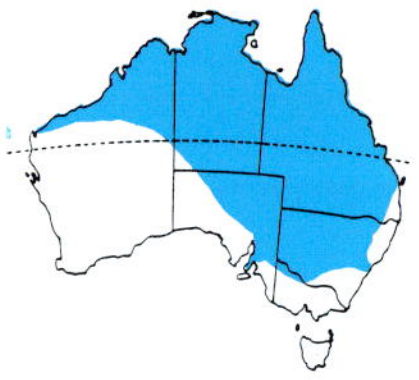

25

26

27

28

29

30

31

32

33

34

35

36

37 EASTERN CURLEW

Numenius madagascariensis 60 cm

The stately long-billed Eastern Curlew breeds in eastern Siberia and is seen all around the coasts of Australia between September and April/May as a non-breeding migrant. A few stay all year. Common along northern and eastern coasts, it is usually alone or in small numbers, seldom in large flocks, on beaches and muddy estuaries. Wary. Probes for crabs and worms. Often it roosts in mangrove trees at high tide. **Sexes and immatures**: *alike in non-breeding plumage; breeding male is tinged cinnamon above.* **Voice**: *'ker-loo' or 'kor-lew' including in flight.*

Non-breeding

38 WHIMBREL

Numenius phaeopus 46 cm

Breeds all around the high Arctic latitudes and overwinters in Australia from September to March/April. It is seen more often along the northern and eastern coasts than in the south and west. It frequents muddy estuaries, sandflats, salt marshes and coral cays, alone or in small flocks, often with other waders. Normally wary but has become quite tame at some Great Barrier Reef resorts. Some may remain all year in northern Australia. **Sexes and immatures**: *alike.* **Voice**: *high clear repeated whistles.*

Non-breeding

39 LITTLE WHIMBREL (Curlew)

Numenius minutus 35 cm

A non-breeding visitor from north-eastern Siberia and seen sometimes in very large flocks in the north west of the continent from September to May but rarely south of the Tropic of Capricorn. Usually in small groups, it frequents paddocks, playing fields, airfields and flood plains; and feeds on insects, their larvae, and possibly seeds. Individual birds or small flocks may move southwards to avoid especially heavy monsoonal rains. 'Freezes' when flushed. May become tame near habitation. **Sexes and immatures**: *alike.* **Voice**: *variations on 'chee-chee'.*

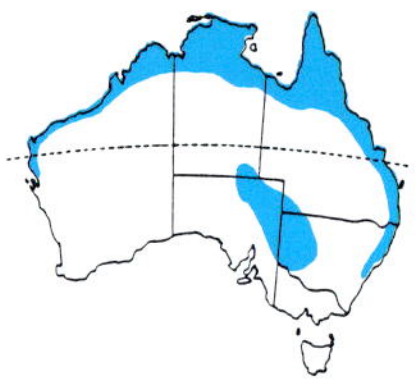

40 BLACK-TAILED GODWIT

Limosa limosa 36-43 cm

An uncommon non-breeding visitor from northern Europe and Asia, arriving in Australia during September and departing in May. Solitary, in small flocks, sometimes in hundreds, it frequents coastal tidal estuaries, fresh and brackish lagoons and swamps often with Bar-tailed Godwits. Most common along northern and north-eastern coasts. Flight is swift, sometimes spectacularly acrobatic. **Recognition** (*cf Bar-tailed*): *smaller, less mottled, broad black tip to tail on white rump, white underwing, bill usually straight.* **Sexes**: *alike; breeding plumage striking red-brown.* **Immatures**: *some wing coverts edged with buff.* **Voice**: *sharp 'witta-wit' and 'kip-kip'.*

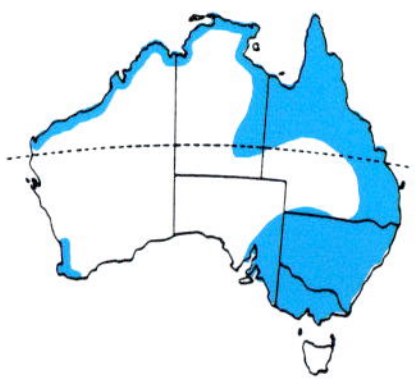

Non-breeding

41, 42 BAR-TAILED GODWIT

Limosa lapponica ♂ 38-42 cm ♀ 42-45 cm

The familiar Bar-tailed Godwit arrives in great numbers in August/September from its breeding grounds in north-east Siberia and spreads right around the coast for its summer visit until take-off time for breeding in April/May. Concentrations of tens of thousands have been recorded on Eight Mile Beach between Broome and Port Hedland in WA. It is seen on coastal sandy shores, mud-flats and marshes, probing, sweeping and jabbing into mud or sand between the tides for small crustaceans and worms. Large flocks may form to fly swiftly in tight wheeling formation. Departing birds fatten for their long trans-world journey and may have assumed their breeding plumage before leaving. Many birds remain behind to overwinter. **Recognition** (*cf Black-tailed Godwit*): *bill usually slightly up-curved, plumage lighter and more mottled.* **Female**: *larger.* **Immatures**: *more buff edging on feathers.* **Voice**: *variations on 'witta-wit', 'ki-kip'.*

Non-breeding left, breeding male right

43, 44 [Ruddy] TURNSTONE

Arenaria interpres 23 cm

The Turnstone visits Australian coastal regions from September to April, from its breeding grounds in eastern Siberia and Alaska. It reaches Tasmania and is especially abundant in the north of the continent. Some birds, probably non-breeding immatures, stay to overwinter. It frequents rocky shores, shingly beaches, estuaries and coral cays, and characteristically runs about overturning stones, shells and beach debris, or pushing up heaps of seaweed to expose small prey with its sharp short bill. It also skilfully uses it to open hard-shelled mussels and barnacles. At high tide it roosts on exposed rocks, or retreats to sandspits with other waders. In flight it shows a panel of white in the wing, and a black-barred undertail. **Sexes**: *male brighter than female.* **Immatures**: *more buff edging to feathers.* **Voice**: *'kut' or 'kitititit'.*

Non-breeding left, breeding right

45 RUFF (Reeve)

Philomachus pugnax 23-30 cm

The Ruff takes its name from the very large ruff fluffed up by the male in courtship and territorial displays (but not likely to be seen in Australia). The female of the species is known as a Reeve. It was only in the early 1960s that the ruff was first noted here, possibly because it had previously been overlooked as a Sharp-tailed Sandpiper, which the female somewhat resembles. It frequents swamps, lagoons, sewage farms and tidal mudflats, often with other waders, but it remains as yet a rare summer visitor. Breeds in Eurasia. **Female**: *smaller.* **Immatures**: *buff wash, neck darker.* **Voice**: *normally silent; very rarely muted notes.*

Non-breeding

46 PIED OYSTERCATCHER

Haematopus longirostris 48-52 cm

The handsomely-plumaged and gregarious Pied Oystercatcher, common in Tasmania, is also found right around the Australian coast on islands, sandy beaches, mudflats and estuaries. Its stout bill, flattened at the sides, is used to dig into the sand for crustaceans and molluscs, and for opening hard-shelled bi-valves and prising limpets from the rocks. Wary. Sedentary. Breeds August-January. Nests in a scrape in the sand. Related to a world-wide superspecies. **Sexes**: *similar; female slightly larger.* **Immatures**: *duller, bill and legs paler.* **Voice**: *penetrating high-pitched 'kleep-kleep' or 'peepapeepapeep'.*

47, 48 SOOTY OYSTERCATCHER

Haematopus fuliginosus 48-52 cm

(A) Race *fulginosus* (southern form)
(B) Race *ophthalmicus* (northern form)

For most of the year, the Sooty Oystercatcher, or Spectacled Oystercatcher in the north, lives close inshore on rocky coasts and coral reefs around the coast, but its distribution is patchy compared with that of the Pied Oystercatcher. Where they meet, they more or less keep to their different habitats, but may share nesting sites — though without interbreeding. The Sootys feed on crustaceans, sea worms and molluscs, opening them in the same way as the Pied Oystercatcher. Generally nests on islands, and does so in a shallow scrape in the sand, or in a rock cleft. Endemic. **Sexes**: *similar; female slightly larger.* **Immatures**: *wing coverts edged buff, legs grey.*

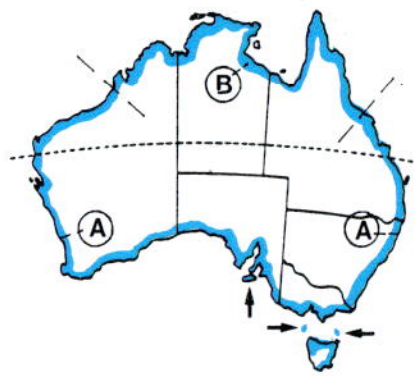

Race (A) left, Race (B) right

37

38

39

40

41

42

43

44

45

46

47

48

21

Seabirds

THE MARINE ENVIRONMENT (Also see map on inside back cover)

The birds in this Section are those that freqent the seas over Australia's continental shelf. This covers an area of over 2.45 million square kilometres, extending from a coastline of some 20 000 km to a 200 metres depth-line, and varying in width from about 50 km to over 500 km. It links Australia to New Guinea in the north — where the surface temperature reaches 29°C in the summer, one of the warmest open sea surfaces in the world after the Red Sea — and reaches southwards over some 2000 nautical miles of latitude to just south of Tasmania in the vicinity of the Subtropical Convergence. Here the surface water temperature drops abruptly from 16° to 12°C in summer and from 12° to 8°C in winter, and marks the limits of the Antarctic Ocean. Two natural marine regions are recognised: a northern tropical region known as the Dampierian Region and a southern temperate region known as the Flindersian Region. They meet at Shark Bay (latitude 26°S) on the west and Fraser Island (latitude 25°S) on the east; approximately at the southern limits of the coral reefs. Australia is surrounded by a large number of islands — 1267 named (866 in the northern tropical region and 401 in the southern temperate region) and many smaller nameless ones. Most are 'continental' islands which have been detached from the mainland by erosion.

THE SEABIRDS

Worldwide, there are four orders of seabirds comprising twenty families and 312 species. Over 100 species representing thirteen families occur in Australian waters. Forty-two are breeding species, and over sixty are regular non-breeding visitors. A number of others have also been recorded as vagrants.

The breeding species
Most species breed on the islands; only a few on the mainland. Cliff rookeries are uncommon and are confined to a few rocky islands in southern waters. The distribution of the breeding species is conditioned by food supply and the availability of nesting sites, and while several seabirds breed in Australia generally, many show marked preferences either for tropical or temperate seas. They separate into three categories:

- **Species that breed in the northern tropical region** (17) — six terns, three noddys, three boobies, two frigatebirds, a shearwater, a petrel and a tropicbird. The highest concentrations are off the northwest coast, but species like the Sooty Tern also nest in large colonies on the Great Barrier Reef.
- **Species that breed in the southern temperate region** (14) — four shearwaters, two gulls, Fairy Penguin, Shy Albatross, Australasian Gannet, Black-faced Cormorant, White-faced Storm Petrel, Great-winged Petrel, Common Diving-Petrel, Fairy Prion. Notable nesting islands are in Bass Strait and around Tasmania, and include those used by the Short-tailed Shearwater (Muttonbird) which breeds in prodigious numbers.
- **Species that breed in Australia generally** (11) — Pelican, Silver Gull, five terns and four cormorants.

The visiting species
During the winter, great numbers of birds, especially shearwaters, petrels and albatrosses, arrive from their breeding grounds in Antarctic waters. Many of them gather to feed in the nutrient rich waters off south-eastern Australia. Visitors in the summer include terns and skuas from the northern hemisphere, some, such as the Arctic Tern, coming from as far as the Arctic Circle.

TUBE-NOSED SEABIRDS Order Procellariiformes

The four families of this order comprising ninety-three species spend most of their lives wandering over the high seas, and coming to land only to breed. They vary greatly in size from the Wandering Albatross to the tiny storm-petrels. All have twin external nostrils in the form of tubes on top of or at the sides of their upper mandibles.

ALBATROSSES *Family Diomedediae*

Of thirteen species of albatrosses, four are confined to the north Pacific, and nine frequent southern seas and visit Australian waters — two only infrequently, the Royal Albatross and Buller's Albatross (these two are not illustrated). The southern birds fall into three groups: two white-backed (great) albatrosses; five dark-backed smaller mollymawks; and two sooty albatrosses. Albatrosses are essentially birds of the winds, and are able to sustain their prolonged and graceful gliding flight on motionless wings by utilising the updrafts from wave-fronts and cover great distances without landing. The Wandering Albatross and others cover prodigious distances, and may circumnavigate the world many times in a lifetime. They feed on fish, squid, crustacea, krill and offal, and most follow in the wake of ships ready to drop on jettisoned galley waste, sometimes keeping up with even the fastest vessels for days and nights on end. Most of the southern species breed within the Antarctic Circle in the summer months and disperse northwards in winter, in some cases as far as the Equator. Albatrosses mature slowly and in some cases undergo many phases of plumage development from the juvenile state to adulthood — which may take as long as ten years for the great albatrosses and seven for the mollymawks. Pairs mate for life, which may extend for forty years or more. Superstitious beliefs regarding albatrosses, epitomised in Coleridge's famous ghostly tale of the *Ancient Mariner*, have been strongly held by mariners for centuries. 1-8

FULMARS, GADFLY PETRELS, PRIONS, SHEARWATERS Family Procellariidae

Some fifty-five species worldwide make up this large family of four natural groups, ranging from the giant petrels to the dainty fairy prions. Thirty-six of the species are known with varying degrees of familiarity in Australian seas, and nine nest here, mostly on oceanic islands.

Fulmars A collective title for a small group of five genera linked by a preference for the colder waters and the habit of nesting on the surface instead of in burrows. 9-11, 13, 23

Gadfly and other Petrels The so-called gadfly petrels belong to the genus *Pterodroma* and constitute a diverse group of about twenty-six small to medium species with short, stout black bills. They are distributed widely in the Pacific from the tropics to the sub-Antarctic. Thirteen species have been recorded in Australian seas and two breed here:

Shy Albatross

Gould's Petrel on Cabbage Tree Island north of Sydney and the Great-winged Petrel off the south-west coast of Western Australia. The five species illustrated are regular visitors. 14-18

Prions Wide hooked bills equipped at the sides with hairlike combs (lamellae) for filtering zoo plankton from the water distinguish these small dove-grey tube-nosed petrels, also the prominent M-marking across the upper wing. The six species have all been recorded in Australian waters, but are indistinguishable except in the hand. 19-22

Shearwaters Shearwaters are small to medium petrels which range widely over all oceans. Of seventeen species, eleven occur in Australian waters, eight regularly. Shearwaters skim close to the surface with straight and stiff wings in swift gliding flight, appearing to cut the water when wheeling — hence the name. They nest in vast colonies in burrows on oceanic islands, and many species congregate in dense feeding rafts offshore during the day, returning to land at dusk. The Short-tailed Shearwater — the best known of several species popularly known as 'muttonbirds' — is one of five shearwaters that breed in Australia. It nests in millions on islands in Bass Strait, with smaller colonies off south-east New South Wales. In December 1798, during Flinders' historic voyage with George Bass in the *Norfolk* through the strait that bears the latter's name, first contact was made with these birds. The breeding cycle of these shearwaters involves annual migrations over prodigious distances. The parents desert the single chick before it has fully fledged — leaving it entirely to fend for itself to reach maturity. It then launches itself at night over the sea to find its own way over the migratory journey northwards, relying entirely on the body fat acquired from the last meal given by its parents. Muttonbirds were harvested in great numbers by sealers for their oil, flesh and feathers — a tradition still maintained, but under strict controls. The name 'muttonbird' comes from the flavour of the flesh. 24-31

STORM-PETRELS Family Oceanitidae

The Storm-Petrels, the size of a swallow and the smallest of the oceanic birds, are renowned for appearing to walk on water while feeding — an illusion created by their habit of flying close to the waves with legs dangling and patting the surface. And it is from this that the name petrel is believed to have come from an allusion to St Peter — recalling his walk on the sea of Galilee. When Storm-Petrels appear suddenly at sea, sailors used to regard this as a sign of an impending storm, and called them Mother Carey's Chicken, derived from *Mater Cara* (Beloved Mother) their traditional guardian. Storm-Petrels range widely over all oceans. Of about twenty species, seven visit Australian waters regularly. 32-36

DIVING-PETRELS Family Pelecanoididae

Diving-Petrels are confined to the colder latitudes of the southern seas. They characteristically 'fly' underwater like penguins. They tend to keep close to the coast near their breeding grounds and are rarely seen far out at sea. Because of the danger of predation of nesting sites by skuas, they come to land only after dark. Of the four species, one breeds in Australia. 46

PENGUINS Order Sphenisciformes

PENGUINS Family Spheniscidae

The sixteen species of penguins live almost entirely in the southern hemisphere, some entirely in Antarctica. The smallest of the family, the Fairy or Little Penguin, breeds in Australia, the Rock-hopper and Fiordland Penguins are not infrequently seen off Tasmania and the southern coast, and eight other species are accidental visitors to Australian waters. They have waterproof and chillproof coats of dense glossy scale-like feathers over a thick layer of down, which traps air, and a body lining of blubber. In 1578, the Welsh members of Sir Francis Drake's ship the *Golden Hind* called the hoards of birds seen during the transit of the Magellan Strait 'pengwyn' (meaning white head) and so established a name which has become a lasting memorial to them. 47-48

SKUAS, GULLS, TERNS AND ALLIES Order Charadriiformes (Part)

SKUAS Family Stercoraiidae

Skuas are piratical birds that force other seabirds to disgorge their prey. There are two genera: *Stercorarius*, comprising three Arctic-breeding species known as jaegers; and *Catharacta*, the skuas, comprising four heavy-bodied species, two of which breed in Antarctica. The jaegers are transequatorial migrants which travel as far as Antarctica to overwinter. All three are seen in Australian waters on passage. The Antarctic-breeding skuas migrate northwards to overwinter and also pass through Australian waters; the Southern Skua being seen as a regular visitor. 37-39, 43-44

GULLS, TERNS AND NODDIES Family Laridae

Gulls Universally known as 'seagulls' gulls are recognised by their similarity of appearance, habit and raucous cries as the most familiar of all shorebirds. They vary in size from 28 cm to 70 cm, and are members of a family of forty-five species, most of which live in the northern hemisphere. They are adapted to almost every niche, except the poles and some desert areas, and are rarely seen far out to sea. 41-42, 45

Terns Compared with gulls, the graceful terns are slimmer and smaller and the wings are longer and more slender. They hover ready to dive or skim over the surface when feeding. There are thirty-five species, most of which are found in tropical and subtropical seas. Seventeen species occur in Australia, of which eleven breed here — mostly in crowded and noisy colonies on coastal and oceanic islands. The others, including the Arctic Tern from the northern polar regions, are mostly regular, seasonal migrants, returning to overwinter during the southern summer. 49-65, 68

Noddies Noddies, closely related to terns, are on the whole darker, and have white caps instead of black. Of the five species, three breed in Australia. Unlike terns, they nest in trees. Breeding White-capped and Lesser Noddies are so tame that they can be lifted from the nest. 67, 70-72

Opposite: Black-faced Cormorants

21

CORMORANTS, PELICANS AND ALLIES Order Pelecaniformes

The six families of this order are found worldwide in temperate and tropical regions, and all are represented in Australia. They comprise the tropicbirds, frigatebirds, gannets and boobies, cormorants, darters and pelicans. They are the only birds that have all four toes joined by a single web (totipalmate). They have sealed nostrils which enables them to dive for food and feathers that require frequent preening for waterproofing. All, except the tropicbirds, have a naked gular or throat patch which is fluttered to assist cooling in heat.

TROPICBIRDS Family Phaethontidae

Tropicbirds, characterised by their long tail streamers, are usually seen far from land when not breeding in tropical and sub-tropical seas. They normally fly high and gracefully with pigeon-like wingbeats and dive for food like gannets. They are often attracted to ships which they circle with loud screams. They are nearly always airborne and their legs are so short that they are almost helpless on land. The snowy beauty of their plumage is reflected in their scientific name, derived from 'shining' in Greek. Two of the three species frequent Australian waters. 66, 67

FRIGATEBIRDS Family Fregatidae

Frigatebirds are instantly recognisable from their silhouettes — they have exceptionally long and slender angled wings, a long and deeply forked scissor-like tail and a long hooked beak. They spend most of the day on the wing. They are known to sailors as the Man-O'-War Bird from their habit of harassing and out-manoeuvering other birds to force them to give up their prey. They also dive to take food from the surface without causing a ripple. They rarely swim and rapidly become waterlogged. They cannot walk on land, only shuffle about, and roost in trees. The male has a prominent scarlet gular pouch which he inflates like a balloon to attract overflying females during the breeding season. But the choice of a mate is entirely that of the female, which is also the larger bird. On some Pacific Islands, frigatebirds have been trained like carrier pigeons to carry messages. Of five species, two are regularly seen in Australian seas. 73-74

GANNETS AND BOOBIES Family Sulidae

This family comprises three species of gannets and six of boobies. The gannets live in cold and temperate seas, and the boobies in all three oceans in tropical and subtropical waters. Flying fish and squid form a large part of their diet and they dive on these from heights of 30 metres or more, entering the water with a huge splash, and sometimes continuing the dive to a considerable depth. The impact of hitting the water is cushioned by protective air sacs in the throat and breast. Some boobies loaf in a ship's rigging and may dive from the forecastle to seize flying fish. One gannet and three boobies frequent Australian waters. 75-79

CORMORANTS Family Phalacrocoracidae

Cormorant or Shag? Traditionally called cormorants in Australia and America, and shags in New Zealand (and by fishermen almost anywhere). Terminology has recently been unified in Australia and New Zealand — with 'shag' applying to the essentially marine cormorants of the genus *Leucocarbo*, and 'cormorant' applying to the genus *Phalacrocorax*. Their plumage is not completely waterproof and they spend much time drying out on land with wings extended, often perched on a rock or a post. They often fly in a V-formation. They breed in large colonies on cliffs, rocky shores and on inland lakes and rivers. Of thirty-three species worldwide, five occur in Australia, one endemic. 80-84

DARTERS Family Anhingidae

This family inhabits warm temperate and tropical regions of the New World, Africa, southern Asia, South America and the Australian Region. The scientific name *Anhinga* is the name used by the Amazonian Indians. Darters are the least marine of the families of this order and most inhabit freshwater lakes and rivers. They are wary and often swim with only the neck and head showing. They forage differently from cormorants by stalking prey slowly underwater, and mostly eat fish. 85-86

PELICANS Family Pelecanidae

Pelicans are very sociable birds and do almost everything together. Adults may develop scarlet on their beaks when courting, which seems the most probable explanation of the ancient belief that brooding parents pierce their breasts to feed the young. There are six or seven species distributed in the tropical and sub-temperate regions of the world; Australia has one. 87

UNCOMMON AND VAGRANT SPECIES (not illustrated)

King Penguin, Gentoo Penguin, Chinstrap Penguin, Adelie Penguin, Fiordland Penguin, Snares Penguin, Erect-crested Penguin, Royal Penguin, Magellanic Penguin, Buller's Albatross, Antarctic Petrel, Kermadec Petrel, Herald Petrel, Tahiti Petrel, Soft-plumaged Petrel, Mottled Petrel, Cook's Petrel, Snow Petrel, Blue Petrel, Grey Petrel, Black Petrel, Westland Petrel, Manx Shearwater, Audubon's Shearwater, Leach's Storm-Petrel, Sooty Storm-Petrel, South Georgian Diving-Petrel, Cape Gannet, Christmas Frigatebird, Black-tailed Gull, Lesser Black-backed Gull, Franklin's Gull, Black Tern, Saunders's Tern.

Opposite: Great Frigatebird

PHOTOGRAPHERS: B.M. Allwright 1, 23; K. Atkinson 67; H. & J. Beste 2, 58, 65, 81; G. Chapman 50, 52, 54, 56, 60, 63, 69, 83; B.J. Coates 34 (NPI): H.J. de S. Disney 34 (NPI); J. Frazier 86; P.J. Fullagar 16, 31, 34 (NPI), 35; T. & P. Gardner 18, 28, 53, 57, 61, 76, 77, 84; P.N. Green 85; G. Harrow 30; C. A. Henley 66; E. & D. Hosking p. 201; 51; K. Ireland 64, 82; G.W. Johnstone 6 (Ant.Div), 7, 10, 43 (Ant.Div) 44; G.R. Jones (Ardea) 38; R. & D. Keller 9, 11, 47, 68; C.R. Knights (Ardea) 37; N. Kolichis 14; T.R. Lindsey 39; T.W. Pescott 20; M. Price (Ant.Div) 36; P.G. Roach 3; G. Robertson p. 197; 22, 24, 70 73,75; A.E.F. Rogers 4, 5, 17; V. Serventy 13, 74, 79; E.C. Slater 25, 62, 80; P. Soleness (APL) p,. 164, 165; M. Soper 12, 19, 21, 29, 32, 48; R. & V. Taylor p. 199; B. Watkins (Ant.Div) 8; J. Warham 15, 26, 27, 33, 46; C. Webster 42, 55, 78; B.A. & A.G. Wells 72; E.E. Zillmann 71; Authors 40, 41, 45, 59, 87.

21

1 WANDERING ALBATROSS

Diomedea exulans

107-135 cm

Wingspan c 340 cm

The wingspan of this magnificent bird, sometimes over 3.5 metres, exceeds that of all other living birds. Its gliding flight is effortless and graceful, and it may follow even the fastest of ships for days and nights on end ready to drop on scraps thrown overboard. It is a true wanderer and its feats of flight over enormous distances around the globe are legendary. It is seen all through the year in Australian waters but mostly in winter on leaving its breeding islands in the southern oceans. Up to 500 birds have been seen feeding at Sydney's sewage outlet at Malabar. **Recognition** *(cf Royal Albatross): difficult to separate, but at close range black in tail of the Wandering is one clue.* **Sexes**: *similar.* **Juveniles**: *brown with white face. Adult plumage reached in 9 years.* **Voice**: *deep throaty 'bleats', mainly silent at sea.*

2 SHY (White-capped) ALBATROSS (Mollymawk)

Diomedea cauta

100 cm

Wingspan c 198-256 cm

The only albatross to breed in Australia, with nesting colonies on Albatross Island off the north-west tip of Tasmania, and on two of the Maatsuyker Islands off its south coast. It also breeds on islands around New Zealand's subantarctic where two subspecies also occur. The Albatross Island colony was discovered by Flinders and Bass during their circumnavigation of Tasmania in 1798. This colony was subsequently decimated by sealers but has largely recovered. Tends to avoid inshore waters and follows ships less than other species, hence the name 'shy'. Mostly seen in Australian seas between July and October. Breeds from September. **Recognition**: *the largest of the mollymawks; the underwing is mainly white with a small black patch at its root.* **Sexes**: *alike.* **Immatures**: *dusky head and breastband.* **Voice**: *cackling and gurgling.*

3, 4 BLACK-BROWED ALBATROSS (Mollymawk)

Diomedea melanophrys

83-93 cm

Wingspan c 230 cm

Seamen traditionally regard albatrosses as stupid and call them mollymawks — based on a Dutch word meaning a stupid gull. The term is also used in New Zealand for all of the smaller species. The circumpolar range of the Black-browed Albatross spans the oceans from the Antarctic pack ice to the equator, and it is possibly the most abundant and widespread of the albatrosses. It is the albatross most often seen in southern Australian waters, especially from May to November after dispersing from its breeding islands in the sub-Antarctic. The Falklands are its stronghold with a breeding population estimated from 500000 to 3 million. It readily follows ships and gathers in large numbers around trawlers. It forages inshore and in harbours, and feeds on fish, squid and cuttlefish. **Recognition**: *bright yellow-orange bill; underwing blackish.* **Sexes**: *alike.* **Immatures**: *dusky about head and narrow breastband; bill dark grey-green with black tip. (As with all the mollymawk group, full adult plumage takes some seven years to develop.)* **Voice**: *mostly silent, harsh grunting.*

5 YELLOW-NOSED ALBATROSS (Mollymawk)

Diomedea chlororhynchos

71-81 cm

Wingspan 180-200 cm

Smallest of the mollymawks and more slender. It is common, and frequents warmer waters than other albatrosses, sometimes as far north as the Tropic of Capricorn. It frequents Australian seas throughout the year, but is most numerous from May to October. Shy, and follows ships only occasionally, but congregates around fishing boats. It feeds at night on the surface. Breeds on various sub-Antarctic islands. **Recognition**: *underwings are mostly white, with narrow black margins.* **Sexes**: *alike.* **Immatures**: *bill all black.* **Voice**: *shrill bleating laugh.*

6 GREY-HEADED ALBATROSS (Mollymawk)

Diomedea chrysostoma

c 81 cm

Wingspan c 220 cm

The Grey-headed Albatross, with a range extending southwards to the ice pack, frequents colder waters than other albatrosses. Visits to Australian waters are at their peak between May and November, and it is then fairly common off the Western Australian coast and Tasmania, but it is rarely seen off the east coast. It breeds on various sub-Antarctic islands including Macquarie Island. **Recognition**: *relatively broad black band along leading underwing edge.* **Sexes**: *alike.* **Immatures**: *browner and darker.* **Voice**: *braying sounds.*

1

2

3

4

5

6

21

7 SOOTY ALBATROSS

Phoebetria fusca

84-89 cm

Wingspan c 203 cm

The dark-plumaged Sooty Albatross has a distinctive tapered tail, and sails on constantly-adjusted long slender wings. It has the reputation of being the most accomplished and graceful flier of all seabirds. It is a regular follower of ships for long periods, and ancient seafarers believed that the souls of men drowned overboard passed into this bird. Some think that this was the 'pious bird of good omen' shot by the Ancient Mariner in Coleridge's famous tale. It visits Australian waters in winter, chiefly along the southern coasts and Tasmania. Breeds on various sub-Antarctic and Atlantic Ocean islands. **Recognition**: *wholly sooty-brown.* **Sexes**: *alike.* **Immatures**: *bill stripe paler.* **Voice**: *sad screaming 'pee-ooo' at the nest.*

9, 11 SOUTHERN GIANT PETREL (Fulmar)

Macronectes giganteus

86-99 cm

Wingspan 185-205 cm

A large, pale bill, stouter body, and a less graceful flight distinguish the giant petrels from the albatrosses. About one in ten birds of this species fledge white and remain so for life. It ranges widely over the southern oceans and is regularly seen off southern Australia, especially in winter. Frequently follows ships. Like all fulmars, it will eject an evil-smelling stomach oil when approached. They also have a persistent musty smell. Giant petrels are the only petrels to come ashore to feed, where they are aggressive scavengers of carrion, eggs and chicks. At sea they eat fish, molluscs and plankton. The species breeds October-November on various sub-Antarctic islands, including Macquarie Island. **Recognition**: *(cf Northern Giant Petrel): green tipped bill; adults pale head; only the Southern Giant Petrel is seen in white plumage.* **Sexes**: *alike.* **Voice**: *whinnying and squawking.*

8 LIGHT-MANTLED [Sooty] ALBATROSS

Phoebetria palpebrata

79-89 cm

Wingspan 1.83-2.18 m

The name is descriptive of this bird's elegant two-toned plumage which sets it apart from the wholly-brown, closely-related Sooty Albatross. Its flight is equally impressive in ease, grace and manoeuverability — it is able to hang motionless in mid-air beside a cliff edge and land on a ship's masthead in a gale. Its range is circumpolar and extends to the Antarctic pack-ice. It is highly pelagic, and less often seen in Australian waters than the Sooty Albatross. Nests on sub-Antarctic islands, including Macquarie Island, in summer. **Recognition**: *distinctive light-grey mantle on back.* **Sexes**: *alike.* **Immatures**: *browner, no blue line on bill.* **Voice**: *'peee-arr' at nest, trumpeting at sea.*

10 NORTHERN GIANT PETREL

Macronectes halli

81-94 cm

Wingspan 183-218 cm

The Northern Giant Petrel has only in recent years been recognised as a separate species distinct from the Southern, but the two are extremely difficult to distinguish. The habits are very similar but the Northern Giant Petrel forages and breeds on more northerly islands. It is a winter visitor to Australia's southern waters, and arrives up to two months ahead of the Southern Giant Petrel. Breeds on various sub-Antarctic islands July to February. **Recognition**: *(cf Southern Giant Petrel): reddish-tipped bill throughout life; crown and upperparts always darker.* **Sexes**: *alike.* **Immatures**: *all brown-black.* **Voice**: *like Southern Giant Petrel.*

12 WHITE-CHINNED PETREL

Procellaria aequinoctialis

51-58 cm

Wingspan c 340 cm

The largest of other all-dark petrels (Black and Westland), but extremely difficult to separate, except for its white chin — but some birds lack this feature. The species is essentially circumpolar in range, but is only occasionally seen in Australian waters. It follows ships and is aggressive when feeding around fishing trawlers. Breeds on various sub-Antarctic islands generally November-January. **Sexes**: *alike.* **Immatures**: *similar to adults.* **Voice**: *croaking and screaming at the nest.*

7

8

9

10

11

12

21

13 CAPE PETREL
Daption capense 41 cm

The beautifully chequered Cape Petrel is commonly seen in Australian waters from late summer to early spring. It is also widely known as Pintados (meaning painted — the scientific name *daption* is an anagram of pintado), or as Cape Pigeon, perhaps from its habit of pecking food from the water. It floats in large aggressive feeding flocks with much squabbling over food. It often follows ships and also dives shallowly for small sea creatures. Nests in Antarctica and nearby islands. **Sexes** and **immatures**: *alike.* **Voice**: *rasping and rattling sounds.*

14 GREAT-WINGED PETREL
Pterodroma macroptera c 44 cm

This long, slender-winged petrel flies at great speed in wild undulations, shooting upwards to 30 metres or more and diving steeply on swept-back wings. Nearly circumpolar in range and highly pelagic, seldom visiting offshore waters except when breeding. Usually solitary, and mainly a nocturnal feeder. Rarely follows ships. Breeds January-September. It nests in a burrow or on the ground on islands off the coast of WA; also in southern latitudes. **Recognition**: *stocky build, short black bill; no distinctive markings.* **Sexes** and **immatures**: *alike.* **Voice**: *Rapid 'kik-kik', sibilant 'si-si-si', deep braying.*

15 WHITE-HEADED PETREL
Pterodroma lessonii 40-46 cm

Seen usually alone in Australian waters all months, but uncommon except around Tasmania. Circumpolar in range and highly pelagic. Closely related to the Great-winged Petrel, and its flight is equally spectacular. Rarely follows ships. Breeds August-May in southern latitudes. Nests in burrows then disperses both north and south. **Recognition**: *head and underparts white; underwings dark.* **Sexes** and **immatures**: *alike.* **Voice**: *variations on 'wik-wik' and 'oooer' at sea.*

16 PROVIDENCE PETREL
Pterodroma solandri 41 cm

This bird saved the convict settlement established on Norfolk Island in 1788 from starvation — a recorded 170 000 being harvested in only 3 months in 1790. Not surprisingly, the species became extinct there. Now it breeds only on the forested slopes of Lord Howe Island, where it is abundant from February to November, before dispersing in a limited area mainly in the north Pacific. It is common along the edge of the continental shelf off the NSW coast from April to October. **Recognition**: *dark grey-brown, dull white flash underwing.* **Sexes** and **immatures**: *alike.* **Voice**: *screeching 'kir-rer-rer kik-kik-kik' in flight.*

17 GOULD'S PETREL
Pterodroma leucoptera c 30 cm

This small gadfly petrel breeds only at Cabbage Tree Island off Port Stephens on the NSW coast from October to April — with a total breeding population estimated at only 250-300 pairs. It stays at sea during the day feeding and comes in to roost after dark. Nests are mainly in crevices between boulders amongst the cabbage tree palms or *Pisonia* trees, whose sticky seeds are a hazard. **Recognition**: *mainly white below; sooty black crown and darker above than related species.* **Sexes** and **immatures**: *alike.* **Voice**: *rapid high-pitched 'ki-ki-ki-ki'.*

18 BLACK-WINGED PETREL
Pterodroma nigripennis c 30 cm

Where the only known breeding ground was previously on the Kermadec Islands, new colonies are now being found off New Zealand, on Norfolk and Lord Howe Islands and elsewhere. It is regularly but not often seen off the eastern coast, including on Heron and Muttonbird Islands. It is migratory and approaches ships. Breeds November-May and nests in burrows. **Recognition**: *mainly white below with narrow black margins on the wings.* **Sexes** and **immatures**: *alike.* **Voice**: *'peet', 'pee-pee', or 'pee-ur'.*

19 BROAD-BILLED PRION
Pachyptila vittata 25-30 cm

The Broad-billed Prion breeds on various islands of the southern oceans including around New Zealand, and disperses widely in winter. It is seldom seen in Australian waters but dead and dying birds are occasionally washed ashore along the southern coasts. It feeds on plankton by skimming the surface with head submerged or by flying under water. Breeds September-January in burrows. **Sexes** and **immatures**: *alike.* **Voice**: *chicken-like 'ku-aku-a-kuk'.*

20 SALVIN'S (Medium-billed) PRION
Pachyptila salvini 25 cm

Also called Lesser Broad-billed Prion, but is exceedingly difficult to separate from the Broad-billed Prion even in museum specimens. It breeds in millions in the Indian Ocean on Marion, Prince Edward and Crozet Islands, and disperses widely in winter. It is a regular visitor to the seas off southern Australia but is most often recorded from casualties washed up on the beach after storms, especially in the southwest. Breeds October-January in colonies and nests in burrows. **Sexes** and **immatures**: *alike.*

21 ANTARCTIC PRION
Pachyptila desolata 27-39 cm

Also known as the Dove Prion, it ranges widely throughout the cooler waters of the Southern Ocean. It breeds in Antarctic islands including Macquarie Island in early summer. Dispersing widely over the southern oceans, it is one of the prions most commonly washed up as derelicts on the southern beaches. Breeds November-February and nests in burrows. **Voice**: *throaty cooing in flight.*

22 FAIRY PRION
Pachyptila turtur 23-28 cm

The smallest of the prions, and the only one to breed in Australian waters, with large colonies on Bass Strait islands and around Tasmania. It also breeds on Marion Island and around New Zealand. Usually in flocks, it flies swiftly when foraging low over the waves. It deserts the nesting colony in March, returning in August, often to the same partner and nest. Breeds in summer. Nests in burrow or rocky crevice. **Sexes** and **immatures**: *alike.* **Voice**: *'kuk-kuk-kuk-oo-er'.*

23 SOUTHERN FULMAR
Fulmarus glacialoides 46-50 cm

The Southern Fulmar lives and breeds mainly within the cold seas of the Antarctic Ocean and is only seen as a rare visitor in Australian waters off the southern coast in winter. It is similar in appearance and habits to its northern namesake of Arctic waters, including nesting on cliff ledges. It feeds on offal, carrion, plankton and fish and gathers in large squabbling flocks around trawlers and whaling stations. Occasionally follows ships. Breeds November-December. **Sexes** and **immatures**: *alike.* **Voice**: *cackling.*

24 FLESH-FOOTED SHEARWATER
Puffinus carneipes 41-45 cm

Usually in flocks and common during summer in southern waters. Breeds on islands off south-west of WA, also on Lord Howe Island and around New Zealand. After breeding it disperses widely, mainly in tropical waters. It dives and skims the surface on stiff wings when feeding mainly on small fish. Breeds November-March. Nests in colonies in burrow among rocks or low dense vegetation. **Recognition**: *stocky; large pale bill; flesh-coloured legs and feet.* **Sexes** and **immatures**: *alike.* **Voice**: *high-pitched 'which-yew'.*

13

14

15

16

17

18

19

20

21

22

23

24

21

25 WEDGE-TAILED SHEARWATER
Puffinus pacificus 41-46 cm

Widespread throughout the tropical Indian and Pacific Oceans, and breeds on Lord Howe Island and numerous islands off Australian coasts such as Heron Island on the Great Barrier Reef. The drama of the nightly return of clouds of birds to their breeding grounds, and the clamour of countless wailing voices during the night is unforgettable. Breeds September-May. Nests in colonies in burrows, rock crevices and on coral sand. **Recognition**: *lazy gliding flight, wings inclined forward.* **Sexes** and **immatures**: *alike.* **Voice**: *wailing 'ka-koo-er' and catlike cries.*

26 BULLER'S (Grey-backed) SHEARWATER
Puffinus bulleri c 46 cm

Breeds only on Poorknights Islands off New Zealand's north island. Disperses extensively in the Pacific after breeding and is seen relatively frequently at sea off the coast of eastern Australia. Flight, mostly near the surface, is lazy and graceful. Breeds November-January. Nests in burrows and rock crevices. **Recognition**: *dark head; distinctive grey and black upperwing with lateral M-mark; white below.* **Sexes**: *alike.* **Voice**: *mewing calls at nest.*

27 SOOTY SHEARWATER
Puffinus griseus 40-46 cm

The Sooty Shearwater is the common muttonbird in New Zealand and breeds in millions on its numerous off-shore islands. It migrates over enormous distances and returns to its breeding grounds in November. Some breed on islands off the south-eastern coast and in Tasmania and are seen mostly in spring and early summer. Nests in burrows. **Recognition** *(cf Short-tailed): a little larger and darker; white underwing but difficult to separate.* **Sexes** and **immatures**: *alike.* **Voice**: *wheezing, crooning and cackling.*

28 SHORT-TAILED SHEARWATER (Muttonbird)
Puffinus tenuirostris 40-43 cm

The only petrel to breed solely in Australia with millions congregating each year on islands around Tasmania and off southern and south-eastern mainland coasts. Breeds November-March and nests in burrows. Dispersal after breeding begins in mid-April on a vast figure-of-eight 20000 nautical mile migratory route over the Pacific to the Aleutian Islands and back via the central Pacific. It was an important source of food for the early sealers and has since been exploited commercially. **Recognition**: *see Sooty Shearwater.* **Sexes** and **immatures**: *alike.* **Voice**: *very noisy spluttering, wheezing and wailing.*

29 FLUTTERING SHEARWATER
Puffinus gavia 31-36 cm

A common winter visitor to waters off south-eastern Australia after breeding on islands around New Zealand. Some remain throughout the year. Usually in flocks, sometimes large. Flight is characteristically low, direct, brisk, and busy with bursts of rapid wingbeats. Dives for food. Breeds August-March. It is easily confused with Hutton's Shearwater and extremely difficult to distinguish the differences at sea. **Sexes** and **immatures**: *alike.* **Voice**: *cackling sounds.*

30 HUTTON'S SHEARWATER
Puffinus huttoni c 38 cm

This species breeds in New Zealand's south island and is an occasional visitor to Australian seas mainly off the south-eastern coast — sometimes in flocks. Breeds August-April. **Recognition** *(cf Fluttering Shearwater): much alike.* **Sexes** *alike.* **Voice**: *deeper than the Fluttering Shearwater, with repetitive cackling 'kow-wow-haw'.*

31 LITTLE SHEARWATER
Puffinus assimilis 25-30 cm

This species occurs in the Atlantic, Pacific and Indian Oceans. In the Australian region, it breeds on Norfolk and Lord Howe Islands, and islands off the coast of Western Australia. It flies with rapid wingbeats, interspersed with short glides, close to the surface, and frequently dives for food. Relatively sedentary, absent from the nesting sites only in November and December. Breeds in winter and nests in a burrow or under a rock. **Recognition**: *the smallest black and white shearwater in Australian waters. Flight more flapping than close relatives.* **Sexes** and **immatures**: *alike.* **Voice**: *high-pitched wheezing and crooning.*

32 WHITE-FACED STORM-PETREL
Pelagodroma marina 21 cm

The White-faced is the only storm-petrel to breed in Australian waters, and nests on many off-shore islands in the southeast and southwest — the best known colony being on Mud Island in Victoria. Also breeds around New Zealand and on islands in the North Atlantic. Visits nesting burrows after dark. Breeds October-February and spreads northwards, some as far as the Arabian Sea to over-winter. Mostly pelagic and seldom seen inshore except when breeding. In flight, it skips over surface with legs dangling. **Recognition**: *white face and underparts.* **Sexes** and **immatures**: *alike.* **Voice**: *oft-repeated moaning 'pee-oo' at the nest.*

33 BLACK-BELLIED STORM-PETREL
Fregetta tropica 22 cm

An infrequent visitor to seas off the southern coast and Tasmania from breeding grounds on sub-Antarctic islands and New Zealand. Usually alone, sometimes in groups. Flies close to waves swiftly and erratically with legs dangling. Breeds in sub-Antarctic in summer. **Recognition**: *white underparts and underwing (black belly often difficult to see).* **Sexes** and **immatures**: *alike.* **Voice**: *shrill piping in flight.*

34 WHITE-BELLIED STORM-PETREL
Fregetta grallaria 22 cm

Only occasionally seen in Australian waters off the south-eastern coast as a rare visitor from breeding islands in the Atlantic and Pacific Oceans, including Lord Howe Island, where a dark form also occurs. Such black birds cannot easily be distinguished from the black-bellied species, and even the white-bellied form cannot be clearly discerned when the bird is in the water. Breeds December-May dispersing northwards to the tropics. It follows ships. **Recognition**: *black except for white rump and underparts (but some dark below).* **Sexes** and **immatures**: *alike.* **Voice**: *shrill whistling.*

35 GREY-BACKED STORM-PETREL
Oceanites nereis 19 cm

An infrequent but regular winter visitor to Bass Strait and the Tasman Sea from breeding grounds in Antarctic and sub-Antarctic seas. Little is known of its behaviour and movements. It feeds on plankton and is attracted to the lights of ships and lighthouses. **Recognition**: *black head, white underparts, grey rump.* **Sexes** and **immatures**: *alike.* **Voice**: *twittering at the nest site.*

36 WILSON'S STORM-PETREL
Oceanites oceanicus 19 cm

Breeds on Antarctic and sub-Antarctic islands and is possibly the world's most abundant bird. It is widespread as a winter migrant in all oceans. Common in Australian seas, including inshore, from April-June, especially off the east and west coasts. As dainty as a butterfly, it skips and flutters close to the waves while feeding, with legs dangling as though walking on water. Breeds in Antarctica in summer. **Recognition**: *conspicuous white rump, square-cut tail.* **Sexes** and **immatures**: *alike.* **Voice**: *sparrowlike chattering while feeding at sea.*

25

26

27

28

29

30

31

32

33

34

35

36

21

37 ARCTIC JAEGER (Skua)

Stercorarius parasiticus 67 cm (including tail)

Breeds in the Arctic tundra and a regular trans-equatorial summer migrant. It is seen in Australian inshore waters mainly from September to April, more commonly in the east. It has several plumage phases ranging from light to dark. It lives on fish, molluscs, carrion, scraps from ships, and pirates food and eggs from other birds. **Recognition**: *long pointed tail feathers (absent in moulting birds as often seen in the Southern Hemisphere).* **Sexes**: *alike.* **Voice**: *gull-like shrieks, shrill staccato calls.*

38 LONG-TAILED JAEGER (Skua)

Stercorarius longicauda 58 cm (including tail)

Breeds in the Arctic tundra and a regular transequatorial summer migrant. It is the most pelagic of the Jaegers and is rarely seen in Australian waters. Like other skuas, it is a piratical feeder. The long tail feathers, held cocked when the bird is on the water, are distinctive. **Recognition** *(cf Arctic Skua): smaller and slimmer, less conspicuous white wing flashes.* **Sexes**: *alike.* **Immatures**: *mottled and barred.* **Voice**: *shrill yelping.*

39 POMARINE JAEGER (Skua)

Stercorarius pomarinus 78 cm (including tail)

Breeds in the Arctic tundra and is a regular transcontinental migrant to overwinter in the southern waters from September to April. More frequently seen off southeastern Australian coasts than elsewhere, but tends to prefer off-shore waters. Groups of 40 or 50 may follow ships. Plumage very variable, the species occurring in light, intermediate and dark forms. Often harasses gulls and terns to take their food. **Recognition**: *in non-breeding plumage, the twisted tail feathers form blunt 'spoon'.* **Sexes**: *alike.* **Immatures**: *buff mottling and blotching.* **Voice**: *squeaking, harsh chattering.*

40, 41 PACIFIC GULL

Larus pacificus 66 cm

The endemic Pacific Gull is the largest of the gulls found in Australia, and its massive bill is the largest of any gull. It is usually seen singly, in pairs or sometimes loose groups patrolling the water's edge. It is an aggressive feeder, sometimes killing young birds up to the size of a Muttonbird, and eating nestlings and eggs. Occasionally dives for food. There is a belief, though not universally held, that Pacific and Kelp Gulls break open the shells of hard-shelled prey, such as molluscs, crabs and sea-urchins, by dropping them onto rocks from a height — often using the same 'anvil' continually and creating a midden of broken shells. Rarely seen inland, its habitat is on headlands and high places or low sandy spits on the coast and on offshore islands. Adults are sedentary, young birds disperse widely. Breeds August-November. **Recognition** *(cf Kelp Gull): larger size and bigger bill; black band on white tail.* **Sexes**: *alike.* **Voice**: *harsh 'kiaw' and variants of 'ow-ow' and 'yow-yow'.*

Adult left, immatures right

42 KELP GULL

Larus dominicanus 59 cm

A recent colonist, perhaps from New Zealand, the Kelp Gull was first noted in Australia at Botany Bay in 1943 and later found breeding on Moon Island, NSW, in 1958. It is now most common in Tasmania and Bass Strait. It frequents beaches, reefs, and jetties, feeding on fish, small birds and eggs, shellfish and carrion, and may harry terns. Sedentary or dispersive after breeding September-December; nests on the ground. **Sexes**: *alike.* **Voice**: *hoarse 'gorah', sobbing staccato sounds.*

43 SOUTH POLAR SKUA

Catharacta maccormickii 53 cm

This skua, which lives and breeds on the Antarctic Continent was once seen by Scott within 150 miles (c.240km) of the South Pole during his epic and tragic expedition of 1912. It regularly migrates across the equator to the high northern latitudes of the Pacific, Indian and Atlantic Oceans and has even been recorded in Greenland. Only rarely seen in Australian waters. Attracted to fishing trawlers. **Recognition** *(cf Southern Skua): smaller, variable plumage from almost white-bodied to blackish.*

44 SOUTHERN SKUA

Catharacta antarctica 61 cm

Believed by some as a different species from the Great Skua (which breeds in the North Atlantic and is not seen in Australian waters). A marauding, quarrelsome, fearsome and fearless scavenger which breeds on sub-Antarctic islands (including Macquarie). It is a regular winter visitor to southern Australian seas. Sometimes visits harbours and follows ships. It eats other birds, rats, mice, rabbits, carrion and offal as well as piratical feeding. Aggressive, particularly at the nest. **Sexes** and **immatures**: *alike.* **Voice**: *shrieking and crowing.*

45 SILVER GULL

Larus novahollandiae 43 cm

The familiar and ubiquitous Silver Gull lives right around the coast in a variety of habitats, and is also found inland where there are lakes and rivers — sometimes in large flocks. It feeds naturally on small fish and other aquatic life, but will readily gather to scavenge for offal jettisoned by fishing boats, sewage outfalls and refuse dumps. Breeds in colonies usually on small islands, sometimes on headlands or lakes. Also breeds in New Zealand and New Caledonia. **Sexes**: *alike.* **Immatures**: *bill and legs grey.* **Voice**: *raucous cries.*

46 COMMON DIVING-PETREL

Pelecanoides urinatrix 26 cm

This small and stocky petrel is the most wide spread of all diving petrels and is fairly common in southern Australian waters. It is seen singly or in small loose flocks resting on the surface or diving to feed submerged on small fish and other marine life. It breeds on islands in the Bass Strait and south of Tasmania, also in New Zealand and on various sub-Antarctic islands, including Macquarie, and nests in burrows. Relatively sedentary. **Sexes** and **immatures**: *alike.* **Voice**: *variations on 'ku-ku-ah'.*

47 FAIRY (Little) PENGUIN

Eudyptula minor 43 cm

The smallest of the penguins and the only one to breed in Australia, with substantial rookeries all along the southern and Tasmanian coasts and islands. Also breeds in New Zealand. It goes to sea before dawn and returns after dark, a spectacle watched by thousands of people every year at Phillip Island near Melbourne, as the colony tumbles out of the surf to roost. It remains ashore without food for two-three weeks when moulting, living on its fat reserves. Breeds July-March. Nests in a rock crevice or burrow underneath tussock-grass or other vegetation. **Sexes**: *alike.* **Immatures**: *bill smaller.* **Voice**: *'yap-yap' calls.*

48 ROCKHOPPER PENGUIN

Eudyptes chrysocome 60 cm

The Rockhopper Penguin breeds on sub-Antarctic islands, including Macquarie Island, and is a rare non-breeding visitor to the most southerly of Australia's shores during winter. They are aptly named as they hop with great agility over rocks with feet together, and jump into the sea feet first instead of diving like other penguins. Breeds spring to early summer. Nests in rocky areas. **Sexes**: *similar.* **Immatures**: *similar but duller.* **Voice**: *barking and growling.*

37

38

39

40

41

42

43

44

45

46

47

48

21

49, 50 WHITE-WINGED BLACK TERN
Chlidonias leucoptera 23 cm

A regular non-breeding summer visitor to northern Australia from Asia and southern Europe, with occasional incursions southwards, with stragglers reaching Tasmania and South Australia. Most of the birds seen in Australia are in various stages of mottled non-breeding plumage. It frequents swamps, estuaries and coastal lagoons. It often associates with Whiskered Terns in hovering flocks over shallow water, and they are not unalike when both are in non-breeding plumage. It takes insects on the wing or from the surface. **Voice**: *harsh 'kreek'.*

Breeding left, non-breeding right

51 ARCTIC TERN
Sterna paradisaea 38 cm

Breeds in the Arctic and sub-Arctic, and makes a remarkable migratory flight across the world, southwards to the Antarctic pack ice. By moving with the summer between the two hemispheres, it experiences more hours of daylight than any other bird. It is a rare visitor to Australia, mainly in the south-west and south-east, from September to May. **Recognition** *(cf Common Tern): shorter bill, smaller head and long tail streamers.* **Sexes**: *alike.* **Immatures**: *forehead white, crown white instead of all-black.* **Voice**: *like Common Tern.*

52 GULL-BILLED TERN
Gelochelidon nilotica 43 cm

Found throughout the world, and widespread in Australia both inland and around the coast. It frequents swamps, lakes and estuaries, also ploughed land. It eats fish and small reptiles, and skims above the surface hawking for insects. Breeds October-March (May in the north-west). Nests on the ground in colonies. **Recognition**: *stout black bill and longer thicker legs than other terns.* **Sexes**: *alike.* **Juveniles**: *similar to non-breeding adult.* **Voice**: *harsh cries.*

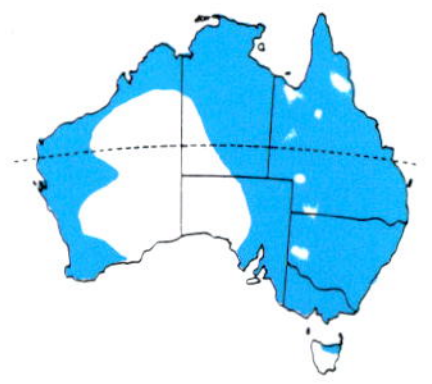

53 BLACK-NAPED TERN
Sterna sumatrana 36 cm

Distributed widely in tropical waters of the Indian and Pacific Oceans, the Black-naped Tern also frequents the coral cays and shallow lagoons of coastal and inshore waters of north-eastern Australia. It roosts and nests colonially on sand-cays of the Great Barrier Reef, and fishes in large flocks, sometimes with other terns and noddies by diving or hovering to pluck prey from the surface. Breeds September-December. **Sexes**: *alike.* **Voice**: *barking and sharp shrill chattering.*

54 COMMON TERN
Sterna hirundo 38 cm

A regular non-breeding visitor from northern Asia. It migrates as far south as Antarctica and is seen in Australia, usually in flocks, between October and March. Frequents coastal estuaries, reefs, bays and salt-fields. It fishes by skimming the surface with bill down, sometimes submerging. **Recognition** *(cf Arctic Tern): bill black or red with black tip.* **Sexes**: *alike.* **Voice**: *grating 'keee-uh' or 'kik-kik'.*

Non-breeding.

55 BRIDLED TERN
Sterna anaethetus 42 cm

Widespread in tropical seas and also breeds in Australia. Usually in flocks, it frequents islands off the northern coasts. It roosts on the ground and in trees or bushes, and feeds at sea by plunging and dipping, often nocturnally. Northern birds are probably sedentary, southern ones migrate northwards after breeding. Breeds September-January. Nests colonially on off-shore islands on sand, shingle and rock ledges unusually under cover. **Recognition** *(cf Sooty Tern): smaller, browner plumage and white eyebrow line continues past and behind eye.* **Sexes**: *alike.* **Voice**: *'yap-yapping'.*

56 WHITE-FRONTED TERN
Sterna striata 42 cm

The White-fronted Tern breeds in New Zealand and crosses the Tasman Sea to Australian waters, as young birds. It is seen mostly in the summer off the south-east coast and around Tasmania, where there have been some recent breeding records. It is mainly coastal, and takes fish in the surf. Sometimes follows fishing boats. Often seen in flocks, sometimes in the company of Crested Terns. **Recognition** *(cf Common Terns): almost white underwing.* **Sexes**: *alike.* **Voice**: *whistling 'tsit', and 'kee-uk'.*

57 ROSEATE TERN
Sterna dougallii 41 cm

The subtle rose-coloured tinge usually present on the breast and underparts gives this graceful tern its name. Usually in flocks, it breeds in colonies on several sand-cays and off-shore islands on the Great Barrier Reef and off the west coast of Australia but is generally uncommon. Widespread in the northern hemisphere, and non-breeding birds reach northern Australia. Breeds September-May. **Recognition** *(cf Common Tern): tail streamers extend well beyond folded wings.* **Sexes**: *alike.* **Voice**: *rattling and rasping.*

58 CASPIAN TERN
Hydroprogne caspia 56 cm

Of worldwide distribution, and the largest of the terns. It frequents coastal habitats, inshore islands, rivers and lakes, but though widespread, it is generally uncommon. It fishes by hovering and by high spectacular dives. Unsociable and aggressive. Breeds October-February (south), all months (north). Nests singly and in colonies on sandy beaches, shingle and headlands. **Sexes**: *alike.* **Voice**: *crow-like 'kraa-uh'.*

59 CRESTED TERN
Sterna bergii 48 cm

The Crested Tern raises its shaggy crown plumes when it is excited or alarmed, and is one of the largest of the terns. It is the commonest tern in Australia, and is found in coastal and near-coastal waters. It dives for fish and may flock around shoals; occasionally takes young turtles and eggs. Sedentary, but young birds disperse. Breeds March-December. Nests in large colonies on off-shore islands. Widespread in the Indian and Pacific oceans. **Sexes**: *alike.* **Immatures**: *dusky, crown mottled.* **Voice**: *cawing and 'wep-wep'.*

60 LESSER CRESTED TERN
Sterna bengalensis 42 cm

Found mainly in northern tropical waters, it is less common than the Crested Tern. It is noisy and restless and dives for fish in deep sea as well as in shallower waters. Often harried by frigatebirds. Sedentary, but young birds may disperse. Breeds May-November. Nests colonially on off-shore islands. Widespread from the Mediterranean to the Southern Hemisphere. **Recognition** *(cf Crested Tern): smaller, and orange-red bill.* **Sexes**: *alike.* **Immatures**: *bill grey-yellow.* **Voice**: *like Crested Tern.*

49

50

51

52

53

54

55

56

57

58

59

60

21

61 WHITE TERN
Gygis alba 36 cm

This dainty pure white bird breeds in colonies in tropical and subtropical seas — on islands which include Norfolk and Lord Howe — and has only been seen rarely in Australian waters off the coast of NSW. A surface feeder on small fish. Its flight is light and its outstretched wings appear to be translucent when seen from below. Breeds in colonies October-March. Nests in a tree fork in pine groves on Norfolk Island. **Sexes**: *alike.* **Voice**: *soft resonating 'tung', quiet chatter.*

62 LITTLE TERN
Sterna albifrons 25 cm

Worldwide distribution. Breeds in Australia along much of the coast except in the west and south-west. Frequents coastal inlets, beaches and islands. Dives for fish. Aggressive during nesting, with diving attacks on intruders. Breeds October—January. Nests on sandy beaches on coasts and islands close inshore. **Sexes:** *alike.* **Voice:** *high-pitched rasping.*

63 FAIRY TERN
Sterna nereis 27 cm

The Fairy Tern, whose scientific name means water nymph, breeds in Australia and is common along much of Australia's coast where the Little Tern is absent. Also breeds in New Zealand and New Caledonia. In shallow coastal waters and estuaries it dives for fish and crustaceans in noisy and sometimes very large flocks. Breeds November-February. Nests colonially on offshore islands and sandy coasts. **Recognition** *(cf Little Tern): slightly larger, paler above, less black before eye, bright yellow bill without black tip.* **Sexes**: *alike.* **Immatures**: *crown mottled.* **Voice**: *high-pitched alarm calls.*

64 WHISKERED (Marsh) TERN
Chlidonias hybrida 26 cm

Breeds from the Mediterranean through Indonesia to Australia, also in Africa. In Australia it is wide spread across the continent, particularly in freshwater lakes and rivers, usually in flocks; rarely on the coast. It hawks for insects over water, scoops fish and other aquatic prey from the surface, or partly submerges. Migratory northwards to winter. Breeds October-December. Nests in colonies in water vegetation, sometimes with other waterbirds. **Sexes**: *alike. (Non-breeding plumage like non-breeding White-winged Black Tern).* **Immatures**: *similar to adults in winter.* **Voice**: *harsh piping.*

Breeding

65, 68 SOOTY TERN
Sterna fuscata 47 cm

Immense colonies of this tern are to be seen on islands of the Great Barrier Reef, Torres Strait and off the northern and north-western coasts of Australia. It is also widespread in tropical and sub-tropical seas of the Atlantic, Indian and Pacific Oceans, including Lord Howe and Norfolk Islands, and is perhaps the most pelagic of all the terns. It feeds by snatching food from just below the surface. In Australia it breeds in spring in the south and in autumn in the north. It nests on a scrape of sandy ground or under a bush. A mass of wheeling and screaming Sooties returning to their breeding islands makes an unforgettable spectacle. **Recognition** *(cf Bridled Tern): slightly larger, forehead patch all forward of the eye, blacker plumage above.* **Sexes**: *alike.* **Immatures**: *upper parts speckled.* **Voice**: *calls like 'wideawake'; barking and growling.*

Adult above, immature below

66 WHITE-TAILED TROPICBIRD
Phaethon lepturus 82 cm including tail

Known to sailors as the Bos'nbird. Widespread in the world's tropical seas except eastern Pacific. Closest breeding is on Cocos Island (where an apricot race occurs) and Walpole Island, a coral inlet 240km south-east of New Caledonia. A rare but regular visitor to eastern coasts, most sightings being of young birds including casualties after storms. A vagrant elsewhere. Follows ships. Food includes flying fish. **Sexes**: *alike.* **Immatures**: *lack streamers, upperparts barred black.* **Voice**: *harsh screams.*

67 WHITE-CAPPED (Black) NODDY
Anous minutus 36 cm

Usually seen in large flocks on islands and reefs of the Great Barrier Reef, Torres Strait and the Coral Sea. Also widespread in tropical Pacific and Atlantic Oceans. It roosts in trees at night and flies to open sea beyond the reefs at daybreak. Sedentary. Breeds September-December. Nests in leafy trees. **Recognition** *(cf Common Noddy): blacker, smaller, slighter tail, whiter cap, longer bill.* **Sexes** and **immatures**: *alike.* **Voice**: *'krik-krik-krik'.*

69 RED-TAILED TROPICBIRD
Phaethon rubricaudus 86 cm including tail

Widespread in the Pacific and Indian Oceans. Breeding areas include Norfolk and Lord Howe Islands and Raine Island off Australia's east coast and Rottnest Island off the west coast, and ranges far out to sea. It dives for fish and squid. Rarely follows ships. Breeds mainly September-December in Australia. In pairs or loose colonies, nests in rock crevice or under a bush. **Recognition** *(cf White-tailed Tropicbird): pure white wings lacking black markings.* **Sexes**: *alike.* **Immatures**: *upperparts mottled, no tail streamers.* **Voice**: *harsh clanging rattle.*

70 GREY NODDY
Procelsterna cerulea 30 cm

This delicate small noddy breeds on various islands in the South Pacific, including Norfolk and Lord Howe. A vagrant to eastern Australia from December to May, most often in January. A fairy-like flight. It paddles in the water and feeds on small fish and crustaceans picked from the surface. **Sexes**: *alike.* **Voice**: *continuous purring note.*

71 COMMON (Brown) NODDY
Anous stolidus 40-45 cm

This large, heavily-built brown noddy is wide spread in tropical and temperate seas, and breeds in huge colonies on islands around the shores of northern Australia. It settles on the water in densely-packed rafts, and rests on any convenient floating perch, including the backs of turtles and even other larger birds. Feeds on fish mainly from surface beyond reefs. Roosts at night ashore. Breeds July-January. Nests on the ground or low bush. **Recognition**: *sharp separation of white cap from black lores.* **Sexes**: *alike.* **Immatures**: *cap absent or indistinct.* **Voice**: *'kraa-kraa-kraa'.*

72 LESSER NODDY
Anous tenuirostris 34 cm

Breeds only on the Seychelles in the western Indian Ocean and the Abrolhos group of Islands off the coast of WA in large numbers. In the morning it flies out to sea to feed, and takes fish near the surface. It returns in the late afternoon to roost and nest in mangroves. Immature birds gather in 'creches' on the beach. Breeds September-December. Nests in vast colonies in mangroves. **Sexes**: *alike.* **Voice**: *rattling, piping, whispering.*

61

62

63

64

65

66

67

68

69

70

71

72

21

73 LEAST FRIGATEBIRD
Fregata ariel 71-81 cm

Widespread in tropical seas and overlaps the range of the Great Frigatebird. Breeds in the Coral Sea, on the Great Barrier Reef and along the northern coasts. Piratical, and often robs boobies of their catches. The adult preens a great deal. During stormy weather, it may shelter in mangroves and spread its wings on the leaves to dry. Breeds May-December. Nests on the ground or in trees. **Recognition**: *male all black with white flank; female has black chin.* **Voice**: *'kukukuk'.*

Female

74 GREAT FRIGATEBIRD
Fregata minor 86-100 cm

Widespread in the tropical seas of the Atlantic, Indian and Pacific Oceans. Breeds on islands in the Coral Sea, and a regular non-breeding visitor to northern and eastern coasts. During the day it is usually seen soaring above its breeding island, often at a great height where it is only a speck. Its flight is swift and graceful. It dives on other seabirds to snatch prey and it also takes flying fish, young turtles and squid from the surface, but does not swim or walk. **Recognition**: *adult male entirely black; female has white chin and throat.* **Voice**: *gurgling and chuckling.*

Male

75 RED-FOOTED BOOBY
Sula sula 75 cm

The smallest booby. Distributed in tropical seas around the world, but seen in Australian waters only off the Queensland and WA coasts. It breeds in colonies on islands in the Coral Sea, on Raine Island and on the Ashmore Reef WA. Often roosts and nests in trees, and feeds mostly at night. Breeds May-September. Nests in bush well off the ground, sometimes with frigatebirds. **Recognition**: *bewildering plumage variations from white to grey-brown. Red legs; sometimes long white tail.* **Sexes**: *alike.* **Voice**: *'krerk' or 'krok'.*

76, 77 MASKED BOOBY
Sula dactylatra 86 cm

Widely distributed in tropical oceans. Breeds on islands of the Great Barrier Reef, Torres Strait, Coral Sea, Lord Howe, Norfolk, Adele and Bedout (WA) Islands. Highly pelagic like all boobies and forages far out to sea — diving mainly for flying fish and squid. Seen only sporadically inshore around the coastline. Its reaction to human intruders at the nest is only rather mildly intimidatory, with side-to-side pecking movements, bill-lunging, or flicking bits of stick or coral. Boobies are so-called by sailors because of their reputation for stupidity of allowing themselves to be easily captured while perched on ships. Disperses widely after breeding. Breeding is continuous, but mostly July-December. Nests on ground in small colonies. **Recognition**: *head pure white, tail black.* **Sexes**: *similar.* **Voice**: *male, 'pseep-pseep'; female 'kerk-kerk'.*

Adult left, juvenile right

78 BROWN BOOBY
Sula leucogaster 75 cm

Widespread in all tropical oceans. Least pelagic of boobies and commonly seen around Australia's northern coasts and at numerous off-shore breeding islands. Large breeding colonies have been recorded on Raine Island Qld, and Middle Island in the Lacepede Group, WA. It is seen around harbours and river mouths, and loafs on marker buoys. Breeds all months; nests mostly on the ground. **Recognition**: *deep plain brown above, white below. (Beware of confusion with Red-footed in dark phase).* **Sexes**: *similar.* **Voice**: *male, wheezing or hissing; female, sonorous 'kruk-kruk'.*

79 AUSTRALASIAN GANNET
Morus serrator 92 cm

Widespread in summer in the coastal and offshore waters of southern Australia, Tasmania and New Zealand. In Australia, it ranges further northward to winter. Its breeding colonies here are on islands in Bass Strait, on Lawrence Rock (Vic) and around Tasmania. Most birds seen around the continent come from breeding colonies in New Zealand. Singly, in small groups or large flocks, it feeds by plunge-diving. Breeds October-December. Nests on a mound on the ground. **Sexes**: *alike.* **Immatures**: *speckled brown.* **Voice**: *variety of harsh loud notes.*

80 BLACK-FACED CORMORANT [Shag]
Leucocarbo fuscescens 69 cm

The endemic Black-faced Cormorant lives entirely in a marine habitat along the offshore islands of southern Australia and Tasmania. Usually singly or in flocks, but it sometimes forages in thousands. Sedentary. Breeds in colonies mainly September-January. Nests on bare rock just above high water mark. **Recognition** *(cf Pied Cormorant): black facial skin, less white on head. Head held low in flight.* **Sexes**: *alike.* **Voice**: *infrequent croaks.*

81 LITTLE BLACK CORMORANT
Phalacrocorax sulcirostris 64 cm

Common and widespread, this small entirely black cormorant lives mainly on inland waters rather than on the coast. It is sociable and often breeds in a large foraging mass. Nomadic. Breeds in colonies all months. Nests very high in swamp trees with other cormorants and waterbirds. Range extends to Indonesia and New Zealand. **Recognition**: *small and entirely black.* **Sexes**: *alike.* **Voice**: *croaking.*

82 LITTLE PIED CORMORANT
Phalacrocorax melanoleucos 64 cm

Most familiar of the cormorants, and often seen close to habitation such as on stock dams and ornamental lakes in city parks, as well as on inland waters, streams, swamps, inlets and rocky islands. It often forages alone, and eats small fish, yabbies, shrimps, tadpoles and larvae. Nomadic. Breeds all months. Nests in colonies in bush or tree. Range extends to Malaysia and New Zealand. **Sexes**: *alike.* **Voice**: *croaking.*

83 PIED CORMORANT
Phalacrocorax varius 66-81 cm

Widespread on mainland except in most arid areas of interior; mostly coastal in south-west. Singly and in flocks sometimes immense, it frequents mangroves, estuaries, rivers, lakes, swamps and rocky islands. Rather uncommon. Breeds most months. Nests colonially on reefs, sand-banks and dead trees. Also breeds in New Zealand. **Recognition**: *conspicuous black trousers; bright orange spot in front of eye.* **Sexes**: *alike.* **Voice**: *threatening 'tok-tok'; other gutteral noises and screaming.*

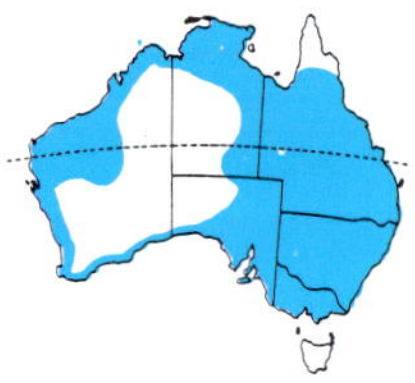

84 GREAT [Black] CORMORANT
Phalacrocorax carbo 92 cm

The world's most widespread cormorant and the largest in Australia. It lives along the coast but more extensively inland on main waterways and freshwater lakes. Sociable and usually in groups, sometimes very large flocks. Often seen resting on sandbanks and mudflats or drying its wings on poles, snags and rocks. Breeds January-September. Usually nests colonially in trees and on ground. **Recognition**: *all black (white flanks when breeding); yellowish chin.* **Sexes**: *alike.* **Voice**: *rather silent; some croaking at nest.*

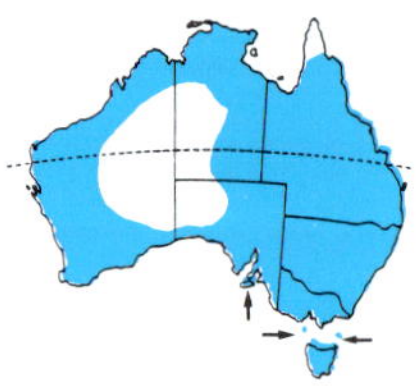

73

74

75

76

77

78

79

80

81

82

83

84

21

85

86

85, 86 DARTER

Anhinga melanogaster 94 cm

Darters are found extensively in freshwater inland habitats of rivers, lakes, swamps and lagoons, particularly in the north. Generally solitary, but sometimes seen in loose parties resting, preening and drying wings (the feathers are not waterproof). Its body is usually submerged when it is swimming with its sinuous snakelike neck and head showing — hence its popular name Snakebird. It is wary and immediately sinks below the surface if approached, or it may dive and resurface well away. It slowly stalks prey underwater; its chief food being fish, speared with a sharp jab of the beak. In flight, it sometimes soars to great heights. Sedentary. Breeds irregularly but mostly October-March. Nests in a tree above water. Chicks over two weeks old will instinctively tumble out into the water if threatened — and scramble back when all is clear. Also found in tropical Asia, Africa and South America. **Immatures**: *like females.* **Voice**: *fearsome witchlike cackle.*

male left, female right

87 AUSTRALIAN PELICAN

Pelecanus conspicillatus 190 cm

The familiar Australian Pelican is relatively common throughout the Australian mainland and Tasmania. It has spread to New Guinea in recent years and is sometimes seen as a vagrant in New Zealand, some Pacific islands and Indonesia. Mostly in flocks, it frequents estuaries, mudflats on the coast, rivers, large inland lakes and lagoons — and fishes in shallow water. Its main food is fish and crustaceans. It chooses remote and secluded sites for nesting. Large breeding colonies are found on the west coast on offshore islands and on the more extensive inland waterways in the east of the continent of which Lake Eyre is a notable example. It is highly nomadic and travels huge distances away from inland Australia in times of drought, as happened in 1977 after one of the most severe and widespread droughts of this century. It often soars at great heights floating on thermals. Its take-off to get airborne is long and involves much paddling and wing flapping. Breeds variably at any time after water levels are sufficient following rain. Nests on the ground or amongst water vegetation. Creches of hundreds of young birds are formed in the breeding colonies. **Sexes and immatures**: *similar.* **Voice**: *mostly silent, grunts or croaks at the nest.*

A wonderful bird is the pelican!
His bill will hold more than his belican.
He can take in his beak
Food enough for a week
But I'm darned if I see how the helican.
Dixon L. Merritt

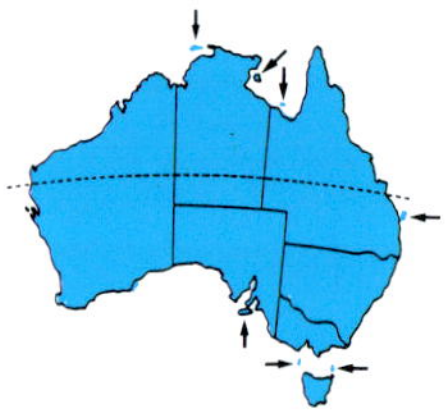

87

1

2

3

4

5

6

7

8

9

10

11

12

13

14

15

16

The Authors and their Photography

We became interested in bird photography in 1965 after some six years wandering through the bush in various parts of New South Wales recording bird calls. We bought a Pentax camera and a 500 mm Takumar lens, and had them mounted on a gunstock fitted with a follow-focus device made by an instrument maker in Sydney to our own design. It worked quite well when used as a gun from the shoulder (see Reef Heron on page 169). When mounted on a tripod, however, the results were disappointingly soft. At that time we neither understood the reason nor the remedy — that was to come in later years — and we sold the equipment. But the excitement of seeing over a hundred species of birds for the first time in close-up through the telephoto lens acted as a spur to explore techniques for producing photographs of birds that would satisfy our ideas of technical excellence and aesthetic appeal. To this end, we became interested in photographing captive and aviary birds just as much as operating in mangrove swamps knee-deep in mud photographing waders 'wild and free in a place of their own choosing'.

STUDIO PHOTOGRAPHY OF BIRDS IN THE FIELD

For two years we worked with the bird-banders photographing birds caught in mist-nets as subjects. For this purpose, we built a series of transparent perspex enclosures so that the bird could be photographed in natural light against a background of the bush in which it lived. Our first unit is illustrated at (11) and the photograph of the yellow-faced Honeyeater on page 135 is an example of the style of photographs taken with the system. The first series of enclosures was eventually abandoned, because of reflections from the transparent end panel, but we eventually solved the problem by the simple expedient of inclining the panel at 45° in a black surround with all sight lines from the camera going to a black surface.

A new design incorporating this principle was built, using clear plastic film for the cladding. A number of other modifications were also introduced, including handling sleeves on both sides and underneath, a gimble-mounted camera platform and a system of variable clamps for holding the prop for the bird. A stand was provided which enabled the unit to be raised, lowered and tilted when composing the background. The bottom of the enclosure was removable so that it could also be operated on the ground. Electronic flash was adopted for illuminating the subject—with the use of two flanking lights and one for modelling. It then became necessary to operate in shade so as to avoid double images from the ambient light as well as flash. To this end, the whole unit, except the front, was enclosed under a canopy, with curtains lined inside with white to avoid a reflected colour cast. When operating under these conditions, the camera aperture is set for correct exposure for the flash on the bird, and the shutter speed is adjusted to bring the exposure of the daylight scene beyond the bird into balance. Additional flashheads are employed to illuminate the background when operating in gloomy conditions in the jungle. A Hasselblad camera is used, and extensive Polaroid tests are made on lighting and composition before shooting begins.

When photographing a captive bird, we always give it ample time to settle into its unaccustomed environment and preen itself, and a whole day might be spent on a single bird. Food and water is always provided. Some birds adapt quickly and become so trusting that they will perch on a finger when free to fly away. It must be emphasised that when working in the field with this system, licenses are always obtained from the authorities to catch birds for the purpose of photography, with an undertaking to release them afterwards.

Photographs were obtained with the new studio of a number of birds never before photographed — under conditions which would have been well nigh impossible by any other means. Examples of photographs taken with the unit are of a Red-backed Kingfisher taken in the Simpson Desert in 1973 on page 83, a Northern Scrub-robin taken at Iron Range in 1975 on page 93, and the Eyrean Grasswren taken in the Simpson Desert in 1977 on page 107. The photographs of the studio at (14) and (15) were taken at Iron Range and the Simpson Desert respectively. We were fortunate on these occasions to be supported by teams of helpers to catch the birds, run the camps and help with the formidable logistical problems. There would have been several gaps in the photographic coverage in this book except for these expeditions, and this is the moment publicly to express our thanks to all those who contributed to their success.

TELEPHOTO LENSES

In no branch of photography is detail more important than when photographing birds, and a degree of sharpness that may be acceptable, say, for sports photography is not necessarily so for bird photography. In the early 1970s, we did a series of comparative tests to find out why the quality of the image obtained with long lenses often fell so short of that seen in the viewfinder. We used a finely calibrated resolution chart which was photographed on Kodak High Contrast Copy Film with up-market cameras at various shutter speeds, and lenses mounted on a heavy tripod with a Miller fluid head. We found that shots taken with lenses over 200 mm all showed evidence of mechanical vibration, and that the degradation of the image was almost entirely vertical, which would seem to confirm that camera vibration is mainly attributable to the mirror action. High shutter speeds did

PHOTOGRAPHERS: D. Keller 7; L.Le Guay 14; remainder by authors

not entirely eliminate it, but the effect was less noticable with slow shutter speeds of 1/8th second or longer — which is predictable as it has been shown that vibration from the mirror is transient and decays very quickly. Vibration was more evident when the camera was unsupported by hand and a cable release used. The breakthrough came during tests with a 300 mm lens, when we noticed that the image obtained on film was much sharper when the camera socket was used instead of the socket on the lens. From this we concluded that by fixing the camera directly on the tripod head, camera shake was effectively insulated from the lens. Accordingly, we mounted the camera and lens on a beam, and provided a front rest in the form of an adjustable traversing bar — as shown in the illustrations opposite. A dramatic improvement in the quality of our long-lens photography resulted. The front support, however, appeared not to be essential, and equally good results were obtained with back-end support only — as demonstrated by the shot of a Papuan Frogmouth on page 21 which was some ten metres up a tree when being photographed, as illustrated at (5).

A vice fitted with a tripod head and turned upside down on a window (7) makes a good support when using a vehicle as a hide.

When stalking without a tripod, every aid has to be used to hold a long lens still. As portability is a major consideration, our favoured combination (2) comprises a camera with a motor winder, a 300 mm *f*4 lens used with a 1.4x extender (equivalent to 420 mm at *f*5.6), a pistol grip, a monopod and a shoulder stock. A shoulder strap and socket for the monopod (4) is sometimes useful, and a small tripod head on the monopod which can be locked one way is preferred to a ball joint. Extenders have been greatly improved from early days, and our resolution tests on the best of them show that the loss of image quality is minimal. We sometimes use a tele-flash when shooting against the light (3); the shot of the Brush Wattlebird on page 125 was taken this way.

SHOREBIRD PHOTOGRAPHY

Among our most pleasant memories are the many days we have spent in the mud at various estuaries, and off the Cairns Esplanade, photographing shorebirds with a series of mobile hides, illustrated at (8), (9) and (10). They were propelled on our knees as shown at (8), and the strategy was to take the unit into the water to approach the birds from the sea. As that is not the direction from which danger normally threatens, timid waders, which would take off if one came within 100 metres of them, could be closely approached. The hides at (9) and (10) are amphibious, with buoyancy provided by sealed 10 gallon oil drums mounted on axles. All the wading birds attributed to us in Part II were taken from these hides.

ZOO AND AVIARY PHOTOGRAPHY

Much of our pleasure from bird photography has come from making pictures of zoo and aviary birds. We built a series of enclosures for this purpose of which (12) and (13) are examples. To this end, we sought to apply conventional portraiture techniques of good composition and lighting. The unit at (12) was set up to photograph a whipbird (the result is at pages 23 and 31). The unit on the trailer (13) included a reflection-free window so that we could use growing vegetation to simulate a natural background. (All the finch shots contributed by us to Section 14 were taken by this means.) A corridor was also incorporated in the unit so that small birds could be photographed in flight (an example is on page 139). One enclosure was built large enough to photograph owls and, through the courtesy of Bruce and Margaret Kubbere, we used it jointly with Eric Hosking at their Featherdale Wildlife Park, Doonside, NSW, in 1982. The Masked Owls (20 and 21 on page 21) were photographed on that occasion. After experimenting with Balcar frontal projection equipment for six months in 1982, we succeeded in adapting it to photographing birds in flight against a natural background, using a budgie borrowed from a neighbour (we kept no birds on our own). Its wings-up portrait at page 68 was taken in our garage. We are indebted to Vic LePla of the Australian Broadcasting Corporation for his technical advice.

TARONGA PARK ZOO

From 1968 to 1982 our work was based at Taronga Zoo, Sydney, where we enjoyed the most generous facilities. Our story might have been very different had it not been for the support of the Director, Ronald Strahan and his Board, the Curator of Birds, Kerry Muller and the Head Bird Keeper, Les Clayton and all members of the Zoo staff. Our photography at the Zoo entered a new phase in 1969 with the construction of a large fixed studio (16), 8 m long, 5 m wide and 3 m high, designed specially to photograph birds in flight and built by John Lysaght in Newcastle. Cruciform in plan, it consisted of three corridors — two diagonal and one transverse. These were wired and louvred at each end and given perches. When opposites louvres were opened, a bird would naturally fly back and forth from light-to-light and, with a photo-electric beam placed mid-way down the track to trigger the camera shutter and set off the flashheads, it would take its own photographs each time it passed through it. Lighting was provided by two heavy duty industrial units, driving six 2500 volt flashheads, made by Ernest Turner Electrical Instruments Ltd of High Wycombe, England. The effective flash duration was rated at 130 microseconds (millionths) at full power and 80 microseconds at half power. Most of the flight shots of ours in this book were taken in this studio.

ACKNOWLEDGEMENTS

The heart of the book lies in its photographs. Behind them is a story of effort, skill, ingenuity, patience and sacrifice on a scale that cannot be over-estimated, and they are in themselves a lasting tribute to the photographers who took them.

Many individuals generously gave their time and expertise in assisting us in the provision of supporting text to complement the photographs and participating in other ways in the making of this book. (They are named in chronological order without implication of priority.)

WAYNE LONGMORE deserves special mention for acting as our consultant from earliest days. He refereed a number of the Sections and proof-read all the galleys before the book went to press.

ARNOLD McGILL, doyen of Australian ornithologists, gave us his full support and encouragement from the beginning, and made major contributions to the final result by providing most of the expertly executed distribution maps, a set of introductory drafts for the Sections and making valuable suggestions as a referee for some of the text.

DARIEL LARKINS provided a sound groundwork on which the Species Notes were gradually built over the ensuing four years.

RAE HENSON wrote drafts for all the Species Notes. She helped throughout with the revisions, and generously gave her time with some of the research.

The text for each Section was read and checked for accuracy through all the drafting stages by our referees: WALTER BOLES, JOHN DISNEY, GRAEME CHAPMAN, TERRY LINDSEY and HAROLD POLLOCK — each according to his expertise and field of interest. Their emendations and suggestions were of great benefit to the final result.

WALTER BOLES checked the identity of all 864 transparencies selected for the book before they were processed. Our special contributors, PENNY AND JERRY OLSEN, and RICHARD DONAGHEY came to the rescue at a time of illness and saw us through a difficult patch. Section 2 (Diurnal Birds of Prey) and Section 15 (Bowerbirds and allies) respectively bear the stamp of their expert knowledge and experience.

ERNEST HOSKIN gave us the benefit of his astonishing gift for bird calls in the descriptions of those of his acquaintance in New South Wales.

TED CARTHEW,a former Executive Trustee of the National Photographic Index of Australian Wildlife supported us with help and advice throughout the project.

ZOË MIDDLETON, a colleague from Index days, and GERALD MIDDLETON, besides assisting with the reading of proofs, were always at hand with help and encouragement.

VINCENT SERVENTY, a Trustee of the National Photographic Index of Austraian Wildlife, gave expert advice whenever needed.

DENSEY CLYNE, writer and film-maker of Mantis Wildlife Films, was ever at hand to answer questions on natural history.

JANE DALBY of the Royal Botanic Gardens, Sydney; DAVID McALPINE of The Australian Museum, Sydney; and DR MARY CARVER of the Department of Entomology, CSIRO, Canberra advised on honeydew, manna and lerps.

JOHN LLOYD, Ranger at Sherbrooke Forest Park, Victoria, and his wife, gave us much valuable literature and information on the lyrebird.

JOHN WATERHOUSE readily dipped into his highly-organised ornithological library for answers to queries whenever asked.

The following gave valuable assistance and advice in connection with research for the special article on pages 72 and 73 'Land of Parrots': GLENYS McIVOR, Curator of Maps at the National Library who cheerfully responded to numerous queries over a period of two years and provided many key references; ALAN TRELOAR, Honorary Fellow at New England University, Armidale, and FATHER GABRIEL PEDRON at the Benedictine Monastery, Arcadia, who generously contributed translations from the Latin of inscriptions on sixteenth century world maps; Associate Professor FRANK BROEZE, Department of History at the University of Western Australia, who encouraged the pursuit of the subject; INGRID DILLA, Department of History at Rijksuniversiteit Te Leiden, The Netherlands, who provided valuable background on the voyage of Hendrick Brouwer across the Southern Indian Ocean in 1611; Dr HANS STRUNDEN, ornithologist of Essen, W. Germany who provided helpful comments; and TONY CAMPBELL, Acting Head of the Map Library of the British Library, who gave assistance during a two-day visit to London in September 1986.

Our thanks and appreciation also go to the many others who helped and encouraged us.

We would like to express our thanks to the members of the following organisations for the facilities provided.

The National Photographic Index of Australian Wildlife
It was very rewarding after the thirteen years associated with the Index project during its formative days from 1968 to 1981 to be able to benefit from its services during the making of this book. In 1983 we looked at all the photographs then represented in the collection and made our first selections. Our thanks and appreciation go to Ronald Strahan, Mary Rose Gordon, Heather Lawrence, Denise Greig and their colleagues for all the help and encouragement given in various ways throughout the course of our project. Photographs obtained from photographers through the Index's agency are identified (NPI).

Bird Observers Club
Frank Stephens, the Club's Slide Librarian, assisted by his wife Rona, gave us valuable help and encouragement and put us on the track of certain photographs. These are identified (BOC).

Photographic Agencies
The following photographic agencies simplified the problems of filling some of the gaps: Animals, Animals, New York
Antarctic Division, Department of Science (Ant. Div.)
Ardea London Limited (Ardea)
Australasian Nature Transparencies (ANT)
Australian Picture Library Pty Ltd (APL)
Photographic Library of Australia (PLA)

Photo Processing Laboratories
The following provided valuable services:
Richard Carter and Bob Smith
Vision Graphics, Sydney
Marinus Deboer
Combined Exposure Pty Ltd
John Palmer
Australian Color Laboratores Pty Ltd
Ian Maher
Adcolor (Sydney) Pty Ltd
Andrzej Opiela
Authentic Print, Sydney

Artwork and Design
The following assisted in the production of pilot pages of text and the assembly of dummy books:
Keith McLean and Denise Curry
Stewart Printing, Hornsby
Brian Troy
Hornsby Printing, Hornsby
Sue Beskin
Ku-ring-gai Typesetting Services
Mt Ku-ring-gai

Equipment
We wish to record the invaluable facility afforded by the HK Vidichrome Colour Analyser, generously made accessible to us by Vision Graphics, for calibrating the adjustments needed to achieve balance in the composite pages between individual photographs for colour and density.
The major problem of scaling 864 photographs for the composite pages was simplified with the acquisition of an Artiscope Rescaler, an invaluable piece of equipment made in Austral[illegible]. With this we were able to project each photograph for tracing in place for the guidance of the platemakers. This task alone took over a year; without the Artiscope, it would have been well nigh impossible.

Libraries
Facilities were afforded by the following libraries:
National Library of Australia, Canberra
Map Department
British Library, London
Map Library
State Library of New South Wales, Sydney
General Reference Library
Mitchell Library
Hornsby Shire Library
including the Galston Branch Library
Hindwood Memorial Library
NSW Gould Leagure Headquarters, Beecroft
Fisher Library, University of Sydney
The Library of the Royal Historical Society, Sydney

REFERENCES

We acknowledge our indebtedness to the many ornithological and associated works consulted in preparing and editing the text for this book. They include:

Ornithology

A Field Guide to the Birds of Australia, G. Pizzey (1983)
A Field Guide to Australian Birds Vols 1 and 2, P. Slater (1979)
A Field Guide to Nests and Eggs of Australian Birds, G. Beruldson (1980)
A Field Guide to the Birds of Britain and Europe, R.T. Peterson et al (1954)
A Field Guide to the Birds of Western Australia, G.M. Storr, R.E. Johnson, M. Thompson (1985)
A Field Guide to the Seabirds of Australia and the World, G. Tuck and H. Heinzel (1980)
A Guide to the Birds of Tasmania, M. Sharland (1958)
A List of the Vertebrates of South Australia, Edited by J.H. Aslin (South Australian Museum) (1985)
An Index of Australian Bird Names, CSIRO (1969)
Australian Finches in Bush and Aviary, K. Immelmann (1982)
Australian Parrots, J.M. Forshaw and and W. Cooper (1981)
Australian Parrots in Bush and Aviary, I. Harman (1981)
Australian Parrots in Field and Aviary, A.L. Lendon (1979)
Australian Warblers, A.R. McGill (1970)
Bird in the Hand, H.J. de S. Disney and others (1974)
Bird Life, I. Rowley (1982)
Birds and Where to Watch Them, R. Wheeler and T.R. Lindsey (1987)
Birds in Bass Strait, K. Simpson (1972)
Birds in the Australian High Country, H.J. Frith (1984)
Birds of Australia, J.D. Macdonald (1973)
Birds of New Guinea and Tropical Australia, W.S. Peckover and L.W.V. Filewood (1976)
Birds of North America, C.S. Robbins, B. Brunn, H.S. Zim and A. Singer (1983)
Birds of the Kimberley Division, Western Australia, G.M. Storr (1980)
Birds of the Northern Territory, G.M. Storr (1977)
Birds of the World, O.L. Austin and A. Singer (1965)
Birds of Western Australia, D.L. Serventy and H.M. Whittell (1976)
Budgerigars Bush and Aviary, N.W. Cayley (1933)
Checklist of the Birds of Australia Part 1, H.T. Condon (1975)
Collins Bird Guide: a Photographic Guide to the Birds of Britain and Europe, S. Keith and J. Gooders (1980)
Collins Concise Atlas of the World (1984)
Collins Junior Australian Atlas (1985)
Common Australian Birds of Towns and Gardens, G. Chapman (1969)
Complete Book of Australian Birds, Readers Digest (1976, 1986)
Complete Book of New Zealand Birds, Readers Digest (1985)
Eagles, Hawks and Falcons of Australia, D. Hollands (1984)
Evolution and Ecological Relationships Between Quail-Thrushes, J. Ford (*Emu*, Vol. 83, pp 152-172 1983)
Handlist of Birds in New South Wales, A.K. Morris, A.R. McGill, G. Holms (NSW Field Ornithologists Club) (1981)
Hawks in Focus, J and L. Cupper (1981)
Interim List of Australian Songbirds, R. Schodde (1975)
Kookaburras, V. A. Parry (1970)
New Light on the Discovery of Australia as Revealed by the Journal of Captain Don Diego de Prado y Tovar, H. N. Stevens, Haklyut Society (1930).
Nightwatchmen of Bush and Plain, D. Fleay (1979)
Nocturnal Birds of Australia, R. Schodde and I.J. Mason (1980)
Owls, J. Sparks, T. Soper and R. Gillmor (1970)
Pigeons and Doves of Australia, H.J. Frith (1982)
Recommended English Names for Australian Birds, RAOU (*Emu*, Vol. 77 1978)
Revised List of Queensland Birds, G.M. Storr (1984)
Seabirds, P. Harrison (1983)
The Atlas of Australian Birds, M. Blakers et al. (1984)
The Birds of Australia, K. Simpson and N. Day (1984)
The Emu, RAOU
The Fairy-Wrens (1982), R. Schodde
The Handbook of Australian Sea-Birds, D.L. Serventy, V. Serventy and J. Warham (1971)

The Lore of the Lyrebird, A. Pratt (1945)
The Official Checklist of the Birds of Australia, RAOU (1926)
The Reader's Digest Complete Atlas of the World (1968)
The Romance of the Lyrebird, A.H. Chisholm (1960)
The Seabirds of Australia, National Photographic Index of Australian Wildlife (1986)
The Slater Field Guide to Australian Birds, P. Slater, P. Slater and R. Slater (1986)
The Times Atlas of the World, Comprehensive Edition (1980)
The Times Concise Atlas of the World (1982)
The Waterbirds of Australia, National Photographic Index of Australian Wildlife (1985)
The Wrens and Warblers of Australia, National Photographic Index of Australian Wildlife (1982)
What Bird is That?, N.W. Cayley (1984)
Waterfowl in Australia, H.J. Frith (1982)

History

A Separate Creation, G. Pizzey (1985)
Australia Unveiled, G. Schilder (1976)
Diary 1844-1845, John Gilbert (Mitchell Library)
Discovery of Australia, G. Collingridge (1895; Fac.1983)
Dutch Discoveries of Australia, J.P. Sigmond/L.H. Zuiderbaan (1976)
Early History of Australian Zoology, G.P. Whitley (1970)
Java la Grande, L. Fitzgerald
Journal of an Overland Expedition in Australia from Moreton Bay to Port Essington 1844-1845, L. Leichhardt (1847)
More Early History of Australian Zoology, G.P. Whitley (1975)
Strange Journey; the Adventures of Ludwig Leichhardt and John Gilbert, A.H. Chisholm (1955)
The Discovery of Australia, A. Sharp (1963)
The Literature of Australian Birds, H.M. Whittell (1954)
The Mapping of the World 1472-1700, R.W. Shirley (1984)
The Secret Discovery of Australia, K.G. McIntyre (1977)
The Voyage of Pedro Alvares Cabral to Brazil and India, Hakluyt Society (1938)
The Voyage of Torres, B. Hilder (1980)
To Sail Beyond the Sunset, C.M. Finney (1984)

RECORDINGS OF BIRD CALLS

Recordings of bird calls can be obtained from a number of sources, including the following:

National Photographic Index of Australian Wildlife
The Australian Museum
College Street
SYDNEY NSW 2000 Cassettes: '*What Bird Song Is That?*' (1987)

Bird Observers Club
P.O. Box 185
NUNAWADING Victoria 3121 Cassettes

The Gould League of NSW
Mary Street
BEECROFT NSW 2119 Disc: '*The Birds Around Us*'

John Hutchinson
Wildlife Sound Studio
Brockman Street
BALINGUP WA 6253 Cassettes

Ray Swabey
79 Cambridge Terrace
MALVERN SA 5061 Cassettes

There are also two reference libraries of recorded bird calls in Canberra that can be consulted. These are administered respectively by CSIRO's Division of Wildlife and Rangelands Research at Lyneham, ACT (P.O. Box 84 Lyneham ACT 2602), and the National Film and Sound Archive at McCoy Crescent, Acton, ACT (GPO Box 2002 Canberra 2601). A list of all recordings of Australian birds known up to December 1986 is available in *A guide to recorded wildlife sound — Birds — by Dr Peter Fullager and Norman Robinson* (1987) which can be obtained from the National Film and Sound Archive (the publisher) or any branch of the Australian Government Printing Service.

Index of English Names

Numbers in *italics* refer to photographs

Index of Scientific Names

Numbers in *italics* refer to photographs

Donald Trounson was born in Redruth, Cornwall, and educated at Mill Hill School, London. After war service in North Africa, Sicily and Italy, he joined the British Diplomatic Service with postings to Rome, the Foreign Office, New York (United Nations), and to Sydney as a member of the British High Commission. In 1967 he founded the National Photographic Index of Australian Wildlife on behalf of the Australian Museum in association with the National Library, and served as its Executive Officer until 1981. His interests include music, motor racing, archaeology, watercolour painting and photography. He was granted Australian citizenship in 1985.

Molly Trounson (née Clampett) was born in Adelaide and educated at Woodlands CEGGS, Adelaide. She was the Deputy Representative for the BBC in their Australian office from 1957 to 1981, and also a voluntary contributor to the work of the National Photographic Index of Australian Wildlife from 1967 to 1981 as well as Honorary Secretary to its Executive Committee. Her interests include music, golf, gardening and bird photography.

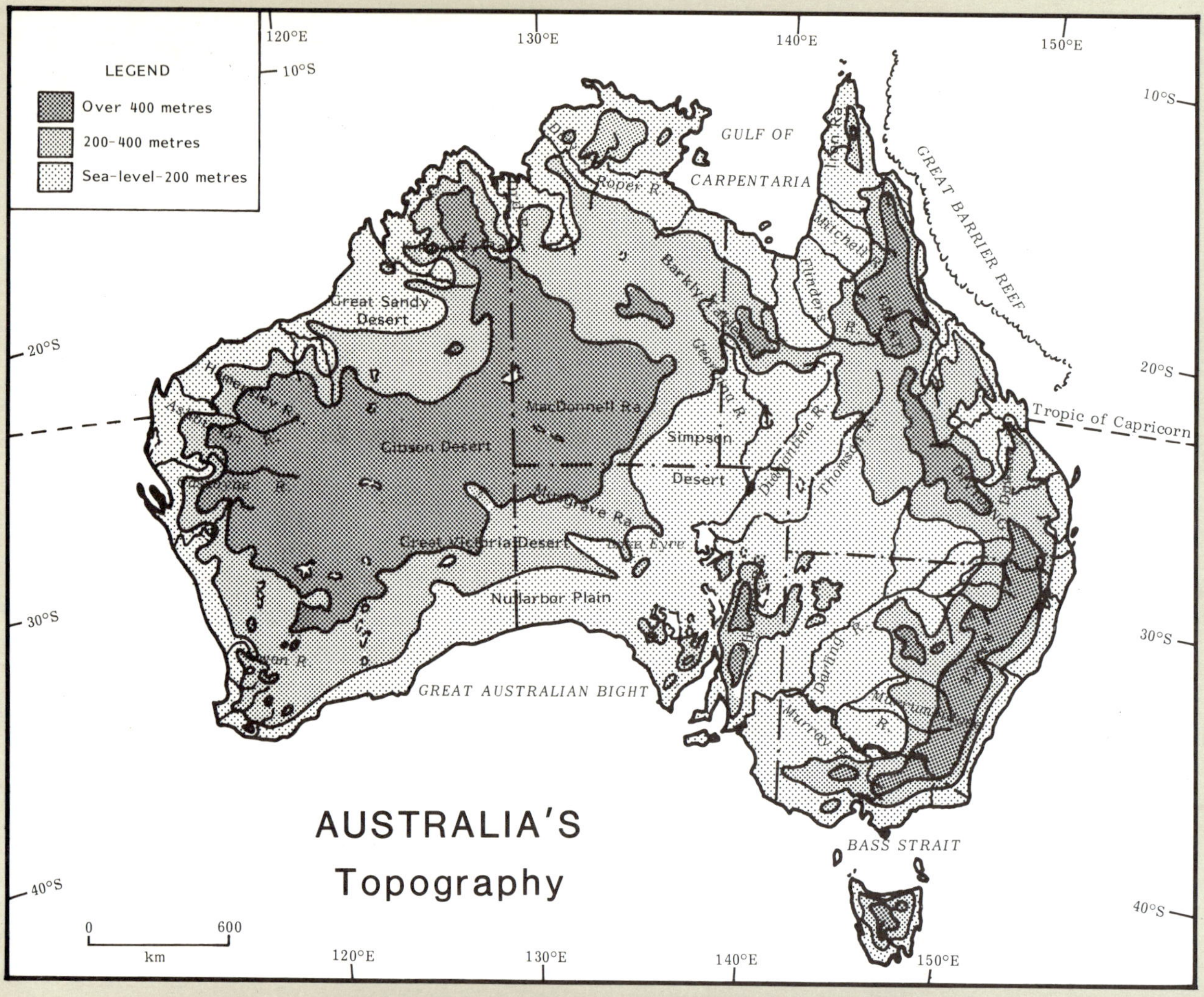

AUSTRALIA'S Topography

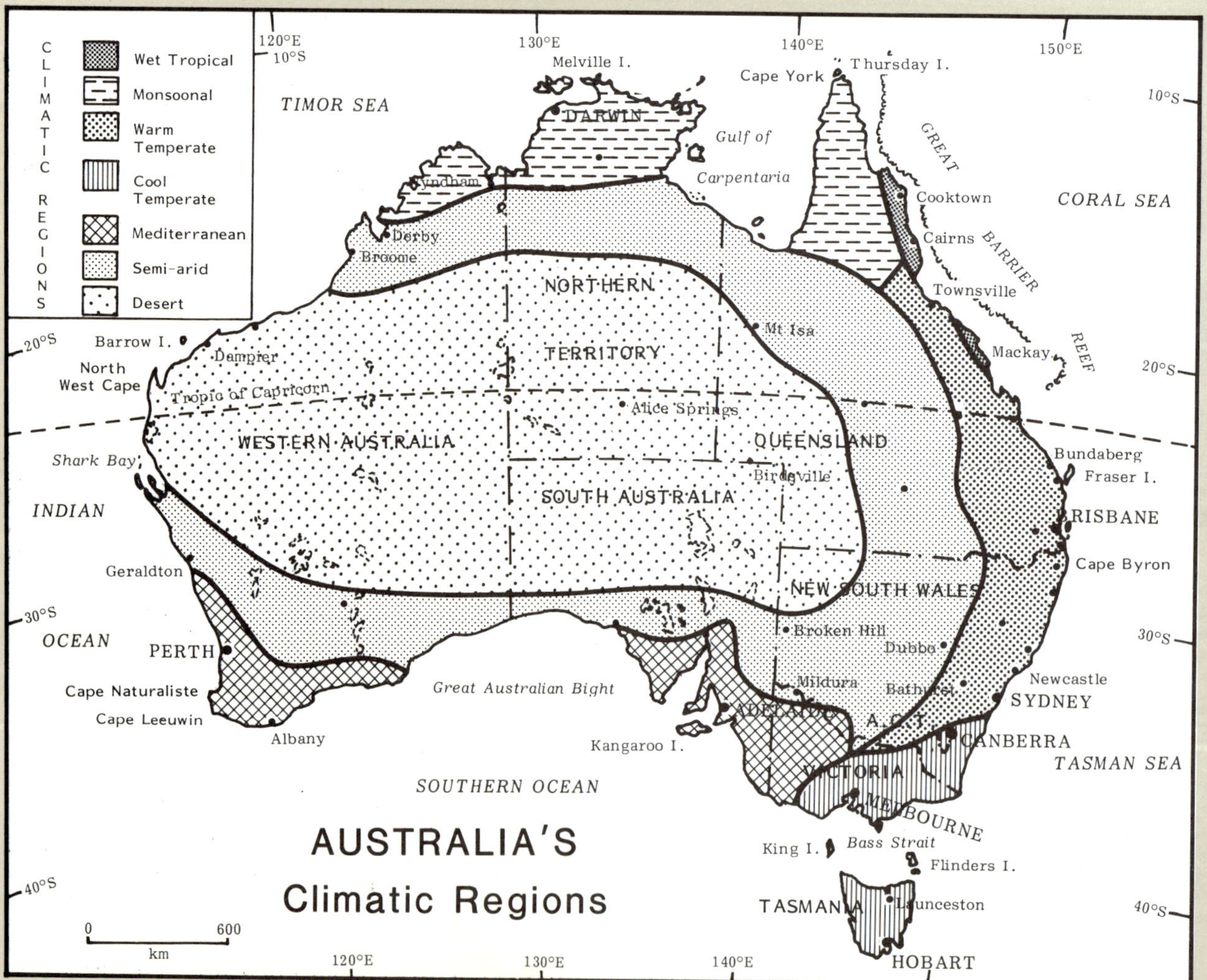

AUSTRALIA'S Climatic Regions